Trainee Guide

Heating, Ventilating, and Air Conditioning

Level 1

National Center for Construction Education and Research

Wheels of Learning
Standardized Craft Training

Prentice Hall

Upper Saddle River, New Jersey   Columbus, Ohio

 Prentice-Hall, Inc.
A Simon & Schuster Company
Upper Saddle River, New Jersey 07458

This information is general in nature and intended for training purposes only. Actual performance of activities described in this manual requires compliance with all applicable operations procedures under the direction of qualified personnel. References in this manual to patented or proprietary devices do not constitute a recommendation for their use.

Printed in the United States of America

10 9 8 7 6

ISBN: 0-13-245002-X

Prentice-Hall International (UK) Limited, *London*
Prentice-Hall of Australia Pty. Limited, *Sydney*
Prentice-Hall of Canada, Inc., *Toronto*
Prentice-Hall Hispanoamericana, S. A., *Mexico*
Prentice-Hall of India Private Limited, *New Delhi*
Prentice-Hall of Japan, Inc., *Tokyo*
Simon & Schuster Asia Pte. Ltd., *Singapore*
Editora Prentice-Hall do Brasil, Ltda., *Rio de Janeiro*

This volume is one of many in the *Wheels of Learning* craft training program. This program, covering more than 20 standardized craft areas, including all major construction skills, was developed over a period of years by industry and education specialists. Sixteen of the largest construction and maintenance firms in the U.S. committed financial and human resources to the teams that wrote the curricula and planned the national accredited training process. These materials are industry-proven and consist of competency-based textbooks and instructor guides.

Wheels of Learning was developed by the National Center for Construction Education and Research in response to the training needs of the construction and maintenance industries. The NCCER is a nonprofit educational entity supported by the following industry and craft associations:

- Associated Builders and Contractors
- American Fire Sprinkler Association
- Carolinas Associated General Contractors
- Metal Building Manufacturers Association
- National Association of Minority Contractors
- National Association of Women in Construction
- National Insulation Association
- Painting and Decorating Contractors of America

Some of the features of the *Wheels of Learning* program include:

- A proven record of success over many years of use by industry companies
- National standardization providing "portability" of learned job skills and educational credits that will be of tremendous value to trainees
- Recognition: Upon successful completion of training with an accredited sponsor, trainees receive an industry-recognized certificate and transcript from NCCER.
- Approved by the U.S. Department of Labor for use in formal apprenticeship programs
- Well illustrated, up to date, and practical. All standardized manuals are reviewed annually in a continuous improvement process.

Acknowledgments

This manual would not exist were it not for the dedication and unselfish energy of those volunteers who served on the Technical Review Committee. A sincere thanks is extended to:

Larry DeFrees	Leslie E. Seigle	David M. Skarphol
Ralph Eisenberger		Mike Van Zeeland

Contents

Introduction to HVAC

Module 03101

INTRODUCTION TO HVAC

Objectives

Upon completion of this module, the trainee will be able to:

1. Explain the basic principles of heating, ventilation, and air conditioning.
2. Identify career opportunities available to people in the HVAC trade.
3. Explain the purpose and objectives of an apprentice training program.
4. Describe how certified apprentice training can start in high school.
5. Describe what the Clean Air Act means to the HVAC trade.

Prerequisites

Successful completion of the following Task Modules is required before beginning study of this Task Module: Common Core Curricula.

Required Student Material

1. Trainee Task Module

Course Map Information

This course map shows all of the *Wheels of Learning* task modules in the first level of the HVAC curricula. The suggested training order begins at the bottom and proceeds up. Skill levels increase as a trainee advances on the course map. The training order may be adjusted by the local Training Program Sponsor.

Course Map: HVAC, Level 1

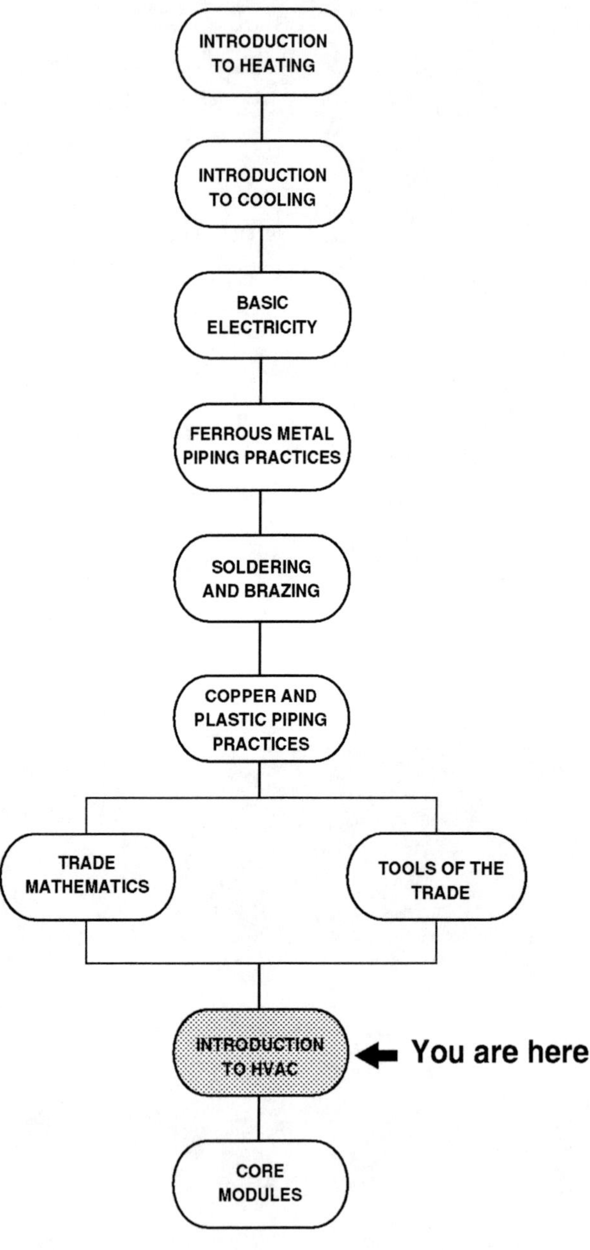

TABLE OF CONTENTS

Trade Terms Introduced In This Module

Chlorofluorocarbon (CFC) refrigerant: A class of refrigerants comprised of chlorine, fluorine, and carbon that have a very adverse affect on the environment.

Compressor: In a refrigeration system, the mechanical device that converts low pressure, low temperature refrigerant gas to high temperature, high pressure refrigerant gas.

Condenser: A heat exchanger that transfers heat from the refrigerant flowing inside it to the air or water flowing over it.

Evaporator: A heat exchanger that transfers heat from the air flowing over it to the cooler refrigerant flowing through it.

Expansion device: Also known as the liquid metering device or metering device. Provides a pressure drop that converts the high temperature, high pressure liquid refrigerant from the condenser into the low temperature, low pressure liquid refrigerant entering the evaporator.

Heat transfer: The transfer of heat from a warmer substance to a cooler substance.

Mechanical refrigeration: The use of machinery to provide cooling.

Noxious: Harmful to health.

Reclamation: Remanufacture of used refrigerant to bring it up to the standards required of new refrigerant.

Recovery: The removal and temporary storage of refrigerant in containers approved for that purpose.

Recycling: Circulating recovered refrigerant through filtering devices that remove moisture, acid, and other contaminants.

Refrigeration cycle: The process by which a circulating refrigerant absorbs heat from one location and transfers the heat to another location.

Toxic: Poisonous.

1.0.0 INTRODUCTION

Since ancient times, people have sought ways to make the buildings in which they live, work, and play more comfortable. Today, the Heating, Ventilation, and Air Conditioning (HVAC) industry provides the means to control the temperature, humidity, and even the cleanliness of the air in our homes, schools, offices, and factories. The members of the HVAC trade are skilled workers who install, maintain, and repair the equipment that makes this possible.

2.0.0 HEATING

Early humans burned fuel as a source of heat. That hasn't changed; what's different between now and then is the way it's done. We no longer need to huddle around a wood fire to keep warm. Instead, a central heating source such as a furnace or boiler does the job using the **heat transfer** principle. That is, heat is created in one place and carried to another place by means of air or water.

For example, in a common household furnace, fuel oil or natural gas is burned to create heat, which warms metal plates known as heat exchangers (*Figure 1*). Air from living spaces is circulated over the heat exchangers and returned to the living spaces as heated air. This type of system is known as a forced-air system.

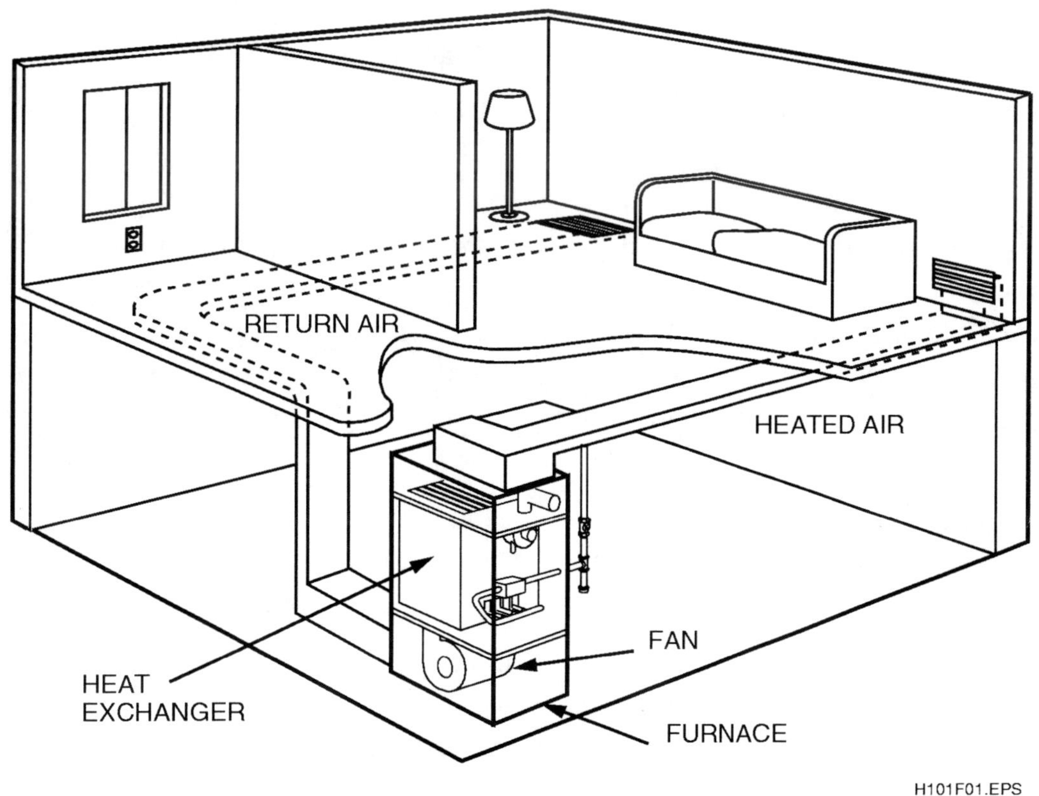

Figure 1. Forced-Air Heat

Water is also used as a heat exchange medium. The water is heated in a boiler (*Figure 2*), then is pumped through pipes to heat exchangers where the heat it contains is transferred to the surrounding air. The heat exchangers are usually baseboard heating elements located in the space to be heated.

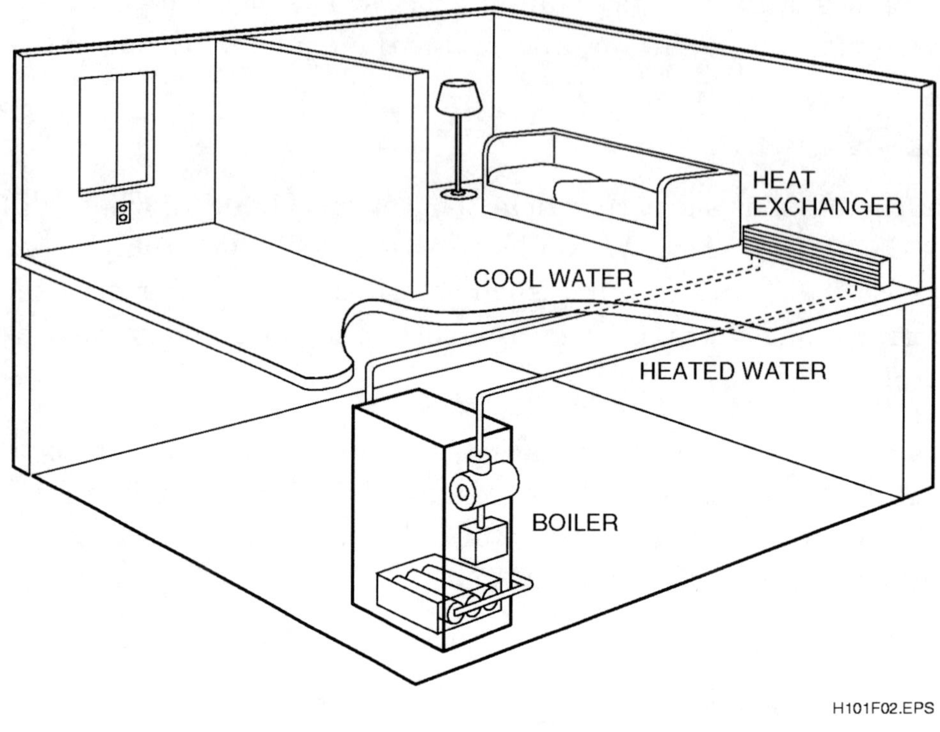

Figure 2. Hot Water Heat

Natural gas and fuel oil are, by far, the most widely used heating fuels. Electricity is also used as a heat source. In an electric heating system, electricity flows through coils of heavy wire, causing the coils to become hot. Air from the conditioned space is passed over the coils and the heat from the coils is transferred to the air. Because electricity is expensive to use for heating, this method is no longer common in cold climates. It is more likely to be used in warm climates where heat is seldom required.

3.0.0 VENTILATION

Ventilation is the introduction of fresh air into a closed space in order to control air quality (*Figure 3*). Fresh air entering a building provides the oxygen we breathe. In addition to fresh air, we want clean air. The air in our homes, schools, and offices contains dust, pollen, and molds, as well as vapors and odors from a variety of sources. Relatively simple air circulation

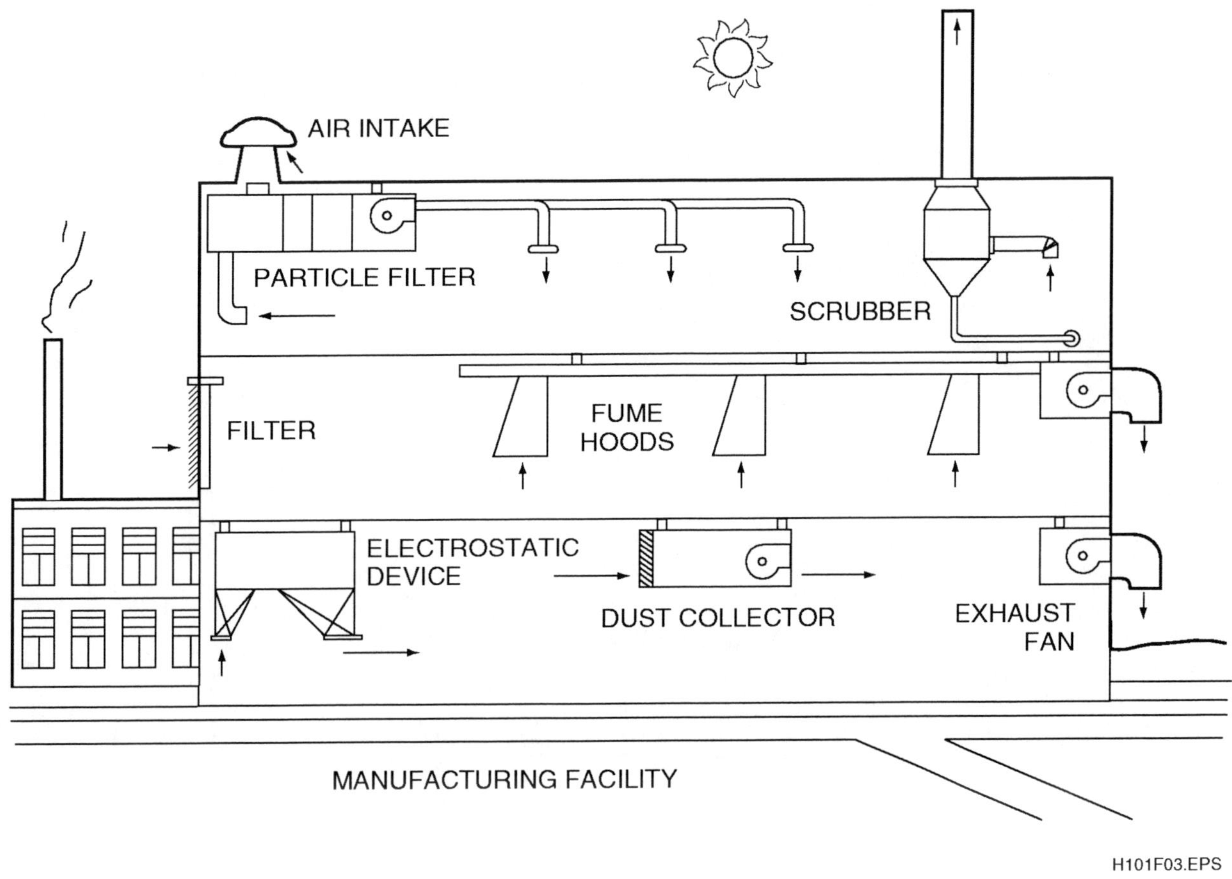

Figure 3. Ventilation System

and filtration methods, including natural ventilation, are used to help keep the air in these environments clean and fresh. Many industrial environments, on the other hand, require special ventilation and air management systems. Such systems are needed to eliminate **noxious** or **toxic** particles and fumes that may be created by the processes and materials used at the facility.

The U.S. Government has strict regulations governing indoor air quality in industrial environments and the release of toxic materials to the outside air. Where noxious or toxic fumes may be present, the indoor air must be constantly replaced with fresh air. Fans and other ventilating devices are normally used for this purpose. Special filtering devices may also be required; these not only protect the health of building occupants, but prevent the release of toxic materials to the outside air.

Through the ages, mankind has used many methods to stay comfortable in hot weather. In this study, however, we will focus on what is known as **mechanical refrigeration**, which came into use in the twentieth century. It is based on a principle known as the **refrigeration cycle** (*Figure 4*).

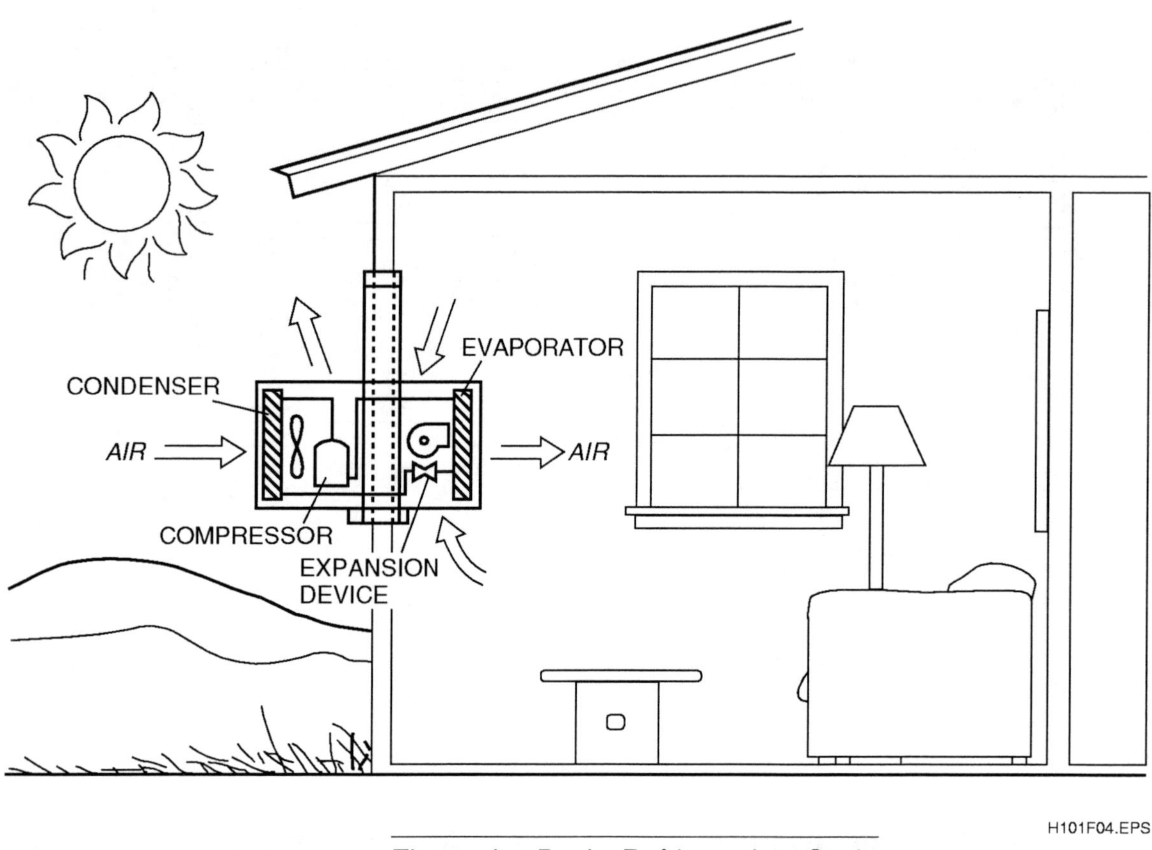

Figure 4.   Basic Refrigeration Cycle

Simply stated, the refrigeration cycle relies on the ability of chemical refrigerants to absorb heat. If a cold refrigerant flows through a warm space, it will absorb heat from the space. Having given up heat to the refrigerant, the space becomes cooler. The colder the refrigerant, the more heat it will absorb, and the cooler the space will become. If the super-hot refrigerant flows to a cooler location, the outdoors for example, the refrigerant will give up the heat it absorbed from the indoors and become cool again.

A mechanical refrigeration system is a sealed system operating under high pressure. The main elements of a mechanical refrigeration system are the:

*   **Compressor** – Provides the force that circulates the refrigerant and creates the high pressure necessary for the refrigeration cycle to work. A special refrigerant compressor is used.

- **Evaporator** – A heat exchanger where the heat in the warm indoor air is transferred to the cold refrigerant.
- **Condenser** – Also a heat exchanger. In the condenser, the heat absorbed by the refrigerant is transferred to the cooler outdoor air.
- **Expansion device** – Provides a pressure drop that lowers the boiling point of the refrigerant as it enters the evaporator. This allows the refrigerant to absorb heat in the evaporator.

The relationship between temperature and pressure is critical to mechanical refrigeration. As you study the process, you will learn that the same refrigerant can be very cold at one point in the system (the evaporator input) and very hot at another (the condenser input). These two points are often only inches apart. This is possible because of pressure changes caused by the compressor and expansion device. In addition to the circulation of refrigerant, air must also circulate. Fans at the condenser and evaporator move air across the condenser and evaporator.

This is a simple explanation of the refrigeration cycle. It is meant to give you a basic idea of how an air conditioner works. Later in the training program, you will explore this subject in greater detail. The relationship between temperature and pressure will also be studied in depth. It is the key to understanding and troubleshooting mechanical refrigeration systems.

The refrigeration cycle is the same in all refrigeration equipment, from the small air conditioner in your car to the huge system that cools the largest office building. The difference is in the size and construction of the components and piping and the amount and type of refrigerant.

5.0.0 CAREERS IN HVAC

Career opportunities in the HVAC trade are many and varied. There is a large existing base of HVAC systems that needs servicing, repair and replacement. In addition, every time a new residential, commercial, or industrial building is constructed, it contains one or more of the HVAC elements.

To get an idea of how vast the HVAC trade is, picture the town you live in. Then think about the fact that almost every building in town contains some form of equipment to provide heating and cooling as well as air circulation and purification. Expand that view to include the entire country and you realize that there are tens of millions of heating, air conditioning, and air management systems. New ones are being added every day; old ones are wearing out and being repaired or replaced. From that perspective, the opportunities in the trade appear limitless.

Figure 5 provides an overview of career opportunities in the HVAC trade. For the purposes of this discussion, it is convenient to view the HVAC industry as having three segments:

1. Community-Based: Companies that sell, install, and service residential and light commercial equipment and systems such as furnaces and packaged air conditioners.
2. Commercial/Industrial: Companies that install and maintain systems for large office buildings, factories, apartment complexes, shopping malls and so forth.
3. Manufacturing: Companies that build and market HVAC systems and equipment.

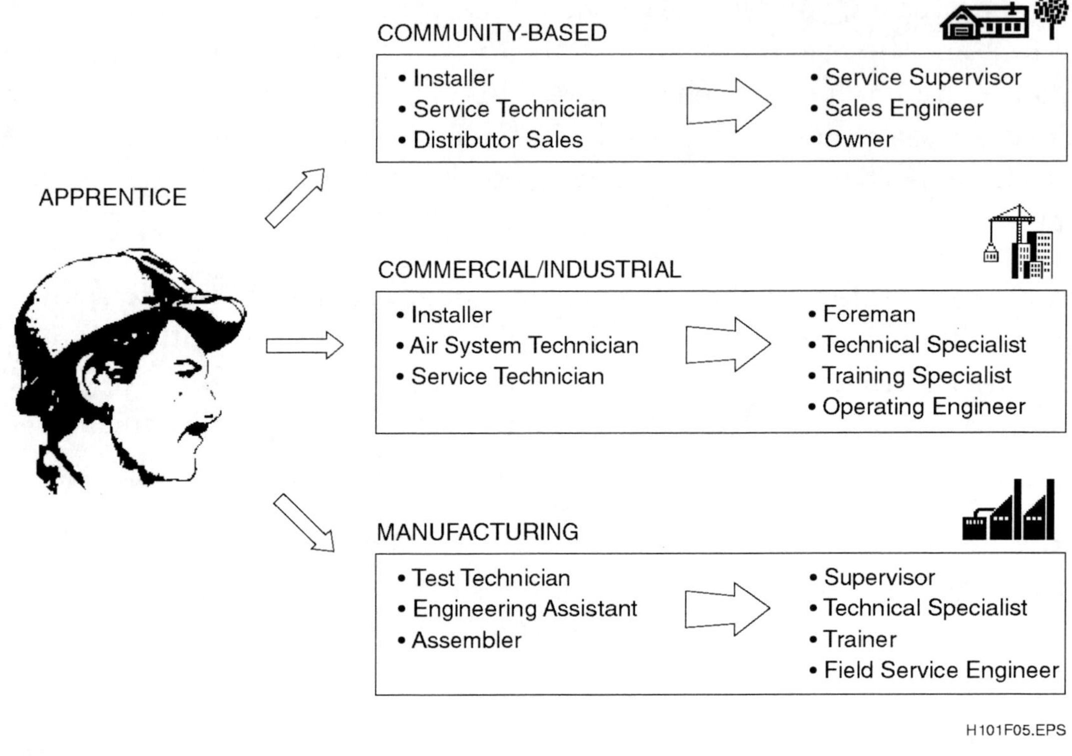

Figure 5. Career Opportunities In The HVAC Trade

5.1.0 COMMUNITY-BASED

At the community level, you might find anything from a one-person installation and service business to a firm with a hundred or more employees, including heating and air conditioning specialists, sheet metal workers, and sales engineers. In such businesses, HVAC specialists may work alone or with a single partner. They typically respond to service calls from homes or small businesses. They may also install furnaces and air conditioning equipment sold by their firm's sales engineer. In other firms, one group may do installations while another group handles troubleshooting and maintenance. At this level, a technician is expected to work with a wide variety of products from many different manufacturers. Systems can consist of anything from a window air conditioner to a two- or three-element heating/air conditioning system. The common thread is that the elements are self-contained units selected by the sales engineer because they meet the heating or cooling requirements of the space in which they

HVAC TRAINEE TASK MODULE 03101

will be used. In a broad sense, these elements could be viewed as off-the-shelf components
for which most of the piping and wiring is done at the factory.

Local or regional distributors provide equipment, parts, special tools, and other services for
the dealers who sell, install, and service HVAC equipment. The distributor needs sales people
who know HVAC equipment. The distributor may also provide engineering support and
service training for its dealers. Distributorships are often affiliated with a single
manufacturer.

5.2.0   COMMERCIAL/INDUSTRIAL

Large commercial and industrial systems have many components and may require thousands
of feet of ductwork and piping. Such systems are designed by engineers and architects. Many
HVAC trade people are required for these projects and an individual is more likely to
specialize. For example, where residential systems are usually controlled by a single wall
thermostat, a large commercial system will often have central computer controls that are
installed and serviced by a system control specialist. Others may specialize in working with
large steam boilers; still others may choose to become experts in installing and servicing high
volume cooling units known as chillers.

Companies that install such equipment are often large construction firms that may work
anywhere in the world. They are likely to do only the installation, and let the building owner
contract with a local firm for maintenance. Many large facilities employ their own HVAC
maintenance people.

5.3.0   MANUFACTURING

There are thousands of HVAC manufacturers. Some of the larger ones cover the entire HVAC
spectrum, while others focus on a particular product or market. For example, one may make
window air conditioners, another gas furnaces, and yet another might make only heavy
commercial equipment. Regardless of their market, they employ HVAC specialists as:

- Test technicians
- Engineering assistants
- Training specialists
- Instruction book writers
- Field service technicians

Working as an HVAC journeyman is satisfying and financially rewarding. If you want to
go beyond the journeyman level, there are many opportunities in supervision and executive
management with dealerships, construction companies, and manufacturers. Companies that
install and service HVAC equipment are usually founded by someone who started as a service
technician. For those who want to teach, there are opportunities to work in vocational schools
and training programs such as the one in which you are now participating.

6.0.0　APPRENTICE PROGRAM

This training program is part of a nationally recognized apprenticeship system approved by the U.S. Labor Department. In combination with your on-the-job training, it will help you learn the skills you need to qualify as a journeyman in the HVAC trade.

Apprentice training goes back thousands of years; its basic principles have not changed in that time. First, it is a means for individuals entering a craft to learn from those who have mastered the craft. Second, it focuses on learning by doing; real skills versus theory. Although some theory is presented in the classroom, it is always presented in a way that helps the student understand the purpose behind the skill that is to be acquired. That is why this program, and others in the *Wheels of Learning* series, are said to be "competency-based."

Apprentice programs as they exist today have some specific characteristics:

* Apprentices must be no less than 16 years old.
* Opportunities must be available to all.
* The program must incorporate both training and on-the-job experience.
* A minimum of 144 hours per year of related training is given.
* There is a progressively increasing schedule of wages.
* There is proper supervision with adequate facilities.
* Job performance and related instruction are periodically evaluated.
* Successful completions are formally recognized.

6.1.0　YOUTH APPRENTICE PROGRAM

The Youth Apprentice Program allows students to begin their apprentice training while still in high school. A student entering the program in eleventh grade will have completed one year of the Wheels of Learning four-year program by graduation. In addition, the program, in cooperation with local craft employers, allows students to work in the trade and earn money while still in school. On graduation, the student can enter the industry at a higher level and with more pay than someone just starting the apprentice program.

This training program is the same one used by Construction Education Foundation learning centers, contractors, and colleges across the country. Students are recognized through an official transcript and can enter the second year of the program wherever it is offered. They may also have the option of applying the credits at a two-year or four-year college that offers an HVAC-related degree or certificate program.

Scientific studies have shown that chemicals known as **chlorofluorocarbons (CFC's)** can damage the ozone layer that protects the earth and its inhabitants from the sun's ultraviolet rays. One of the effects is an increased rate of skin cancer. Many refrigerants used in air conditioning equipment are CFC's.

In the 1980's the nations of the world made a commitment to phase out these chemicals. They agreed to take steps in the interim to prevent their discharge into the atmosphere. In 1990, the U.S. Congress passed the Clean Air Act, which calls for early phaseout of the most toxic refrigerants, eventual elimination of all CFC's, and strict control and labeling of refrigerants. The U.S. Environmental Protection Agency (EPA) is responsible for implementing and enforcing this law, which has a significant impact on the HVAC trade. For example:

- Anyone releasing refrigerants to the atmosphere is subject to a stiff fine and possibly a prison term.
- Anyone handling refrigerants must have EPA-sanctioned certification. Without it, you cannot even buy refrigerants. The training needed to obtain EPA certification is included in this course.
- Records must be kept on all transactions involving refrigerants. This includes purchase, use, reprocessing, and disposal.

Before a sealed refrigeration system can be opened for repair, the refrigerant it contains must be **recovered** and stored in approved containers. When the repair is complete and the system is resealed, the same refrigerant may be returned to the system. It may also be used in another system belonging to the same owner. Recovered refrigerant should, however, be **recycled** before reuse. This removes moisture and impurities that could damage the system. Some recovery units have built-in recycling capability.

If the refrigerant is badly contaminated or no longer needed, it can be **reclaimed**. This is done at remanufacturing centers where the refrigerant is returned to the standards of purity that govern new refrigerants. Reclaimed refrigerant can be resold on the open market, but cannot be classified as "new" refrigerant.

In addition to Federal regulations, there may be state regulations that apply to refrigerants. State regulations may be stricter. It is critical that everyone in the HVAC industry understand and follow the EPA regulations regarding the handling, storage, and labeling of refrigerants. Failure to do so could be very costly to both you and your employer.

SUMMARY

The HVAC trade involves equipment used for heating, cooling, and purifying indoor air. It covers everything from the window air conditioners and furnaces used in our homes to the giant heating and cooling systems used in large office buildings and industrial complexes.

Because of the widespread use of HVAC equipment, there are many career opportunities in the trade. Jobs are available with small local firms, large industrial and commercial contractors, and the manufacturing firms that build and market HVAC equipment. The apprentice program provides an opportunity to learn the trade through a combination of hands-on training and related classroom learning. The Youth Apprentice Program allows students to begin their training while still in high school.

One of the most serious issues affecting the trade is the damage that can be done to the earth's ozone layer by the improper release of refrigerants into the atmosphere. There are severe penalties for improper refrigerant disposal.

References

For advanced study of topics covered in this task module the following books are suggested:

Modern Refrigeration and Air Conditioning, The Goodheart-Willcox Company, Inc., South Holland, Illinois.

Refrigeration & Air Conditioning Technology, Second Edition, Delmar Publishers, Inc., Albany, New York.

1. In a common household forced-air furnace, heat is transferred from:
   a. the heat exchangers to the air.
   b. the natural gas or oil to the conditioned space.
   c. the air to the heat exchangers.
   d. the refrigerant to the outside air.

2. Water is often used in heating systems as a:
   a. fuel.
   b. refrigerant.
   c. heat exchange medium.
   d. heat exchanger.

3. Ventilation is concerned with:
   a. air circulation.
   b. air temperature.
   c. the circulation and cleanliness of indoor air.
   d. the circulation and cleanliness of indoor air and air discharged to the outdoors.

4. The term "refrigeration cycle" refers to:
   a. the process by which circulating refrigerant absorbs heat in one location and moves it to another location.
   b. air conditioned motorcycles.
   c. mobile refrigeration units.
   d. the process by which refrigerant is recycled to remove impurities.

5. Heat is transferred from the indoor air to the refrigerant at the:
   a. compressor.
   b. furnace.
   c. evaporator.
   d. condenser.

6. The expansion device:
   a. raises the boiling point of the refrigerant entering the evaporator.
   b. lowers the boiling point of the refrigerant entering the evaporator.
   c. gets bigger as the temperature increases.
   d. lowers the pressure at the evaporator outlet.

7. The minimum number of formal training hours per year in an approved apprentice program is:
   a. 200.
   b. 180.
   c. there is no minimum.
   d. 144.

8. After two years in the Youth Apprentice Program a student:
   a. receives a college scholarship.
   b. gets no credit toward an apprenticeship.
   c. has completed one year of the four-year apprenticeship.
   d. has completed two years of the four-year apprenticeship.

9. In order to be resold on the open market, a refrigerant must have been:
   a. reclaimed.
   b. recycled.
   c. recovered.
   d. redecorated.

10. Releasing refrigerant to the atmosphere is:
   a. okay providing the refrigerant has been recycled.
   b. okay providing the refrigerant has been reclaimed.
   c. prohibited by federal law.
   d. prohibited in some states.

ANSWERS TO SELF CHECK REVIEW / PRACTICE QUESTIONS

Answer	Section Reference
1. a	2.0.0
2. c	2.0.0
3. d	3.0.0
4. a	4.0.0
5. c	4.0.0
6. b	4.0.0
7. d	6.0.0
8. c	6.1.0
9. a	7.0.0
10. c	7.0.0

The NCCER makes every effort to keep these manuals up-to-date and free of technical errors. We appreciate your help in this process. If you have an idea for improving this manual, or if you find an error, a typographical mistake, or an inaccuracy in the NCCER's Craft Training Manuals, please write us, using this form or a photocopy. Be sure to include the exact module number, page number, a description of the problem, and the correction, if possible. Your input will be brought to the attention of the Technical Review Committee. Thank you for your assistance.

Instructors – If you found that additional materials were necessary in order to teach this module effectively, please let us know so that we may include them in the Equipment/Materials list in the Instructor's Guide.

Write: Curriculum and Revision Department
National Center for Construction Education and Research
P.O. Box 141104
Gainesville, FL 32614-1104
Fax: 352-334-0932

Craft _________________________ Module Name _________________________

Module Number _________________ Page Number(s) _________________

Description of Problem

____

____

____

____

(Optional) Correction of Problem _______________________________

____

____

____

(Optional) Your Name and Address _____________________________

____

Trade Mathematics

Module 03102

NATIONAL
CENTER FOR
CONSTRUCTION
EDUCATION AND
RESEARCH

TRADE MATHEMATICS

Objectives

Upon completion of this module, the trainee will be able to:

1. Solve basic algebraic equations that relate to the HVAC trade.
2. Calculate volume, weight, pressure, vacuum, and temperature.
3. Construct simple geometric figures and solve basic geometry problems that relate to the HVAC trade.

Prerequisites

Successful completion of the following Task Modules is required before beginning study of this Task Module: Common Core Curricula, HVAC Module 03101.

Required Student Material

1. Student Module
2. Calculator
3. Compass
4. Ruler
5. Protractor
6. Pencil and scratch paper

Course Map Information

This course map shows all of the *Wheels of Learning* task modules in the first level of the HVAC curricula. The suggested training order begins at the bottom and proceeds up. Skill levels increase as a trainee advances on the course map. The training order may be adjusted by the local Training Program Sponsor.

Course Map: HVAC, Level 1

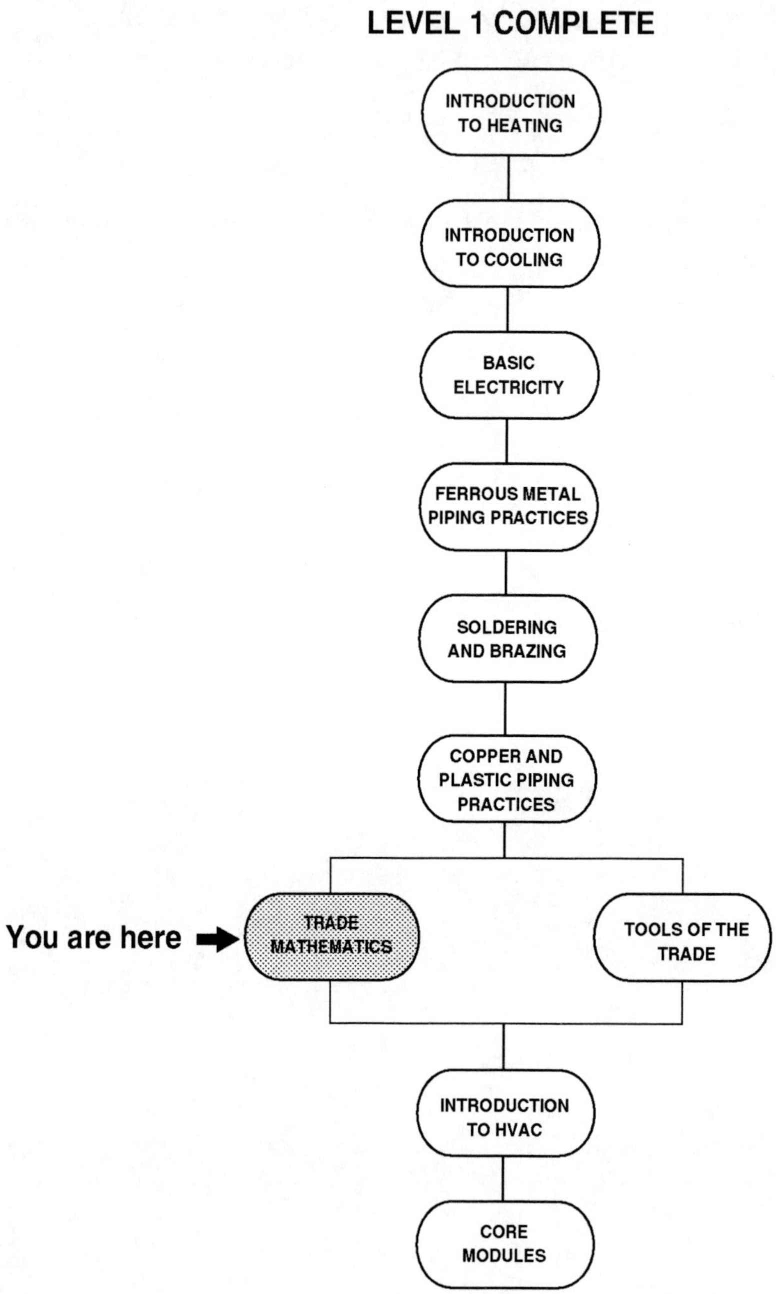

LEVEL 1 COMPLETE

TABLE OF CONTENTS

Trade Terms Introduced In This Module

Absolute pressure: The total pressure that exists in a system. Absolute pressure is expressed in pounds per square inch absolute (psia). Absolute pressure equals gauge pressure plus atmospheric pressure (14.7 lbs. per sq. in.).

Acute angle: An angle that measures less than 90°.

Adjacent angles: The angles formed by two or more rays meeting at the same vertex.

Algebra: The mathematics of defining and manipulating equations containing quantities expressed as symbols rather than numbers.

Atmospheric pressure: The standard pressure exerted on the earth's surface. Normally expressed as 14.7 pounds per square inch (psi) or 29.92 inches of mercury.

Barometric pressure: The actual atmospheric pressure at a given place and time.

Bisect: To divide into two equal portions.

Celsius: Also known as Centigrade. A system for measuring temperature that places the melting point of ice at 0°C and the boiling point of water at 100°C.

Coefficient: A multiplier; i.e., the numeral 2 in the expression 2x.

Complementary angles: Two angles having a combined total measure of 90°.

Constant: An element in an equation with a fixed value.

Ellipse: A figure bounded by a curved line, such that the sum of the distances of any point on the periphery from each of the two fixed points or foci is always a constant.

Equation: A collection of numbers, symbols, and operators connected by an equal sign (=).

Equilateral triangle: A triangle with three sides of equal length.

Fahrenheit: A system for measuring temperature that places the melting point of ice at 32°F and the boiling point of water at 212°F.

Gauge pressure: The difference between the actual pressure in a system being measured and the actual atmospheric pressure. Expressed in pounds per square inch gauge (psig).

Hexagon: A polygon with six equal sides.

Isosceles triangle: A triangle with two sides of equal length.

Obtuse angle: An angle that measures more than 90° but less than 180°.

Octagon: A polygon with eight equal sides.

Parallel: Two or more straight lines that are the same distance apart at all perpendiculars. Parallel lines do not intersect.

Parallelogram: A quadrilateral with opposite sides equal and opposite angles equal.

Pentagon: A polygon with five equal sides.

Perpendicular: A line that forms a right angle (90°) with one or more lines.

Plane geometry: The mathematics of constructing and analyzing two-dimensional representations of three-dimensional objects.

Polygon: Three or more straight lines joined in a regular pattern.

Pressure: Force per unit area. Pressure is expressed in several ways, including pounds per square inch (psi), inches of mercury (in. Hg), and inches water gauge (in. w.g.).

Quadrilateral: A polygon with four sides.

Rays: Two straight lines that meet to form an angle.

Rhombus: A quadrilateral with all sides equal and diagonal angles equal, and no angle measuring exactly 90°.

Right angle: An angle whose rays are perpendicular to one another. The measure of a right angle is always 90°.

Scalene triangle: A triangle with all three sides of different lengths.

Straight angle: An angle whose rays extend outward from a point on a straight line. The measure of a straight angle is always 180°.

Supplementary angles: Two angles having a combined total measure of 180°.

Trapezoid: A quadrilateral with two opposite sides parallel to one another.

Triangle: A polygon with three sides.

Vacuum: Any pressure that is less than the prevailing atmospheric pressure.

Variable: An element of an equation that may change in value.

Vertex: The point at which two rays cross to form an angle.

Volume: The total capacity of a three-dimensional object. Expressed in cubic units, such as cubic inches, cubic feet, or cubic yards.

1.0.0   INTRODUCTION

This module expands on the material learned in Module 00102, *Basic Math,* and applies it directly to the kinds of problems HVAC technicians encounter in the field.

2.0.0   ALGEBRA

Algebra is the mathematics of defining and manipulating equations containing symbols instead of numbers. The symbols may be either constants or variables. They are connected to each other with mathematical operators such as +, -, x and ÷.

As in many fields of mathematics, an understanding of algebra requires the knowledge of some basic rules and definitions. These definitions and rules will be introduced where required to promote the understanding necessary to become proficient in algebra.

2.1.0   DEFINITION OF TERMS

2.1.1   Mathematical Operators

+   Addition
-   Subtraction
x   Multiplication
•   Multiplication
÷   Division

2.1.2   Equations

An **equation** is a collection of numbers, symbols and mathematical operators connected by an equal sign (=). Some examples of equations are:

$$2 + 3 = 5$$

$$P = EI$$

$$\text{Volume} = L \times W \times H$$

2.1.3   Variables

A **variable** is an element of an equation that may change in value. For example, let's look at the simple equation for the area of a rectangle:

$$\text{Area} = L \times W$$

If the area is 12 and the length (L) is 6, the equation would read this way:

$$12 = 6 \times W$$

In this case, it's easy enough to see that the width (W) is equal to 2 (12 = 6 x 2). What is the width if the length is equal to 3?

$$12 = 3 \times W$$

In this case, the width is equal to 4 (12 = 3 x 4). Therefore, in the equations:

$$12 = 6 \times W$$

$$12 = 3 \times W$$

W must be considered a variable.

2.1.4   Constants

A **constant** is an element of an equation that does not change in value. For instance, in the equation:

$$2 + 5 = 7$$

2, 5, and 7 are constants. 2 will always be 2, 5 will always be 5, and 7 will always be 7, no matter what equation they may be in. Constants also refer to accepted values that represent one element of an equation and do not change from situation to situation. One of the most common constants that you will be dealing with is the value of standard atmospheric pressure, 14.7, which may be inserted into any equation that calls for it. Another common constant is Pi (π). It has a value of 3.14 and represents the ratio of the circumference to the diameter in a circle. We will discuss these constants in more detail later in this module.

2.1.5   Coefficients

A **coefficient** is a multiplier. That is, in the equation:

$$\text{Area} = L \times W$$

L is the coefficient of W. It can also be written as LW, without the multiplication sign. No multiplication symbol is required when the intended relationship between symbols and letters is clear. For example:

2x means two times x (2 is the coefficient of x)

IR means I times R (I is the coefficient of R)

Equations often include grouping symbols, such as parentheses ( ), brackets [ ], or braces { } that have a plus (+) or minus (-) coefficient which must be multiplied by each term in the group. If there is no numerical coefficient outside of the grouping symbol, it means that +1 or -1 is the coefficient.

The following rules apply to the removal of grouping symbols:

Plus Coefficient: When the grouping symbol is preceded by a plus (+) sign, simply drop the symbol. This is the same as multiplying each term within the group by +1.

$$3 + (2 - 1) = 3 + 2 - 1$$

Minus Coefficient: When the grouping symbol is preceded by a minus (-) sign, multiply each term within the group by -1, then drop the symbol. Remember that a minus times a minus becomes a plus.

$$3 - (2 - 1) = 3 - 2 + 1$$

Plus Numerical Coefficient: When the grouping symbol is preceded by a plus numerical coefficient, multiply each term within the group by the coefficient and its sign, then drop the symbol:

$$3 + 2(2 - 1) = 3 + 4 - 2$$

Minus Numerical Coefficient: When the grouping symbol is preceded by a minus numerical coefficient, multiply each term within the group by the coefficient and its sign, then drop the symbol:

$$3 - 2(2 - 1) = 3 - 4 + 2$$

2.2.0 Sequence Of Operations

Complicated equations must be solved by doing the indicated arithmetic in a prescribed sequence. This sequence is: **M**ultiply, **D**ivide, **A**dd and **S**ubtract (MDAS). For example, the following equation can result in a number of answers if the MDAS sequence is not followed:

$$3 + 3 \times 2 - 6 \div 3 =$$

To come up with the correct result, this equation must be solved in the following order:

Step 1 Multiply:

$$3 + \mathbf{3 \times 2} - 6 \div 3$$

Step 2 Divide:

$$3 + 6 - \mathbf{6 \div 3}$$

Step 3 Add:

$$\mathbf{3 + 6} - 2$$

Step 4 Subtract:

$$\mathbf{9 - 2}$$

Result:

$$7$$

There are also more complicated equations which may include several variables. "Solving" these equations means to simplify them as much as possible and, if necessary, to separate the desired variable so that it is on one side all by itself, with everything else on the other side. Problems such as these are known as algebraic expressions. When an algebraic expression appears in an equation, the MDAS sequence also applies. For example:

P = R - [5(3A + 4B) + 40L]

The parentheses represent multiplication, so they are worked on first. When working with multiple sets of parentheses or brackets, always begin by eliminating the innermost symbols first, then working your way to the outermost symbols. Thus, in the above equation, the parentheses (3A + 4B) with the coefficient of 5 are multiplied first, giving:

P = R - [15A + 20B + 40L]

The brackets also represent multiplication, so they are worked on next. The minus sign is the same as a coefficient of -1, so each term within the brackets is multiplied by -1, giving:

P = R - 15A - 20B - 40L

At this point, the equation has been simplified as much as possible, and algebra may seem no more useful now than it did in high school. However, let's see what happens when we apply this same equation to a real-life situation. Let's say that you have just installed ductwork in five identical apartments in a complex, and you want to determine the profit on the job. If we wrote this equation out longhand, it would look like this:

Profit (P) is equal to the payment received (R) minus five apartments times three pieces of one type of ductwork in each apartment at a certain cost per piece (A) plus four pieces of a second type of ductwork in each apartment at a certain cost per piece (B) minus forty hours of labor times an hourly rate (L).

It makes a lot more sense to simply write it algebraically:

P = R - [5(3A + 4B) + 40L] *or*

P = R - 15A - 20B - 40L

Now, we'll plug in numbers for the known values. Let's say that R = $1500, A = $10, B = $15, and L = $12. This results in:

P = 1500 - (15 x 10) - (20 x 15) - (40 x 12)

Multiplying, we get:

P = 1500 - 150 - 300 - 480

Result:

P = $570

There are a few simple rules that, once memorized, will help you to simplify and solve almost any equation you encounter as an HVAC technician.

Rule 1: If the same value is added to or subtracted from both sides of an equation, the resulting equation is valid. For instance, given the equation:

$$5 = 5$$

If 3 is added to each side of the equation, the resulting equation remains valid (both sides are still equal to one another).

$$5 + 3 = 5 + 3$$

$$8 = 8$$

In the same way, if we subtract the same number from both sides of an equation, the resulting equation is valid. For instance, given the equation:

$$5 = 5$$

4 may be subtracted from both sides of the equation as follows:

$$5 - 4 = 5 - 4$$

$$1 = 1$$

Moving variables from one side of an equation to another is done in the same way as when moving constants. Recall that when an equation is "solved" for one particular variable, that means that the variable should be on one side of the equation by itself. Consider the following pressure equation that we will study later on in this module:

$$\text{Absolute Pressure} = \text{Gauge Pressure} + 14.7$$

To solve this equation for Gauge Pressure, 14.7 must be moved to the other side of the equation with Absolute Pressure. To do so, we'll subtract 14.7 from both sides of the equation:

$$\text{Absolute Pressure} - 14.7 = \text{Gauge Pressure} + 14.7 - 14.7$$

It should be clear that the +14.7 and -14.7 on the right cancel each other out and we are left with:

$$\text{Absolute Pressure} - 14.7 = \text{Gauge Pressure} \; or$$

$$\text{Gauge Pressure} = \text{Absolute Pressure} - 14.7$$

The equation has been solved for Gauge Pressure. If we wanted to take this new equation, Gauge Pressure = Absolute Pressure - 14.7, and solve it for Absolute Pressure again, we would simply add 14.7 to each side:

$$\text{Gauge Pressure} + 14.7 = \text{Absolute Pressure} - 14.7 + 14.7$$

Again, the +14.7 and -14.7 on the right cancel each other out and we are left with:

Gauge Pressure + 14.7 = Absolute Pressure *or*

Absolute Pressure = Gauge Pressure + 14.7

Rule 2: If both sides of an equation are multiplied or divided by the same value, the resulting equation is valid. For this rule, we'll examine the equation for Ohm's Law, which you will use often in your work as an HVAC technician, and will be studying in detail in *Basic Electricity*. This equation is as follows:

$$E = IR$$

(E = voltage, I = current, and R = resistance.) If you know the voltage (E) and the current (I), but need to find the resistance (R), how do you rearrange the equation? To solve this equation for R, I must be moved to the other side of the equation with E. To do so, we'll divide both sides by I:

$$\frac{E}{I} = \frac{IR}{I}$$

The two I's on the right cancel each other out and we are left with:

$$\frac{E}{I} = R \quad or$$

$$R = \frac{E}{I}$$

The equation has been solved for resistance. If we wanted to take this new equation, R = E/I, and solve it for voltage again, we would simply multiply each side by current:

$$I \times R = \frac{E}{I} \times I$$

The two I's on the right cancel each other out and we are left with:

$$IR = E \quad or$$

$$E = IR$$

Rule 3: Like terms may be added and subtracted in a manner similar to constant numbers. For example, given the equation:

$$2x + 3x = 15$$

The like terms (2x and 3x) may be added directly:

$$5x = 15$$

Dividing both sides of the equation by 5, we have:

$$\frac{5x}{5} = \frac{15}{5}$$

$$x = 3$$

These rules may be used repeatedly until an equation is in the desired form.

Study Example

Let's take a look at one of the pressure equations we'll be using later in the module.

$$P = \frac{hd}{144}$$

(P = pressure, h = height, and d = density.) Solve the equation for density (d).

Step 1   Remove the fraction by multiplying both sides by 144.

$$P \times 144 = \frac{hd}{144} \times 144$$

$$144P = hd$$

Step 2   Divide both sides by h.

$$\frac{144P}{h} = d \ \ or$$

$$d = \frac{144P}{h}$$

The equation has been solved for density (d).

3.0.0   MEASUREMENTS

There are several types of measurements that you will regularly be called upon to perform in your work as an HVAC technician. For example, volume calculations are necessary in air system design, weight calculations are required to compute system charges, and pressure and temperature calculations are needed to troubleshoot heating and air conditioning systems. It is important that you understand and are able to use these measurements and their associated equations. Careful measurements are critical to system maintenance and troubleshooting.

3.1.0   VOLUME MEASURE

Volume is expressed in cubic units, such as cubic inches, cubic feet, or cubic yards. When you compute the volume of a rectangular solid, the volume in cubic units is the product of its length times its width times its height (L x W x H). You may use this basic formula to calculate the volume of a stack of insulation, for example, or the volume of air inside a duct.

Objects such as cones, cylinders, pyramids, and the unusual objects you will find in duct runs all have special formulas for calculating their volumes. These may require more complicated formulas than are necessary for you to learn at this time. Most math, physics, and many technical textbooks will provide these formulas, if you find you need them.

The two most common cubic conversions are:

$$1,728 \text{ cubic inches} = 1 \text{ cubic foot}$$

$$27 \text{ cubic feet} = 1 \text{ cubic yard}$$

Study Examples

1. A container is 1 yard long, 25 inches wide and 30 inches high. How many cubic inches are contained in its volume?

    Volume = 1 yd. x 25" x 30"

    First, convert to a common unit of measure: 1 yard = 36 inches.

    Volume = 36" x 25" x 30"
            = 27,000 cu. in.

2. How many cubic feet are contained in the volume of the above example?

    27,000 cu. in ÷ 1728 cu. in./ft. = 15.625 cu. ft.

3.2.0   WEIGHT MEASURE

Many weights in the English system depend upon the material being weighed. The most common weight measurement uses ounces, pounds and tons as units of measure.

$$16 \text{ ounces} = 1 \text{ pound}$$

$$2000 \text{ pounds} = 1 \text{ ton}$$

Study Examples

1. Convert 12.45 pounds into ounces.

    12.45 lbs. = 12.45 x 16 oz./lb.
               = 199.2 ounces

2. Convert 144 ounces into tons.

    144 oz. = 144 ÷ 16 oz./lb. ÷ 2000 lbs./ton
            = .0045 tons

3.3.0   PRESSURE

Both liquids and gases are capable of exerting a pressure. Gases are compressed under pressure and they expand when the pressure is lowered. Liquids are generally considered to be incompressible.

Pressure is defined as force per unit area. The most common units for expressing pressure include:

> Pounds per square inch: psi
>
> Pounds per square foot: psf (lb./ft.^{2})

The formula for pressure is:

$$\text{Pressure} = \frac{\text{Force}}{\text{Area}}$$

Study Example

A tank of water 1 foot square (144 square inches) is filled with water. (One cubic foot of water = approx. 62.4 lb.). What is the pressure being exerted in pounds per square inch on the bottom of the tank?

$$\begin{aligned}
\text{Pressure} &= \frac{\text{Force}}{\text{Area}} \\[6pt]
&= \frac{62.4}{144} \\[6pt]
&= 0.433 \text{ psi}
\end{aligned}$$

3.3.1   Absolute Pressure

The standard **atmospheric pressure** exerted on the surface of the earth is 14.696 psi taken at sea level with the air at 70°F. For most practical applications, this is often rounded off to 14.7 psi. Atmospheric pressure is also expressed as 29.9213 inches of mercury, which is equal to 14.696 psi. The rounded off value for inches of mercury (in. Hg) is 29.92.

Actual weather conditions usually cause a slight variation in the atmospheric pressure. The actual atmospheric pressure is known as the **barometric pressure**. It is usually ignored in measuring hydraulic machinery pressure, but it cannot be ignored in power plant work and when dealing with the low pressures generated by fans and blowers.

Most gauges measure the difference between the actual pressure in the system being measured and the atmospheric pressure. Pressure measured by gauges is called **gauge pressure** (psig). The total pressure that exists in the system is called the **absolute pressure** (psia). Absolute pressure is equal to gauge pressure plus the atmospheric pressure, either the standard, 14.7, or the actual pressure, if measured. In other words:

Absolute Pressure (psia) = Gauge Pressure (psig) + 14.7

Study Example

A steam boiler pressure gauge reads 295 psig. Find the absolute pressure.

$$Psia = psig + 14.7$$
$$= 295 + 14.7$$
$$= 309.7$$

Boiler pressure is sometimes specified in terms of atmospheres, where one atmosphere is equal to 14.696 (14.7) psi (*see Table 1*).

Multiply	By	To Obtain
Atmospheres	14.7	Pounds per square inch
Pounds per square inch	0.0680	Atmospheres
Pounds per square inch	27.68	Inches of water (H_2O)
Pounds per square inch	2.31	Feet of water (H_2O)
Pounds per square inch	2.04	Inches of mercury (Hg)
Inches of water	0.0361	Pounds per square inch
Feet of water	0.433	Pounds per square inch
Inches of mercury	0.491	Pounds per square inch

Table 1. Pressure Conversions

3.3.2 Static Head Pressure

Municipal water systems usually define pressure in terms of the static head or height in feet from the use point to elevated water reservoirs. Gauge pressure can be converted to static head pressure using the formula:

$$P = \frac{hd}{144}$$

Where:

P = Pressure (psig)

h = Height in feet (head)

d = Density in lb. per cu. ft. (62.43 for water)

144 = Used to convert square feet into square inches

For example, let's say that the water level in a reservoir is maintained at 150 feet above use level. What is the water pressure at use level resulting from this head?

$$P = \frac{hd}{144} = \frac{150 \times 62.43}{144} = 65 \text{ psig}$$

Other terms are necessary to measure extremely low pressures, such as those developed by blowers and fans. These pressures are often measured in inches of water (in. H_2O). Sometimes it is necessary to convert from one measure to another. Consult *Table 1* to solve the following problems.

Study Examples

1. A steam generator operates at a pressure of 320 atmospheres. What is this pressure in terms of psig?

    psig = 320 x 14.7 (from *Table 1*)

      = 4704 psig

2. Blower and fan pressures are usually measured in inches of water because small differences in pressure can be readily detected. Due to the low pressures involved, the exact atmospheric pressure is usually measured to determine the actual discharge pressure. Calculate the discharge pressure in psia if the measured discharge pressure of a blower is 56.55 in. H_2O and the barometric pressure is 28.49.

Step 1 Find the blower discharge pressure in psig:

  psig = in. H_2O x .0361 (from *Table 1*)

    = 56.55 x .0361

    = 2.041455

Step 2 Find the actual atmospheric pressure in psi.

psi = in. Hg x .491 (from *Table 1*)

= 28.49 x .491

= 13.98859

Step 3 Find the absolute pressure of the blower discharge.

Absolute Pressure = Gauge Pressure + Actual Atmospheric Pressure

= 2.041455 + 13.98859

= 16.030055 or approx. 16.03 psia

3.4.0   VACUUM

For computational purposes, a **vacuum** is any pressure that is less than the prevailing atmospheric pressure. It is usually measured in terms of inches of mercury (Hg). Actual barometric pressure must always be determined in measuring a vacuum. Absolute pressure in a vacuum is calculated using the following formula:

Absolute Vacuum Pressure = Barometric Pressure - Vacuum Gauge Reading

Study Example

A vacuum gauge attached to a line reads 17.2 in. Hg. The barometric pressure reads 29.85 in. Hg. What is the absolute pressure in the line?

Absolute Vacuum Pressure = Barometric Pressure - Vacuum Gauge Reading

= 29.85 - 17.2

= 12.65 in. Hg

3.5.0   TEMPERATURE MEASUREMENT

The measurement of temperature is of great importance in the HVAC industry. Unfortunately, there are a number of temperature measurement systems in use. For our purposes, the two most common systems will be discussed.

Most English-speaking countries use the **Fahrenheit** system. This system places the freezing point of water at 32°F (the melting point of ice) and the boiling point of water at 212°F. European countries use the **Celsius** or Centigrade system which places the melting point of ice at 0°C and the boiling point of water at 100°C. Conversions are made using one of the following formulas:

Degrees Celsius = 5/9 x (degrees Fahrenheit - 32 degrees)

Degrees Fahrenheit = (9/5 x degrees Celsius) + 32 degrees

Study Examples

1. Convert 140°F to degrees Celsius.
   Degrees Celsius = 5/9 x (degrees Fahrenheit - 32 degrees)
   = 5/9 x (140 - 32 degrees)
   = 60°C

2. Convert 10°C to degrees Fahrenheit.
   Degrees Fahrenheit = (9/5 x degrees Celsius) + 32 degrees
   = (9/5 x 10) + 32 degrees
   = 50°F

4.0.0   GEOMETRY

Three-dimensional bodies are known as geometric solids. While you naturally work with three-dimensional objects, the usual way of communicating about it on paper is by drawing two-dimensional representations of those objects. For these representations, called "constructions," you'll be using **plane geometry**.

The following sections describe the basic geometric concepts, definitions, and relationships you'll need to understand in order to work with two-dimensional constructions as an HVAC technician. You will find that the geometry used in HVAC work has its greatest applications in the field of sheet metal layout. It is used to determine angles for transition fittings, radius elbows, and many other types of layout challenges.

4.1.0   THE POINT

A point indicates a position only. It has no dimensions, such as length, width, or height. A point is an origin or beginning, such as the starting place for a circle. Two points are necessary to define a line and three points may define a flat surface, or plane, in space. Lines and planes naturally have many other points as well.

4.2.0   THE LINE

A line has only one dimension: length (*Figure 1*). A straight line is defined as the shortest distance between two points. A line with no end points has an infinite length. A line that begins at one point and ends at another has a finite length.

A line having the same direction throughout its length is called a straight line. A broken line is a series of connected straight lines extending in different directions. A line that continuously changes direction is a curved line.

A horizontal line refers to a straight line that is level with the horizon. For practical purposes, this means a line going left-to-right on the page in front of you. A vertical line extends up from the horizon, or in a top-to-bottom direction on your paper.

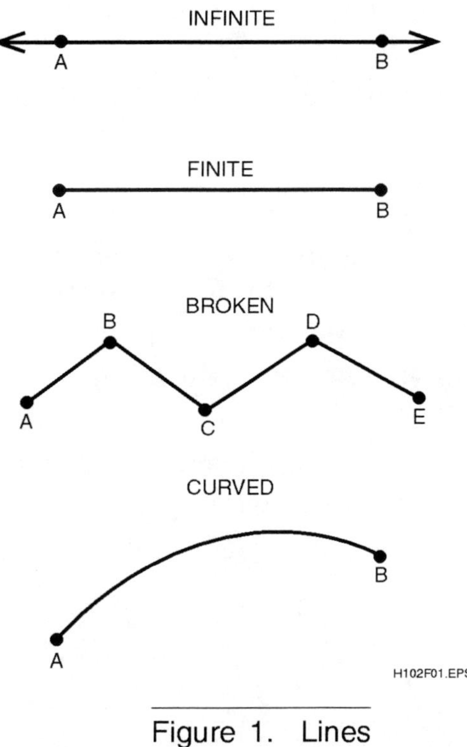

Figure 1. Lines

A line that forms a right angle (90°) with one or more lines is said to be **perpendicular** to those lines (*Figure 2*).

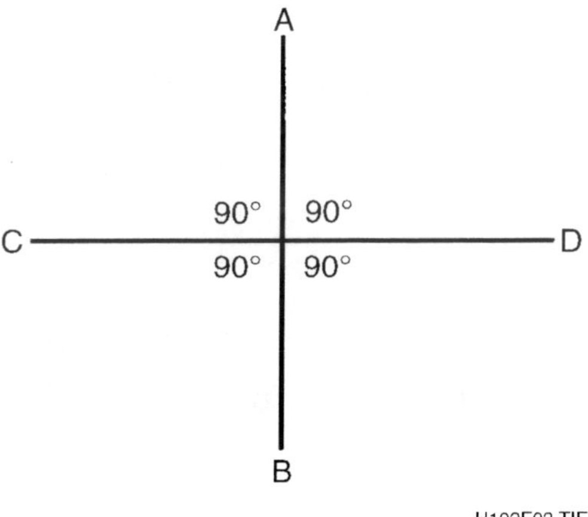

Figure 2. Perpendicular Lines

The distance from a point to a line is the measure of the perpendicular line drawn from that point to the line. Two or more straight lines that are the same distance apart at all perpendiculars are said to be **parallel** (*Figure 3*). Note that parallel lines do not intersect.

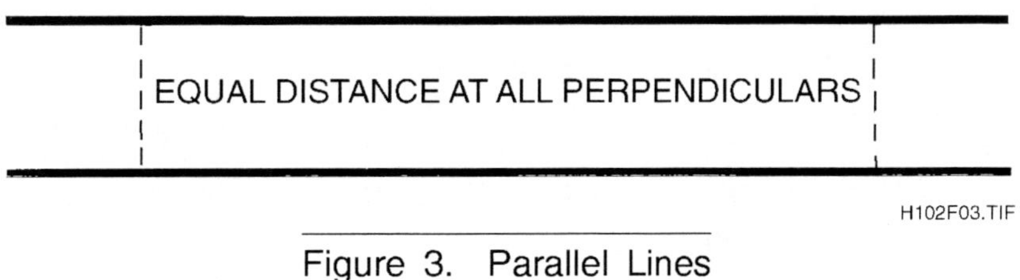

Figure 3. Parallel Lines

4.3.0 THE CIRCLE

As shown in *Figure 4*, a circle is a finite curved line that connects with itself and has these other properties:

1. All points on a circle are the same distance (equidistant) from the point at the center.
2. The distance (radius) from the center to any point on the curved line is always the same.
3. The total measure of all angles formed by all consecutive radii equals 360 degrees.
4. The diameter passes through the center and is equal to twice the distance of the radius. It is also the longest possible straight line that can be constructed within the circle.
5. The circumference of the circle is the distance around the outside of the circle. It can be determined using the equation: Circumference = πd, where π is equal to 3.14 and "d" is the diameter. This equation is helpful when you need to determine the size of a pipe liner or wrap the outside of piping with insulation.

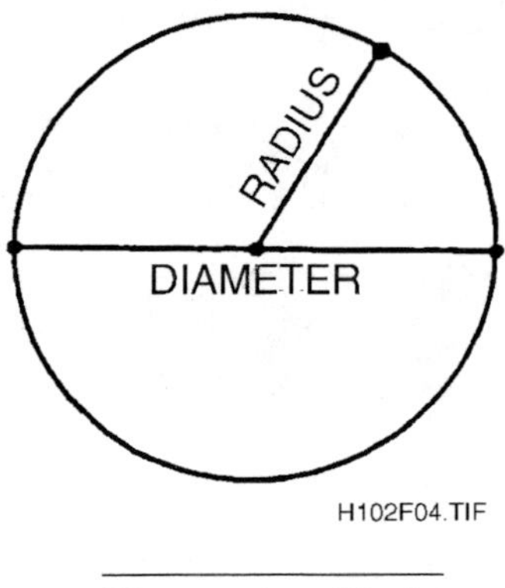

Figure 4.   Circle

4.4.0 ANGLES

Two straight lines meeting at a point, called the **vertex**, form an angle (*Figure 5*). The two lines are the sides of the angle. These lines are also called the **rays** of the angle.

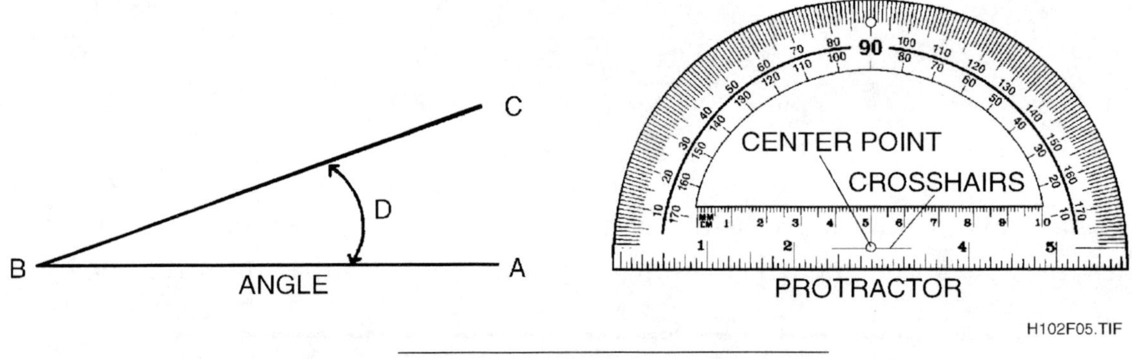

Figure 5.   Angle And Protractor

There are two ways to identify angles. One is to assign a letter to the angle, such as angle D shown in *Figure 5*. This is written: <D. The other way is to name the two end points of the rays and put the vertex letter between them, e.g., <ABC.

When you show the angle measure in degrees, it should be written inside the angle, if possible. If the angle is too small to show the measurement, you may put it outside of the angle and draw an arrow to the inside. Note that degrees can also be expressed using the degree symbol (e.g., 20°).

Angles may be measured using a protractor (*Figure 5*). This small instrument looks like a half-circle with a scale reading along the curve. Place the protractor to align your angle vertex with the protractor's center point and then align one ray with the protractor crosshairs. The other ray will point to an angle measure on the scale along the curve. That measure is the measure of your angle.

Right Angle: This angle has rays that are perpendicular to one another (*Figure 6*). The measure of this angle is always 90°.

Straight Angle: This angle doesn't look like an angle at all (*Figure 6*). The rays of a straight angle lie in a straight line, and the angle measures 180°.

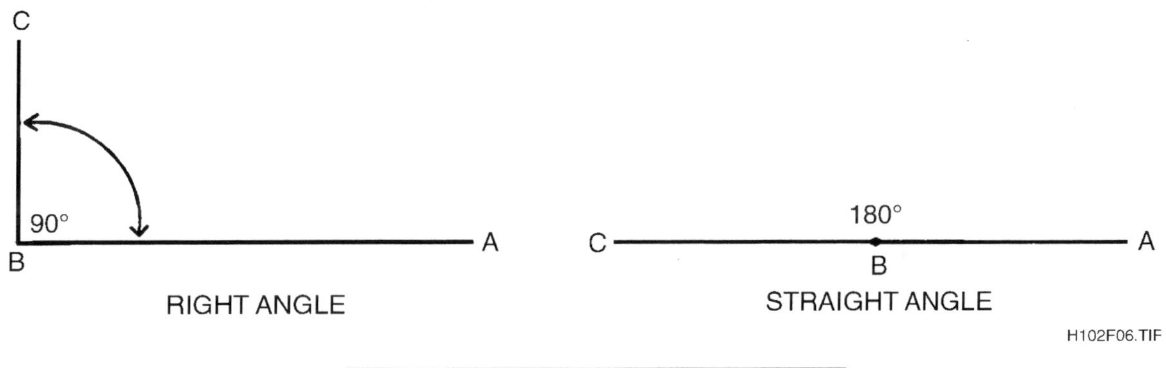

Figure 6.   Right And Straight Angles

Acute Angle: An angle less than 90° (*Figure 7*).

Obtuse Angle: An angle greater than 90° but less than 180° (*Figure 7*).

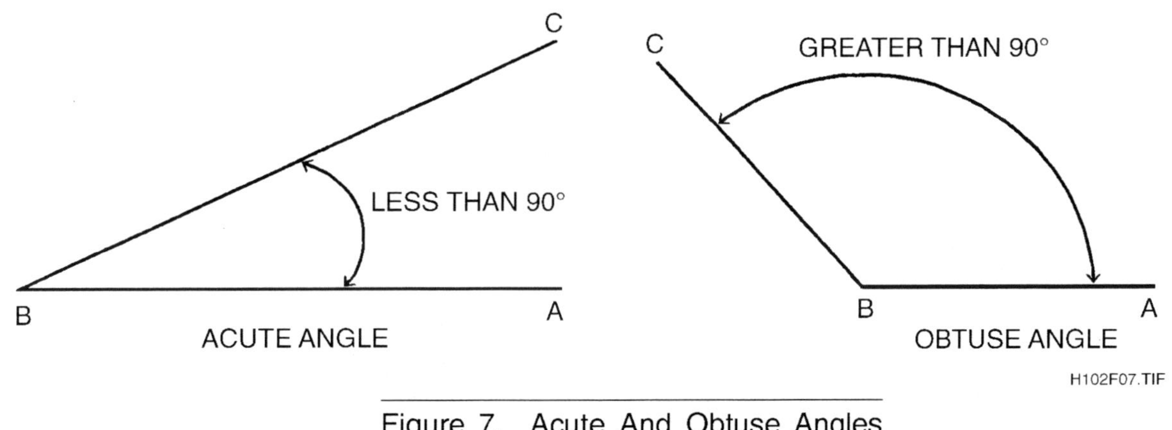

Figure 7.   Acute And Obtuse Angles

Adjacent Angles: When three or more rays meet at the same vertex, the angles formed are said to be adjacent to one another. In *Figure 8,* the angles <ABC and <CBD are adjacent angles. The ray BC is said to be common to both angles.

Complementary Angles: Two adjacent angles that have a combined total measure of 90°. In *Figure 8,* <ABC is complementary to <CBD.

Supplementary Angles: Two adjacent angles that have a combined total measure of 180°. In *Figure 8,* <ABC is supplementary to <CBD.

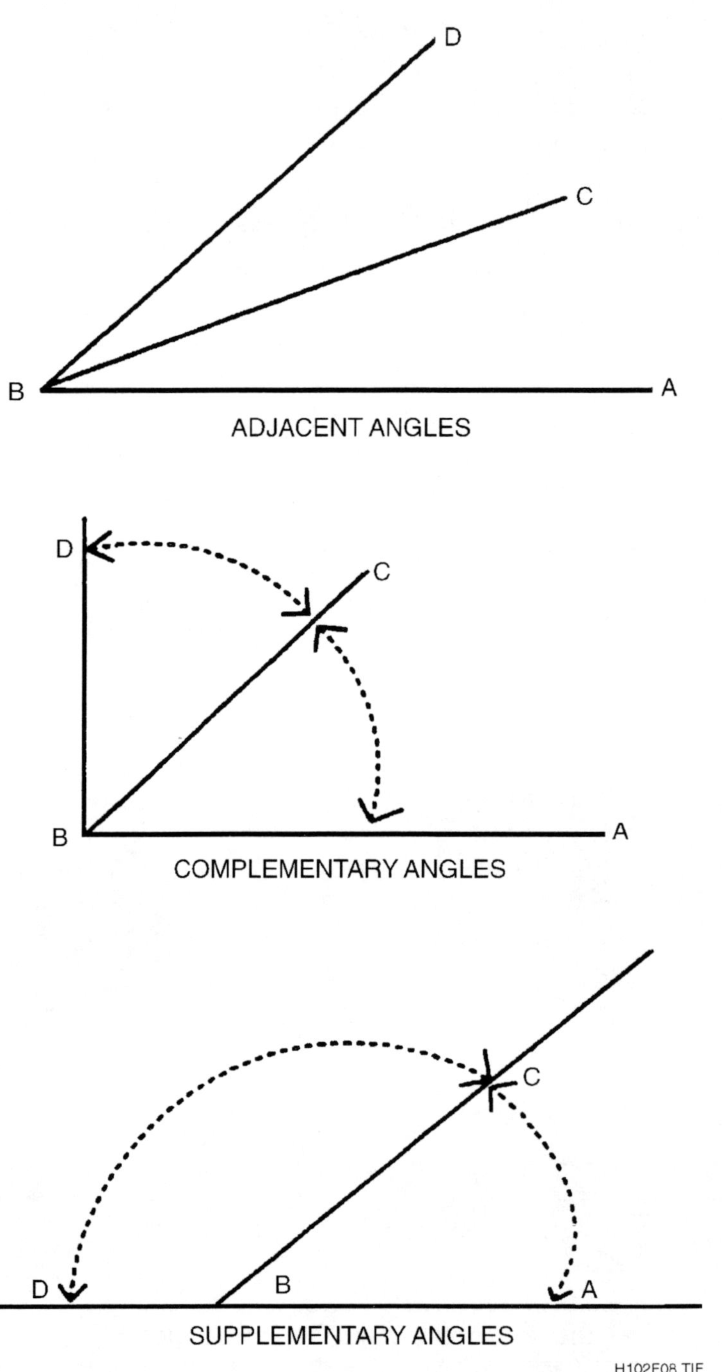

Figure 8.   Adjacent, Complementary, And Supplementary Angles

4.5.0   POLYGONS

Three or more straight lines joined in a regular pattern make up a **polygon**. Some of the most familiar polygons have more common names which generally refer to their number of sides. When all sides have equal length and all internal angles are equal, the figure is said to be a "regular" polygon.

4.5.1   Triangles

A polygon with three sides is called a **triangle**. Three different types of triangles are shown in *Figure 9*. A regular polygon with three equal sides is known as an **equilateral** triangle. Types of irregular triangles are the **isosceles** (having two sides of equal length) and the **scalene** (having all sides of unequal length or $\neq$).

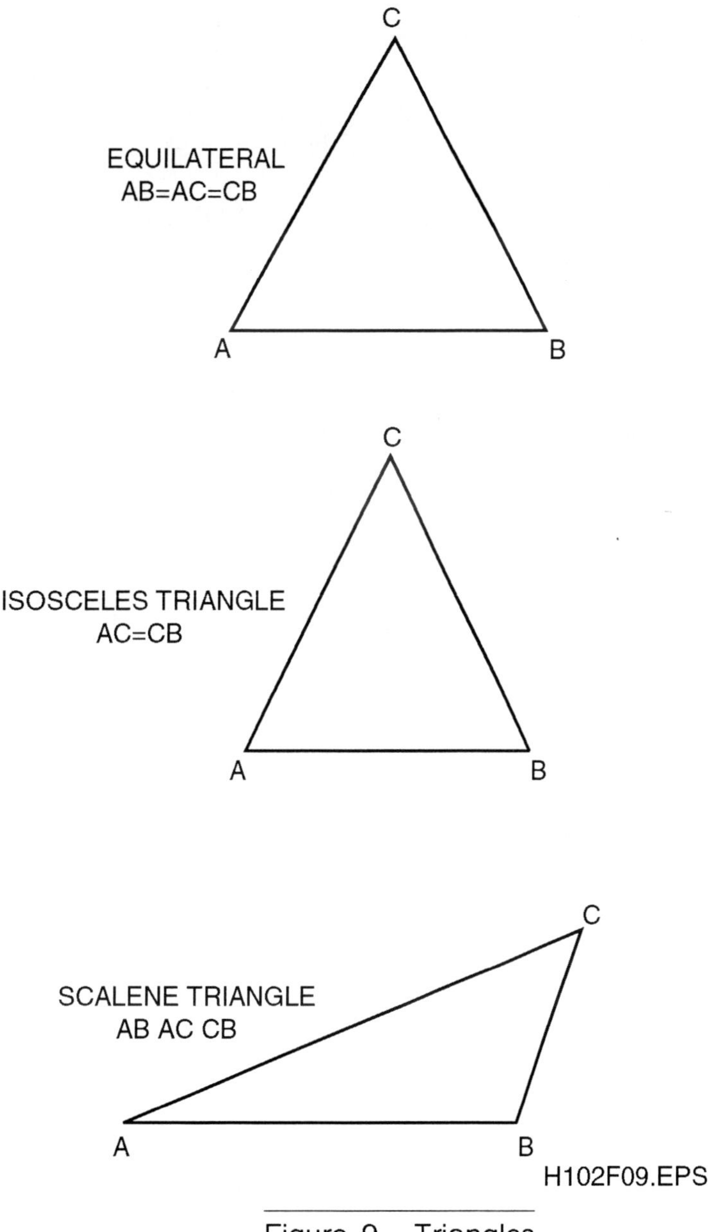

Figure 9.   Triangles

Triangles are also classified according to the angles they contain (*Figure 10*). If one of the three internal angles is 90°, the figure is called a right triangle.

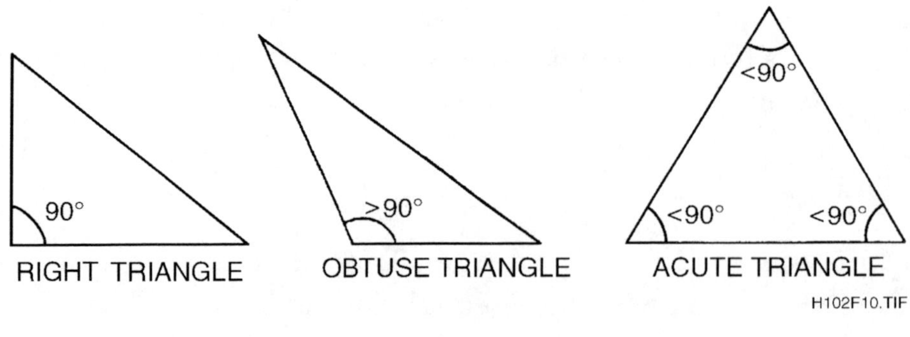

Figure 10.   Right, Obtuse, And Acute Triangles

If one of the three internal angles is greater than 90°, the figure is called an obtuse triangle.

If each of the interior angles is less than 90°, the figure is called an acute triangle. Both acute and obtuse triangles are also known as "oblique" triangles.

The sum of the three interior angles of any triangle is always equal to 180°. This is helpful to remember whenever you know two angles of a triangle and need to calculate the third.

4.5.2   Quadrilaterals

A polygon with four sides is called a **quadrilateral.** Common quadrilaterals are shown in *Figure 11.*

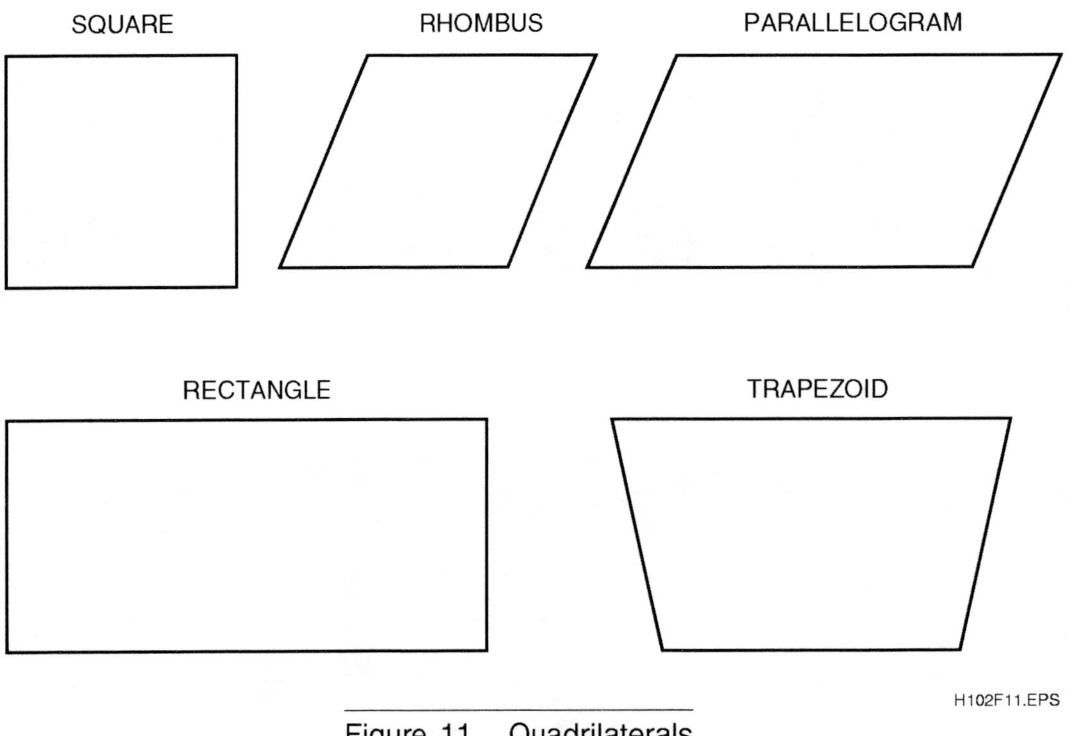

Figure 11.   Quadrilaterals

Squares have all four sides and all interior angles equal. In a square, each interior angle must be 90 degrees.

When all four sides are equal but all interior angles are not, the figure is called a **rhombus**. In a rhombus, each angle must be equal to the one diagonally opposite. None of the angles can be 90 degrees.

Irregular quadrilaterals are many and varied. In a **parallelogram**, opposite sides and opposite angles are equal. Note that the parallelogram is similar to the rhombus. A rhombus is an equilateral parallelogram.

A rectangle has all interior angles equal (90° each) and may or may not have all four sides equal. A square is an equilateral rectangle.

A **trapezoid** has two opposite sides parallel (called "bases") but nothing else needs to be equal. Sometimes, however, the two non-parallel sides might have equal lengths, which could also produce internal sets of equal angles. But a trapezoid is generally an irregular figure.

4.5.3   Pentagon

The **pentagon** is a figure with five equal sides (*Figure 12*). Also note that for this to happen, the internal angles must all be equal as well.

4.5.4   Hexagon

The **hexagon** is a polygon that has been known since ancient times. It has six equal sides, again with all internal angles equal (*Figure 12*). This is the figure you generally associate with honeycombs.

4.5.5   Octagon

An eight-sided polygon is an **octagon** (*Figure 12*). Notice again that all eight sides and internal angles are equal. This figure will be most easily recognized as you drive your car. It's a stop sign!

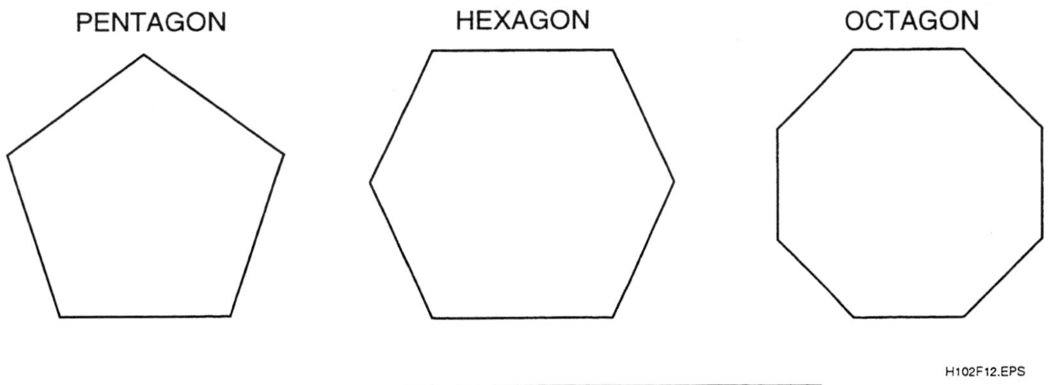

Figure 12.   Pentagon, Hexagon, And Octagon

4.5.6   Ellipse

The **ellipse** (*Figure 13*) is a curved figure. In an ellipse, the sum of the distances of any point on the periphery from each of the two fixed points (called the "foci") is always a constant. This means that every sum measured in this way will be the same.

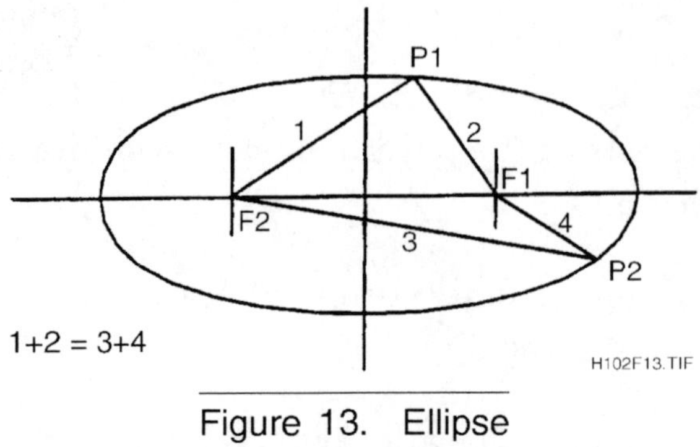

Figure 13.   Ellipse

The ellipse is a regular curve related to the circle. What is of most interest to HVAC technicians is that the shape of a mitered pipe section is an ellipse. Only cuts made perpendicular to the pipe will produce a perfect circle. Any angular cut will make an ellipse.

4.6.0   SIMPLE GEOMETRY PROBLEMS

When you study the following problems, you should draw each required construction. Each step in the construction process should be completed before proceeding to the next.

4.6.1   Erecting A Perpendicular Bisector From A Straight Line

To **bisect** is to divide into two equal portions. A perpendicular bisector devides a line using a second line at a right angle to the first. To erect a perpendicular bisector, follow the sequence described below.

Step 1   Place the point of your compass on point A and extend the compass to a length longer than the distance from A to B. Then swing the compass and strike an arc both above and below the line (*Figure 14*).

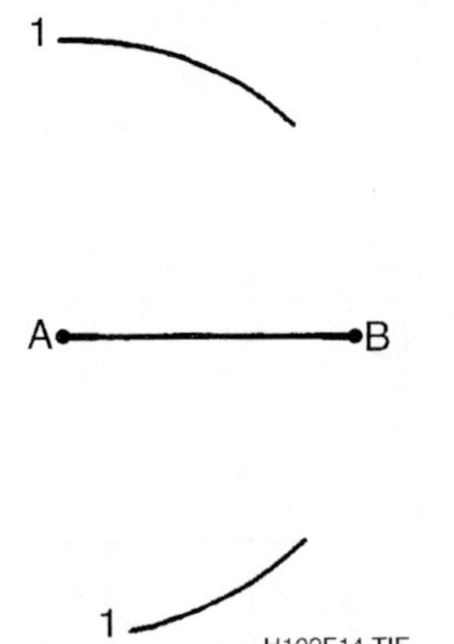

Figure 14.   Erecting A Perpendicular Bisector From A Straight Line (Step 1)

HVAC TRAINEE TASK MODULE 03102

Step 2   Move the point of the compass to point B, making sure you do not change the compass setting. Then strike another arc as before, both above and below the line (*Figure 15*).

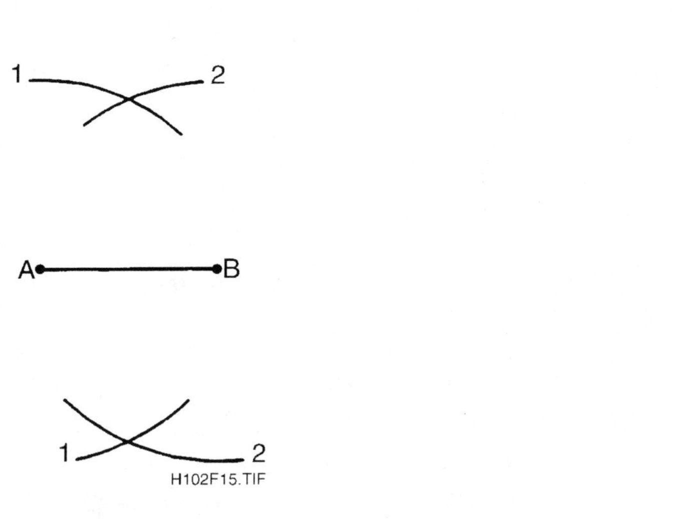

Figure 15.   Erecting A Perpendicular Bisector From A Straight Line (Step 2)

Step 3   Finally, draw a line connecting each point where the arcs intersect (*Figure 16*). The new line, CD, is perpendicular to line AB. It passes through line AB at point E, the center point of AB.

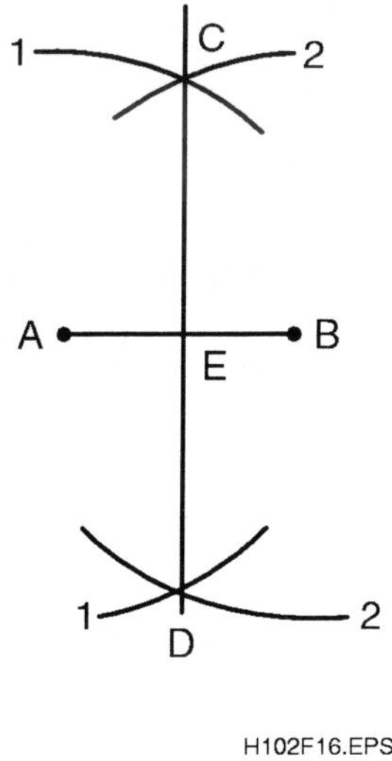

Figure 16.   Erecting A Perpendicular Bisector From A Straight Line (Step 3)

4.6.2 Erecting A Perpendicular Near The End Of A Given Line

To erect a perpendicular near the end of a given line, follow the sequence described below.

Step 1 Near the end of given line AB, locate a point C about $\frac{1}{2}$ inch from B. The perpendicular is to be erected from point C (*Figure 17*).

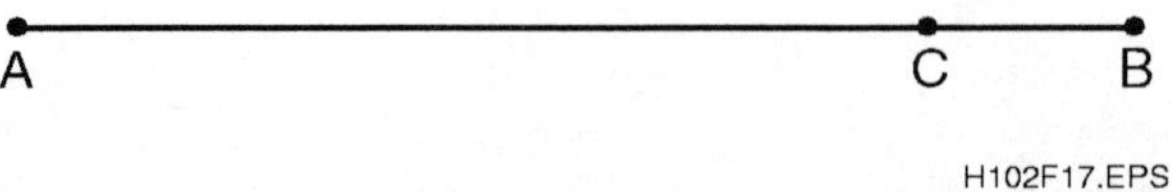

Figure 17.   Erecting A Perpendicular Near The End Of A Given Line (Step 1)

Step 2 With C as your center and using any convenient radius, use your compass to draw the arc from point 1 to point 2 (*Figure 18*).

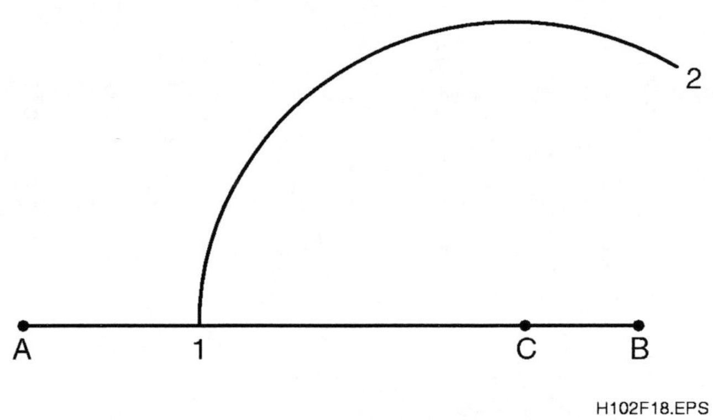

Figure 18.   Erecting A Perpendicular Near The End Of A Given Line (Step 2)

Step 3 Using this same radius, place your compass point on 1, and mark point 3. Then place the compass on 3 and mark point 4 (*Figure 19*).

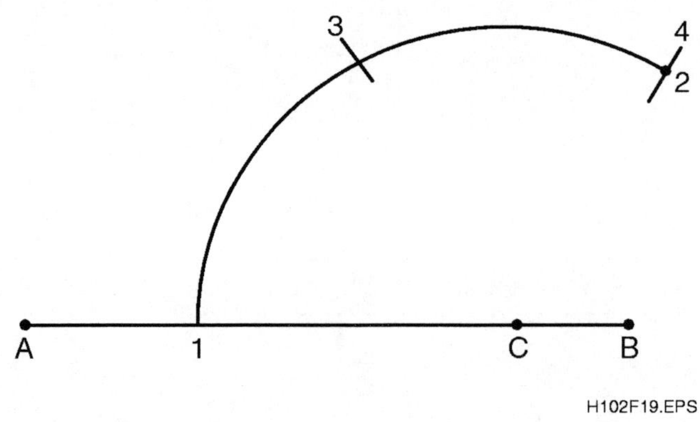

Figure 19.   Erecting A Perpendicular Near The End Of A Given Line (Step 3)

HVAC TRAINEE TASK MODULE 03102

Step 4   Using any radius with points 3 and 4 as centers, describe arcs 5 and 6 intersecting each other at point 7 (*Figure 20*).

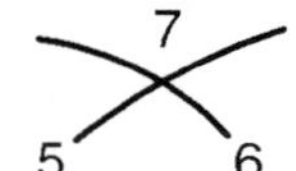
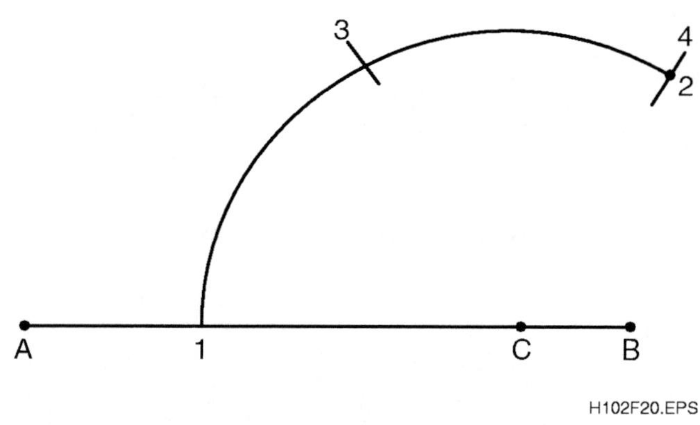

Figure 20.   Erecting A Perpendicular Near The End Of A Given Line (Step 4)

Step 5   Draw a line from C through point 7. This produces the perpendicular at given point C (*Figure 21*).

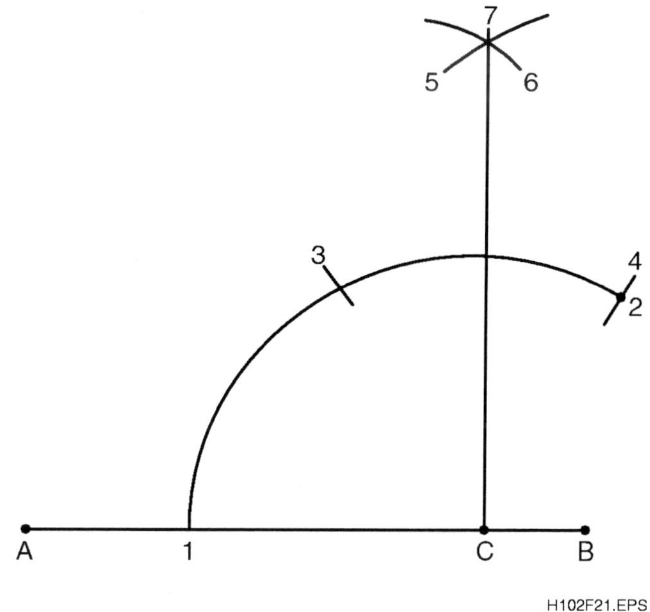

Figure 21.   Erecting A Perpendicular Near The End Of A Given Line (Step 5)

4.6.3   Erecting A Perpendicular At The End Of A Given Line

To erect a perpendicular at the end of a given line, follow the sequence described below.

Step 1   Set the compass point on A of the given line AC (*Figure 22*).

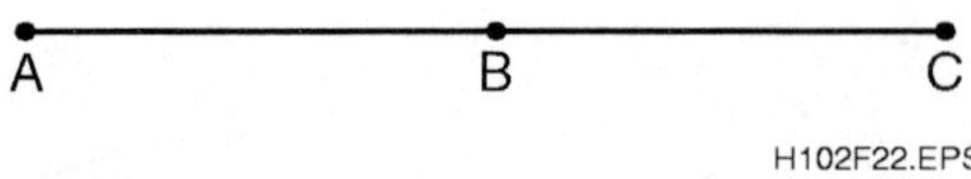

Figure 22.   Erecting A Perpendicular At The End Of A Given Line (Step 1)

Step 2   With any radius, describe arc 2 coming up from the line at any point, such as B (*Figure 23*).

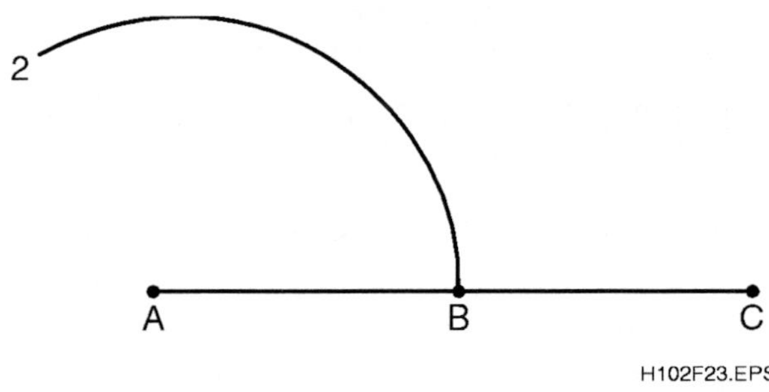

Figure 23.   Erecting A Perpendicular At The End Of A Given Line (Step 2)

Step 3   On B, use the radius AB to describe arc 3 intersecting arc 2 at point D (*Figure 24*).

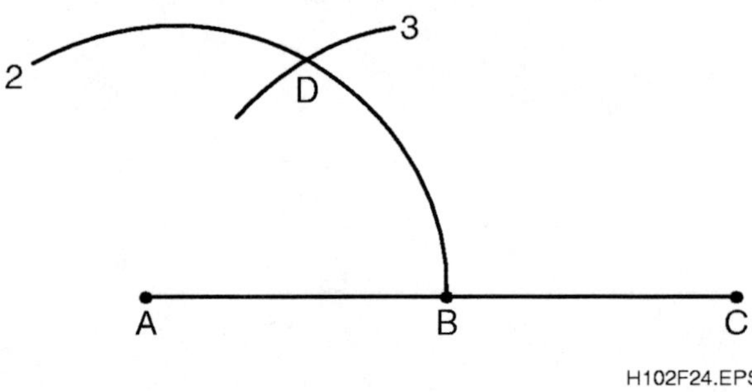

Figure 24.   Erecting A Perpendicular At The End Of A Given Line (Step 3)

Step 4   Now connect D and B with a straight line as shown in *Figure 25*.

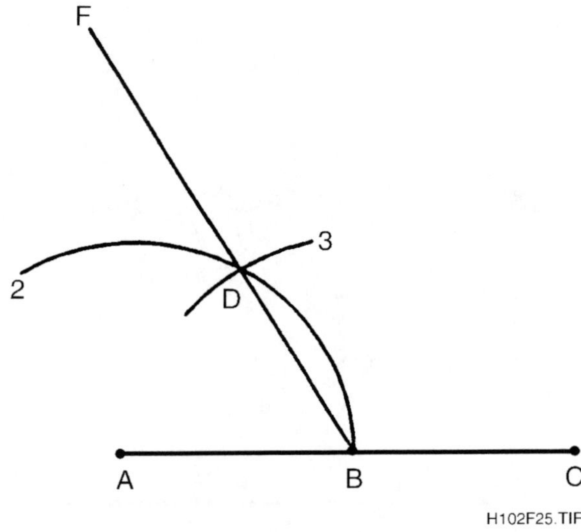

Figure 25.   Erecting A Perpendicular At The End Of A Given Line (Step 4)

Step 5   Using the radius AB, strike another arc, arc 4, from point D to intersect the straight line at point E (*Figure 26*). Note that radius AB = BD = DE.

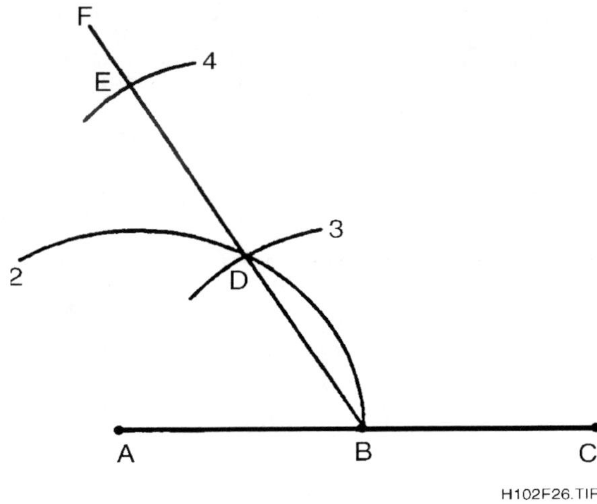

Figure 26.   Erecting A Perpendicular At The End Of A Given Line (Step 5)

Step 6 Finally, connect E to end point A, which will give you the perpendicular required (*Figure 27*).

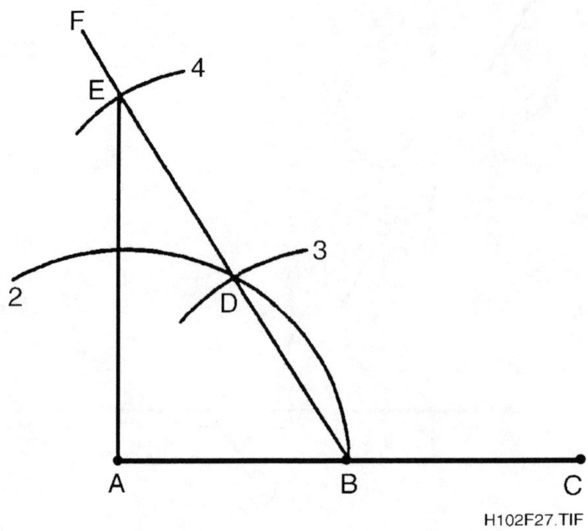

Figure 27. Erecting A Perpendicular At The End Of A Given Line (Step 6)

4.6.4 Bisecting Lines

To bisect a line, follow the sequence described below.

Step 1 To bisect a straight line AB, or the arc AB, let A and B be the center points for your compass (*Figure 28*).

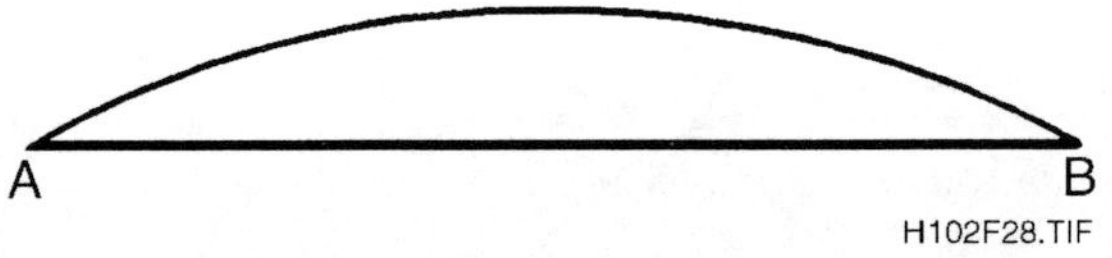

Figure 28. Bisecting Lines (Step 1)

Step 2 Using any radius greater than one-half of the line or arc, first place the compass point at A and describe arc 1 above and below the line/arc as shown in *Figure 29*.

HVAC TRAINEE TASK MODULE 03102

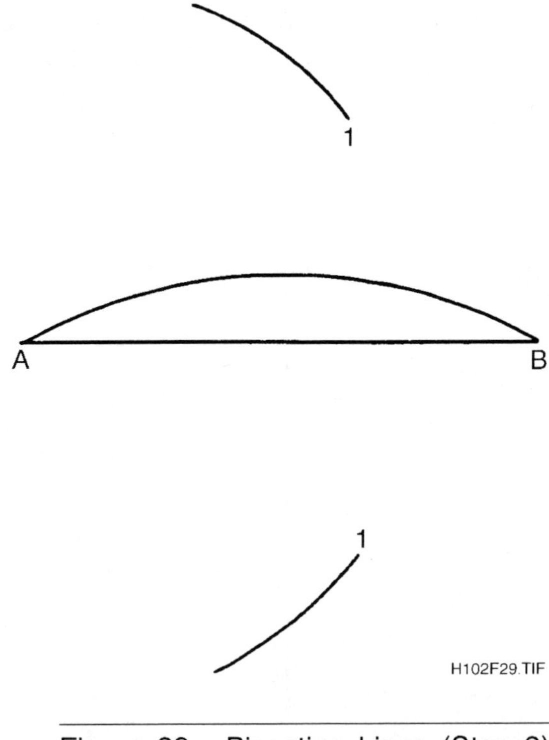

Figure 29.   Bisecting Lines (Step 2)

Step 3   Without changing the compass setting, place the point at B and describe arc 2 above and below the line/chord as shown in *Figure 30*.

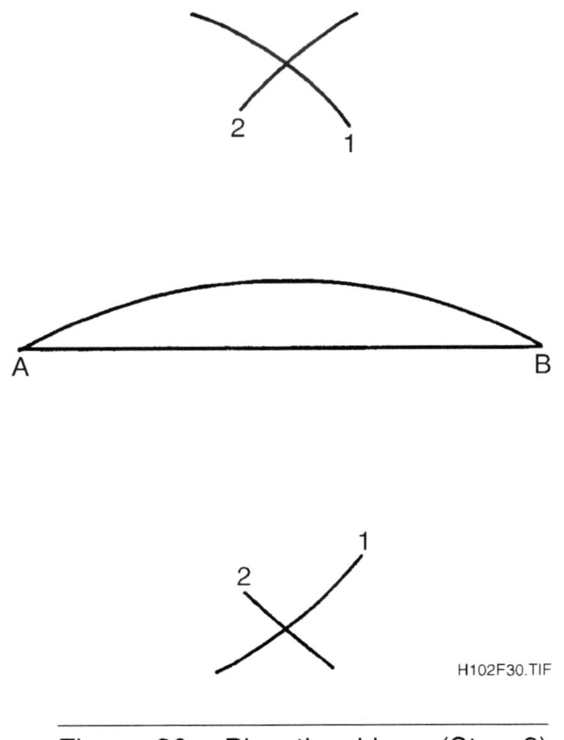

Figure 30.   Bisecting Lines (Step 3)

Step 4 Through both intersecting points of arcs 1 and 2, draw a straight line through line AB or arc AB (*Figure 31*). This line now evenly bisects both the line and the arc. It means, geometrically, that with midpoints C and D, line AC equals line CB or arc AD equals arc DB.

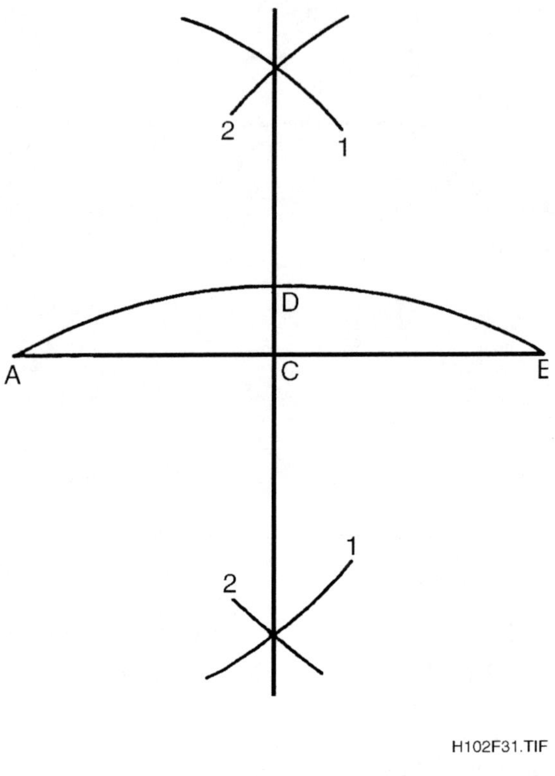

Figure 31. Bisecting Lines (Step 4)

4.6.5 Bisecting Angles

To bisect a 90° angle, 45° angle, or any other angle with rays of equal length, follow the sequence described below. Angle bisection is useful for creating perfect miter cuts.

Step 1 Set the compass point on B and with any radius, describe arc 1 (*Figure 32*).

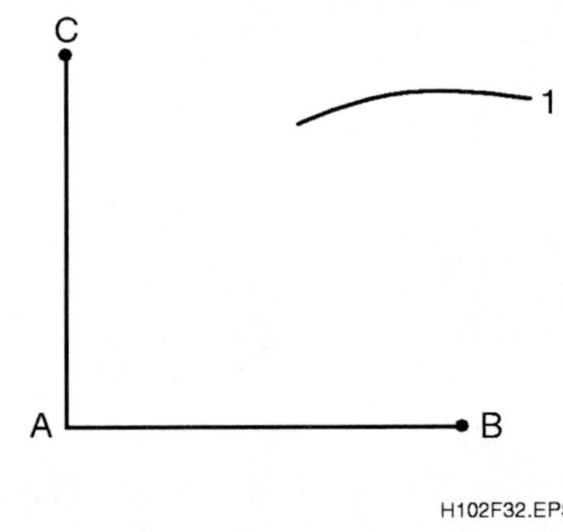

Figure 32. Bisecting Angles (Step 1)

HVAC TRAINEE TASK MODULE 03102

Step 2 Set the compass point on C and with the same radius, describe arc 2, making sure not to change your compass setting (*Figure 33*).

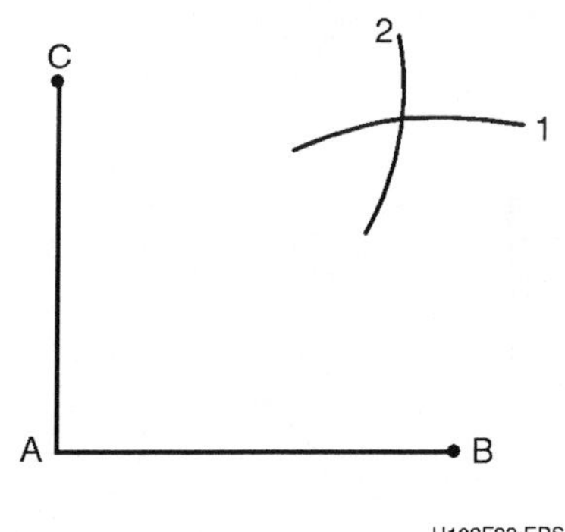

Figure 33.   Bisecting Angles (Step 2)

Step 3 Draw a straight line from their point of intersection to point A. The angle is now divided in half (*Figure 34*). Each of these new angles will be 45°.

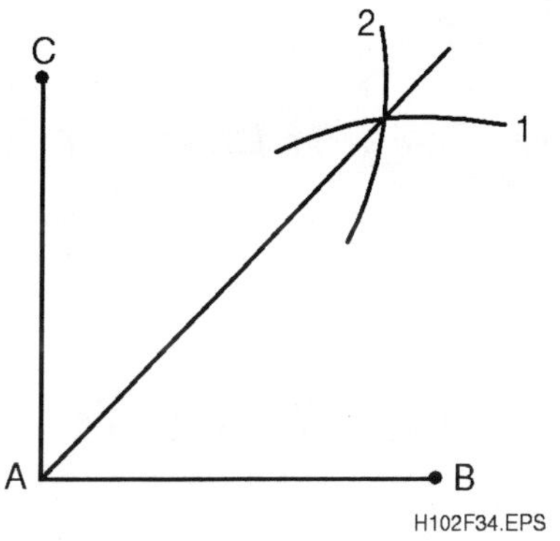

Figure 34.   Bisecting Angles (Step 3)

You may bisect each of these into $22\frac{1}{2}°$ angles by following the same procedure. To do so, we must first find D.

Step 4 To find point D, set your compass at point A, and use radius AC or AB (they are the same) to describe the arc that connects C and B (*Figure 35*). Where that arc intersects your first bisecting line is the location for point D. Now, using D and C or B and D, you may bisect these angles as described above.

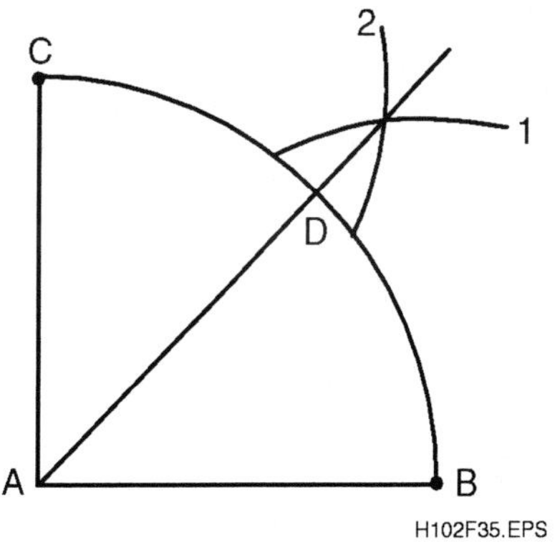

Figure 35.   Bisecting Angles (Step 4)

Step 5   After you've bisected to make the angles illustrated (*Figure 36*), you could continue to bisect each of these to obtain a great variety of angles.

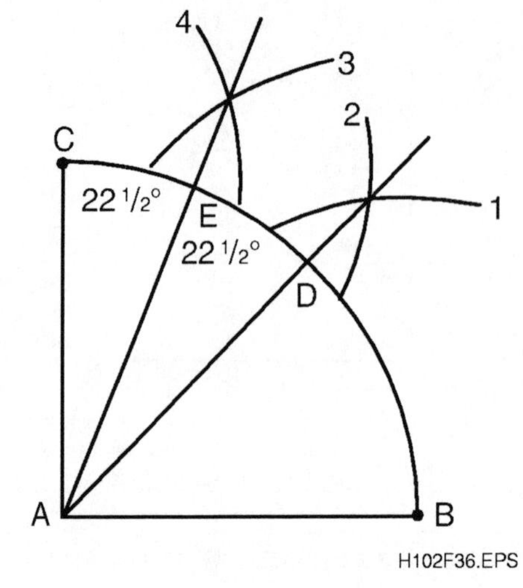

Figure 36.   Bisecting Angles (Step 5)

4.6.6   Bisecting Angles With Rays Of Uncertain Length

To bisect a given angle when you cannot be sure that each ray is of the same length, follow the sequence described below.

Step 1   Let <CBA be the given angle (*Figure 37*).

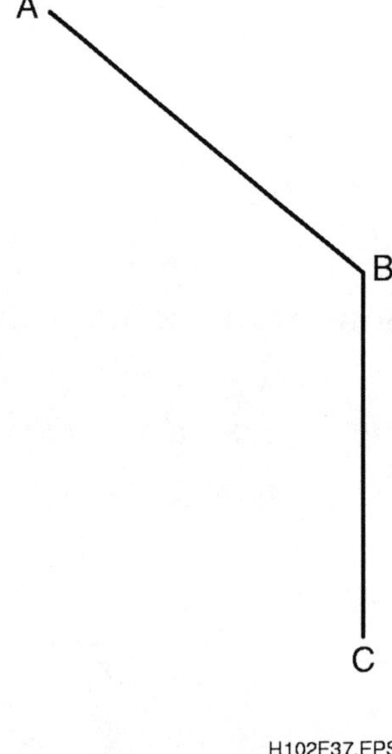

Figure 37.   Bisecting Angles With Rays Of Uncertain Length (Step 1)

Step 2   If you cannot be sure each ray is of equal length, you cannot use the end points A or C for your compass. Instead, place your compass point at B and, using any convenient radius, describe arcs 1 and 2 (*Figure 38*).

HVAC TRAINEE TASK MODULE 03102

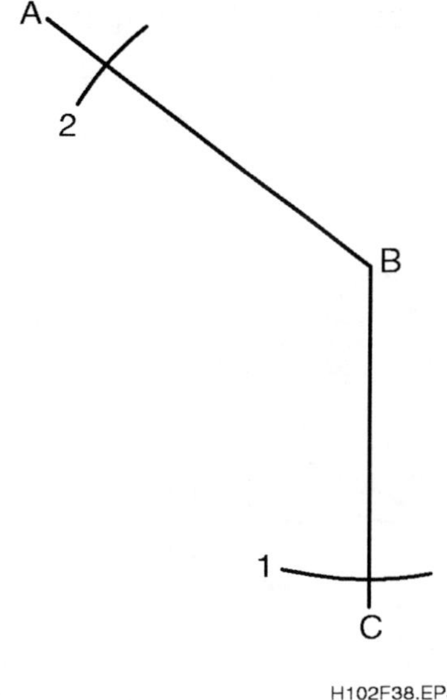

H102F38.EPS

Figure 38. Bisecting Angles With Rays Of Uncertain Length (Step 2)

Where arcs 1 and 2 intersect each ray of the angle is where you will place your compass point to bisect the given angle.

Step 3 With the same or a larger radius and using points 1 and 2 as centers, describe arcs 3 and 4 (*Figure 39*).

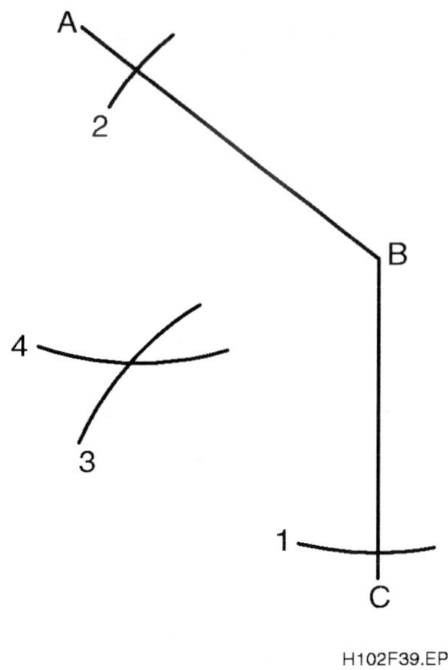

H102F39.EPS

Figure 39. Bisecting Angles With Rays Of Uncertain Length (Step 3)

Step 4 At their point of intersection, draw a straight line to B, which divides <CBA into two equal parts (*Figure 40*).

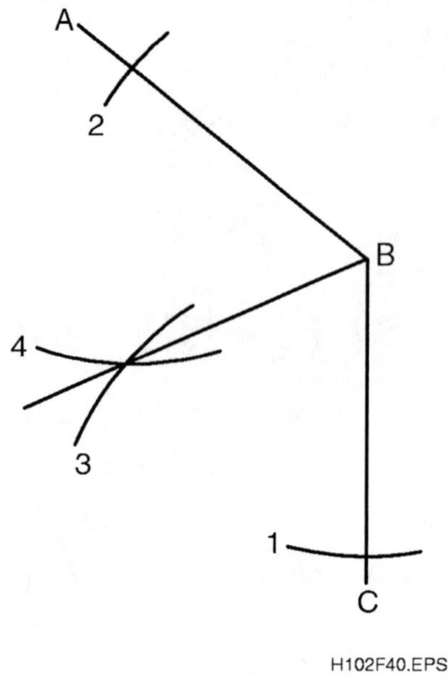

Figure 40. Bisecting Angles With Rays Of Uncertain Length (Step 4)

Step 5 This process can be used to obtain the miter line for dividing a sheet metal elbow, as shown by the dashed lines in *Figure 41*. Simply extend the bisecting line through B to the end point of the metal angle, and the piece can be cut in half.

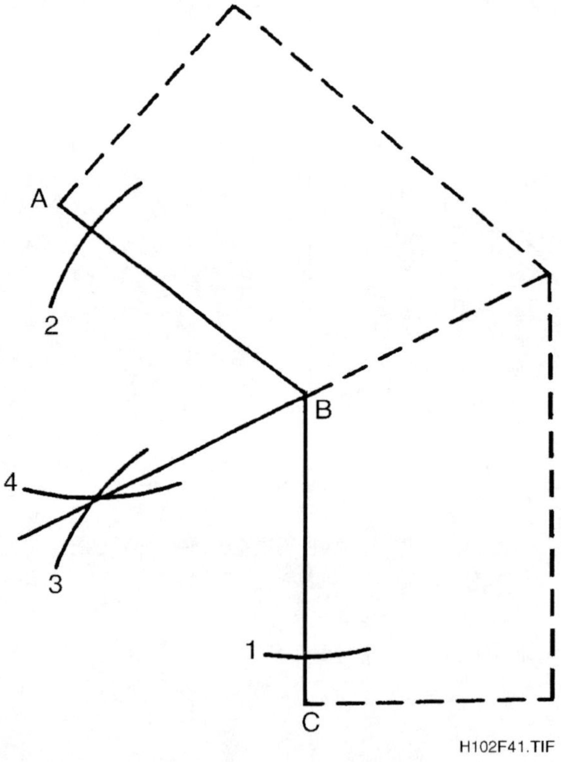

Figure 41. Bisecting Angles With Rays Of Uncertain Length (Step 5)

HVAC TRAINEE TASK MODULE 03102

4.6.7 Constructing Angles

To construct an angle equal to a given angle, follow the sequence described below. Accurate constructions are crucial when drafting piping layouts that slope at a particular angle.

Step 1   Let <BAC be the given angle (*Figure 42*).

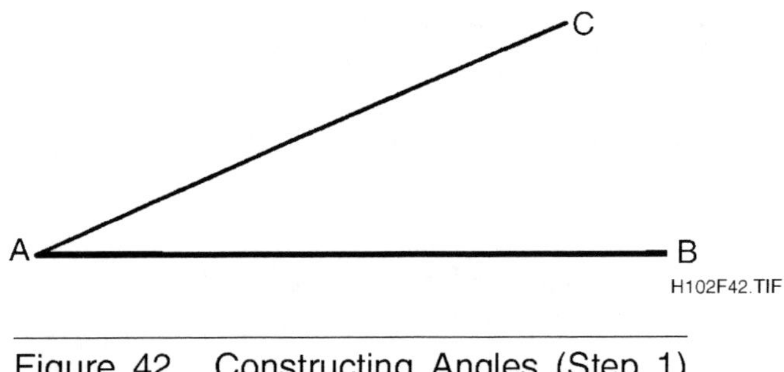

Figure 42. Constructing Angles (Step 1)

Step 2   With point A as the center and using any convenient radius less than either ray, describe arc 1-2 (*Figure 43*).

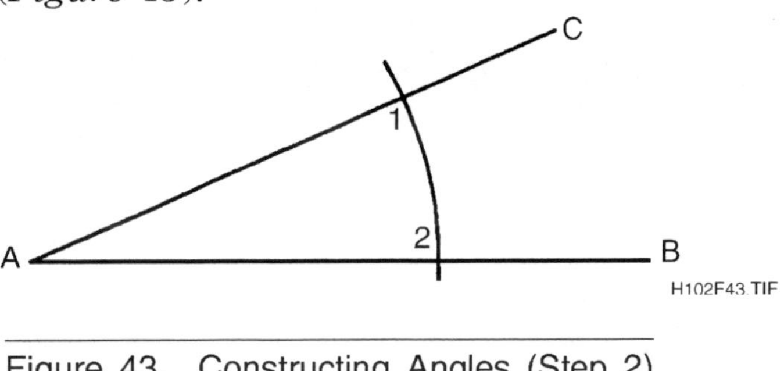

Figure 43. Constructing Angles (Step 2)

Step 3   At a new point, draw a line DE to equal line AB (*Figure 44*).

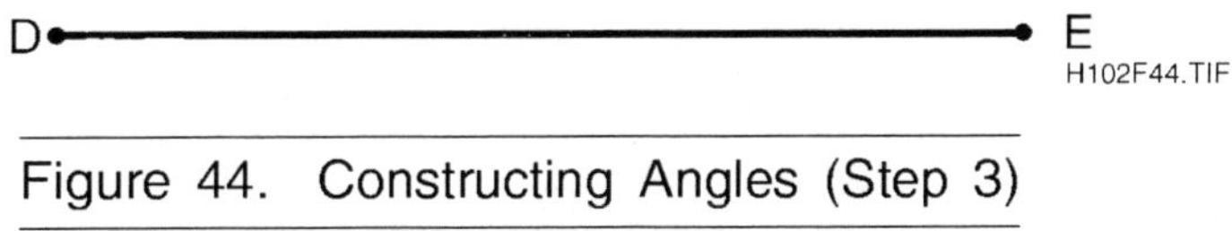

Figure 44. Constructing Angles (Step 3)

Step 4   Using D as the center, take the same radius that produced arc 1-2 and describe arc 3-4 (*Figure 45*). Now, place your compass point at 2 of the given angle and carefully adjust the compass radius to the exact distance from point 2 to point 1.

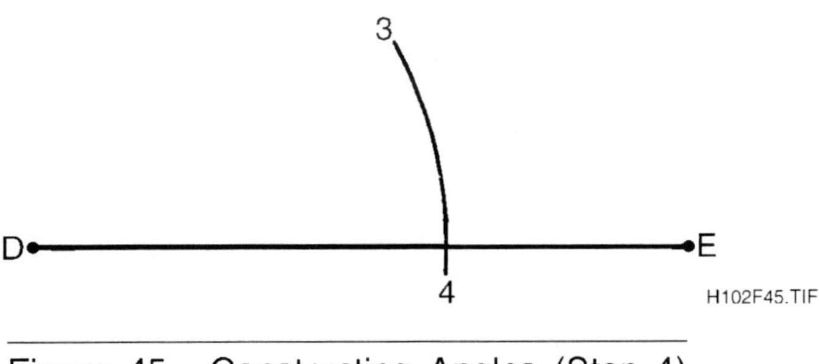

Figure 45. Constructing Angles (Step 4)

Step 5   Set your compass point at 4 on the new angle and describe arc 5 (*Figure 46*).

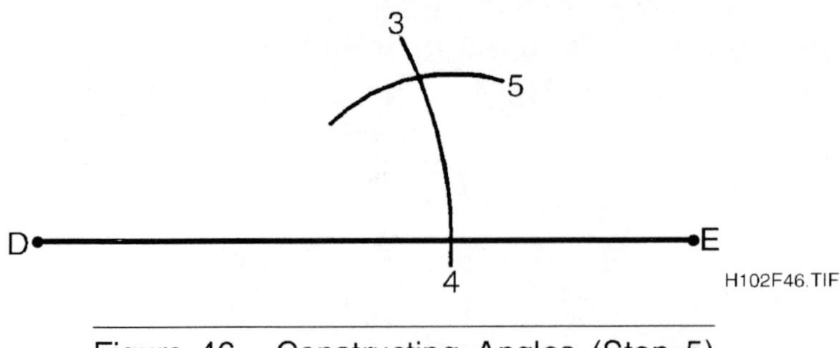

Figure 46.   Constructing Angles (Step 5)

Step 6   Where arc 5 intersects arc 3-4 is point F. Draw a line from D through F and you have copied the original given angle (*Figure 47*).

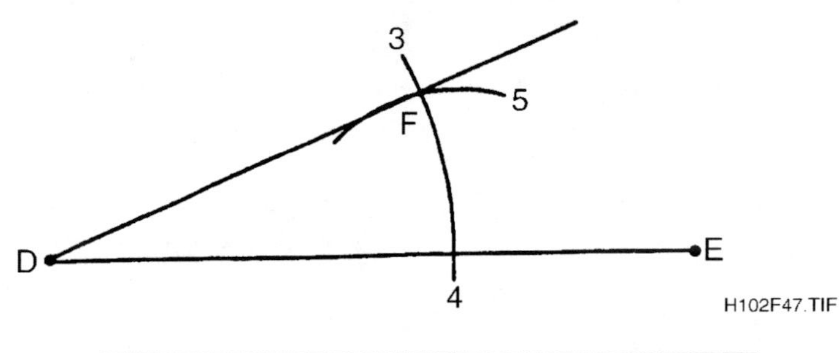

Figure 47.   Constructing Angles (Step 6)

4.6.8   Constructing Circle Parts

This exercise will familiarize you with the various properties of a circle (see *Figure 48*).

Step 1   First, draw the diameter line AB on your paper 3.75 inches long.

Step 2   Bisect diameter AB to locate point C. This will be the center of your circle.

Step 3   With your compass point at C, measure radius CA and then describe the entire circle. Note that your circumference line will pass through point B.

Step 4   Mark off any arc on the circumference of the circle. (We used arc F.) Use the method for copying an angle to reproduce <ABD in the circle on your paper.

Step 5   Once you have an arc, you can draw a straight line to connect the end points of the arc, and then you've described a chord. The area within the circle and between the chord and the arc is called a segment. When you reproduced <ABD, you actually copied arc DB and chord DB, and so your segment should also be equal to the one shown here.

Step 6   By drawing any two radii, you describe an area within a circle known as a sector. This, for all practical purposes, might be most easily remembered as geometrical language for "a piece of pie." In *Figure 48,* the sector ACE is described, which you should also copy to your paper using the method for copying angles.

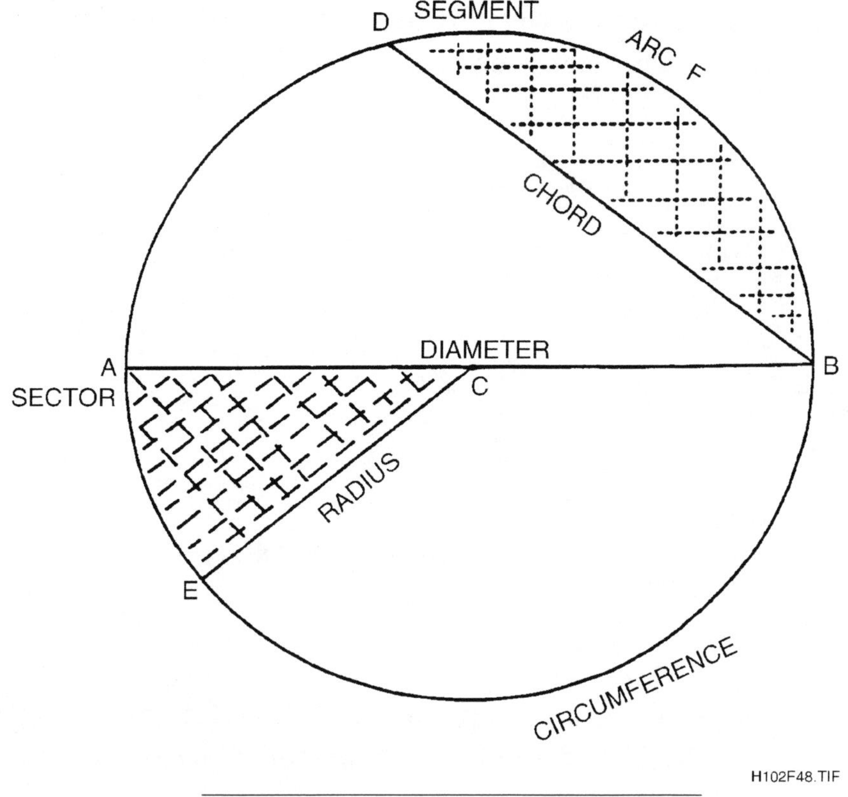

Figure 48.   Constructing Circle Parts

4.6.9   Constructing Parallel Lines

To construct parallel lines, follow the methods described below. Method 1 is faster, but it is not as precise as Method 2. You may use whichever method best suits your purpose. The ability to construct parallel lines is useful in sheet metal work.

Method 1

In order to draw a line parallel to a given line, you only need to locate two points on the new line, both of which are the same distance from the given line.

Step 1   Set your compass for the same radius and use each end point of the given line as centers. Describe arcs 1 and 2 (*Figure 49*).

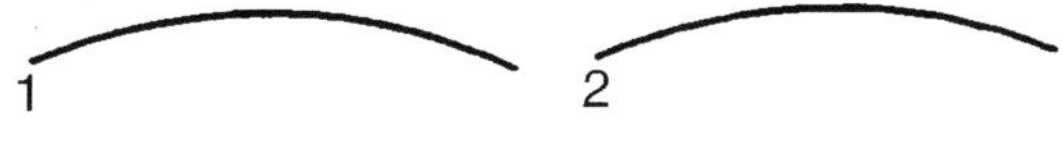

Figure 49.   Constructing Parallel Lines (Method 1 - Step 1)

Step 2 Connect arcs 1 and 2 with a straight line tangent to both. That is, the line should touch the outermost points of each arc. Your new line CD should be parallel to AB (*Figure 50*).

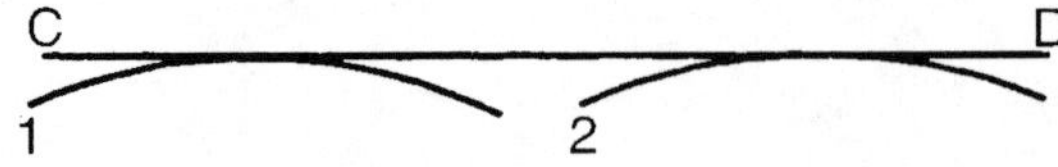

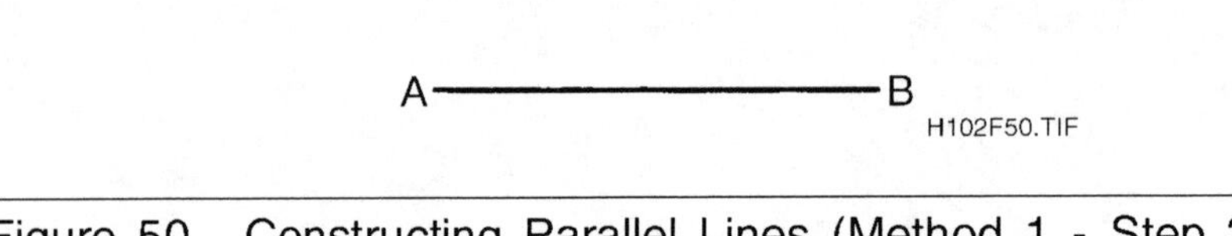

Figure 50. Constructing Parallel Lines (Method 1 - Step 2)

Method 2

This is a far more precise method for drawing parallel lines. Here you are erecting two equal perpendiculars to given line AB. This builds on what you learned earlier about erecting a perpendicular at the end of a given line.

Step 1 We will begin with a single perpendicular at the end of a given line (*Figure 51*). Set your compass to measure arc 2.

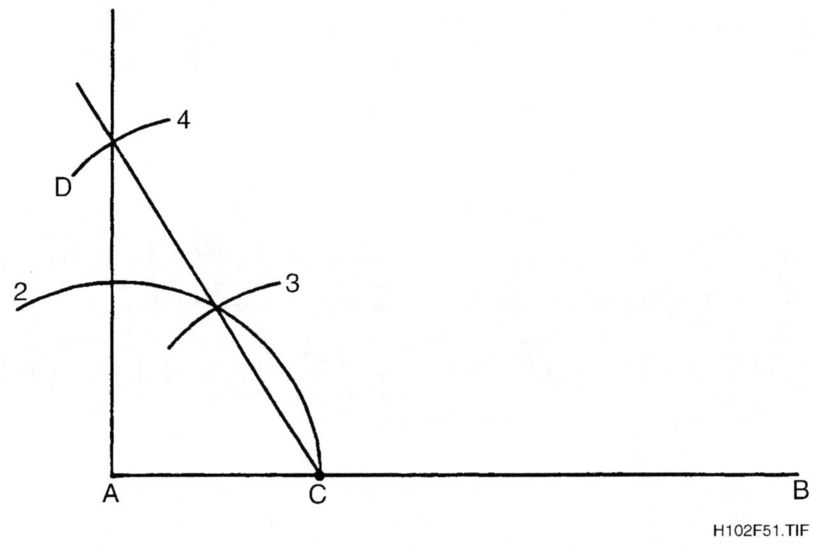

Figure 51. Constructing Parallel Lines (Method 2 - Step 1)

HVAC TRAINEE TASK MODULE 03102

Step 2   Using the same radius as arc 2, describe arc 5 (*Figure 52*).

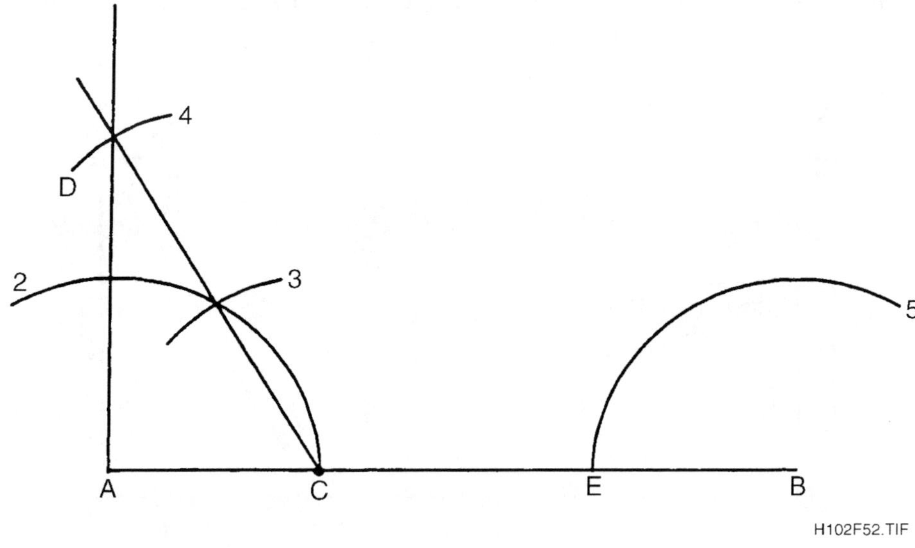

Figure 52.   Constructing Parallel Lines (Method 2 - Step 2)

Step 3   Complete the construction of the second perpendicular as described previously (*Figure 53*).

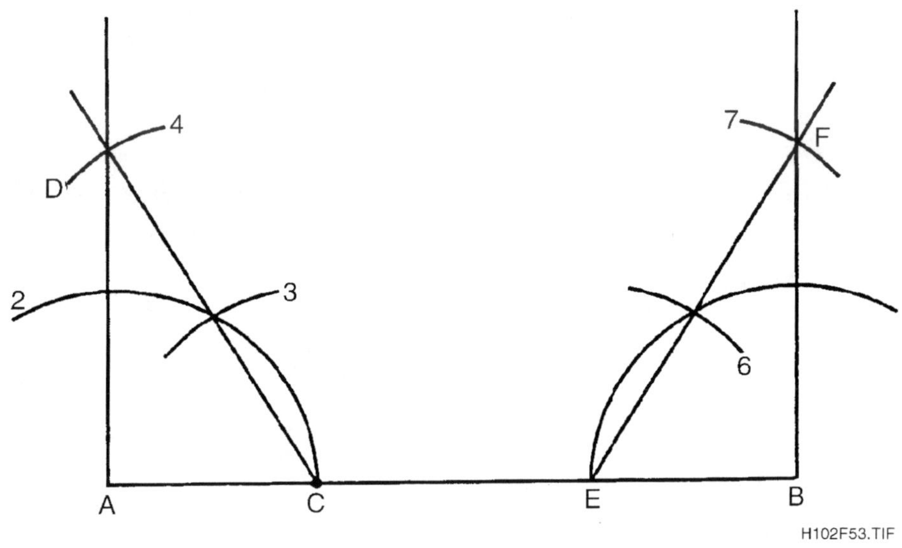

Figure 53.   Constructing Parallel Lines (Method 2 - Step 3)

Step 4 Draw line DF to pass through the specific end points of each equal perpendicular (*Figure 54*). This produces a new line that is parallel to the original given line.

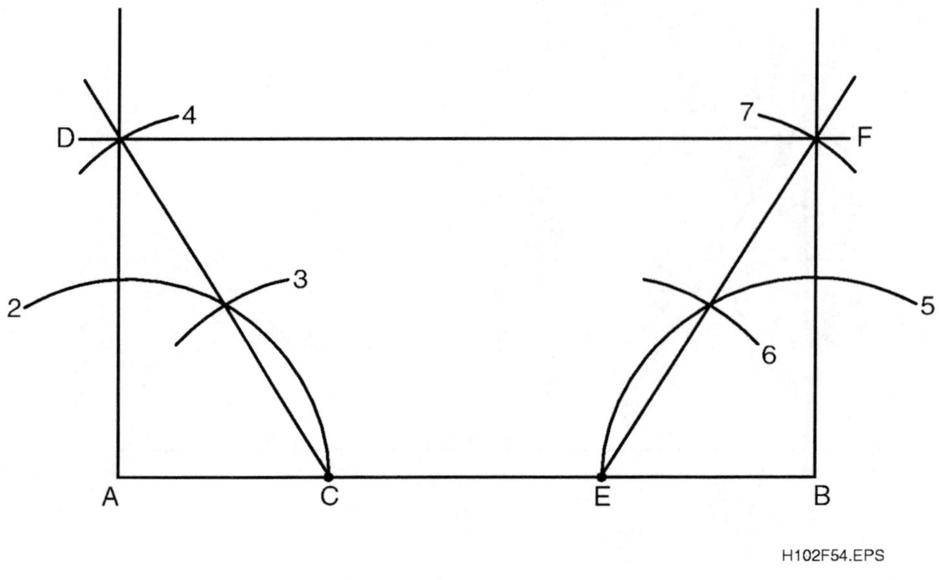

Figure 54. Constructing Parallel Lines (Method 2 - Step 4)

4.6.10 Locating Centers

There are two methods that may be used to establish the center of a given circle. Both methods use the perpendicular bisectors of two chords; you may use whichever method you find most convenient.

Method 1

Step 1 With a compass set at a convenient distance, place it on the circumference of the circle and establish arc 1 (*Figure 55*). Move the compass to a point on the opposite side of the circle and establish arc 2. Draw a line to cross where the arcs intersect to establish the diameter of the circle.

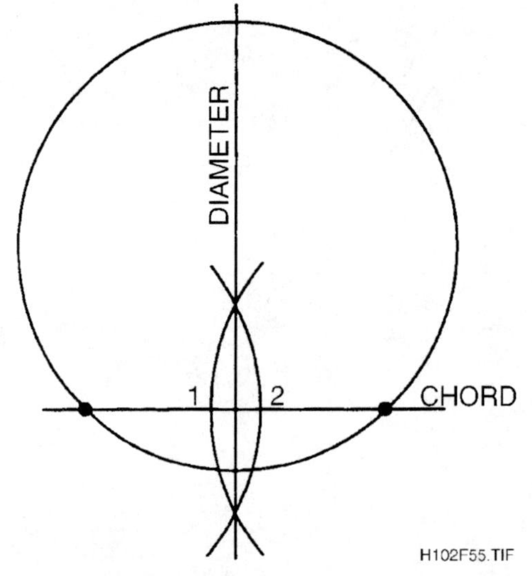

Figure 55. Locating Centers (Method 1 - Step 1)

Step 2   Using the same method as described in Step 1, establish a second chord (*Figure 56*). The perpendicular bisectors of these chords will cross at the center of the circle (point X).

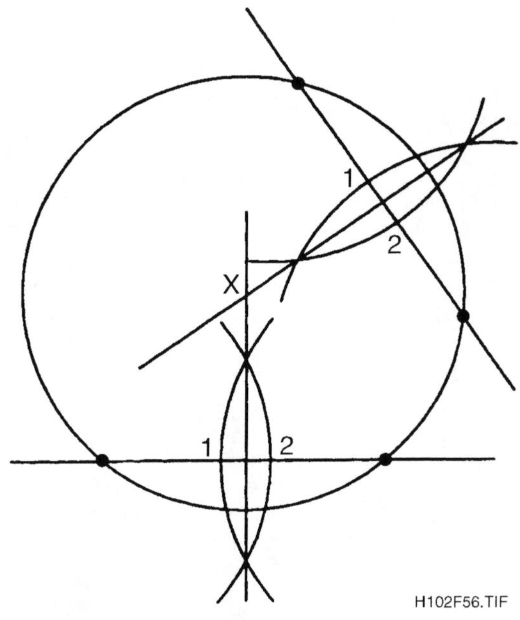

Figure 56.   Locating Centers (Method 1 - Step 2)

Method 2

Step 1   From any point on the circle (A), and with any convenient radius, draw arc 1 to establish points B and C (*Figure 57*).

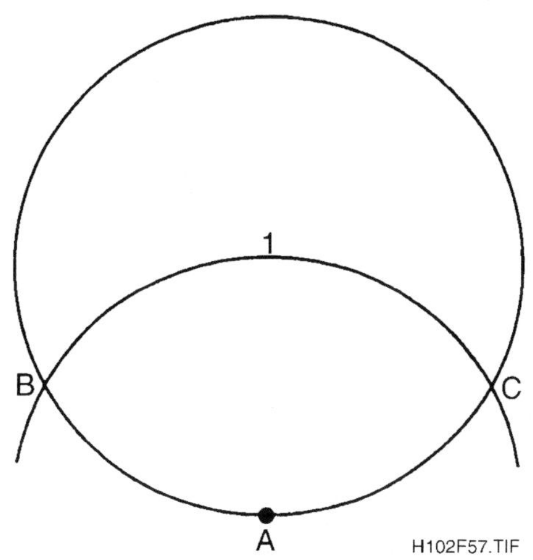

Figure 57.   Locating Centers (Method 2 - Step 1)

Step 2   With the same radius for both, establish point D from B and establish point E from C (*Figure 58*).

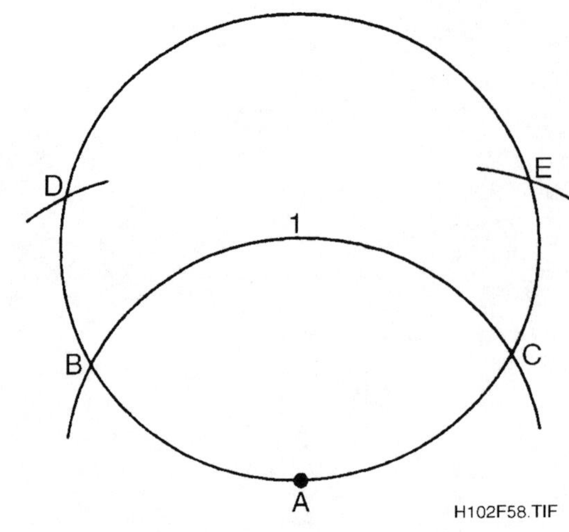

Figure 58.   Locating Centers (Method 2 - Step 2)

Step 3   Keeping the same radius, use point D to establish arc 2. Where arc 2 intersects arc 1, establish point F (*Figure 59*). With the same radius, use point E to establish arc 3. Where arc 3 intersects arc 1, establish point G. Note that arcs 2 and 3 intersect at points B and C.

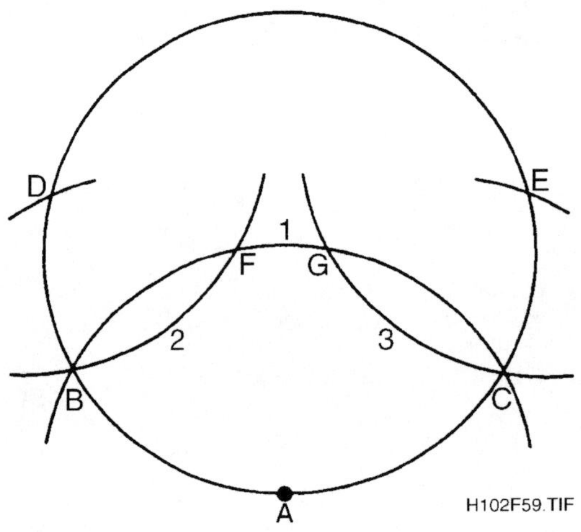

Figure 59.   Locating Centers (Method 2 - Step 3)

HVAC TRAINEE TASK MODULE 03102

Step 4   Draw a straight line extending through the circle circumference that joins points B and F, and an additional line that joins points C and G (*Figure 60*). These lines intersect at the center of the circle (point X).

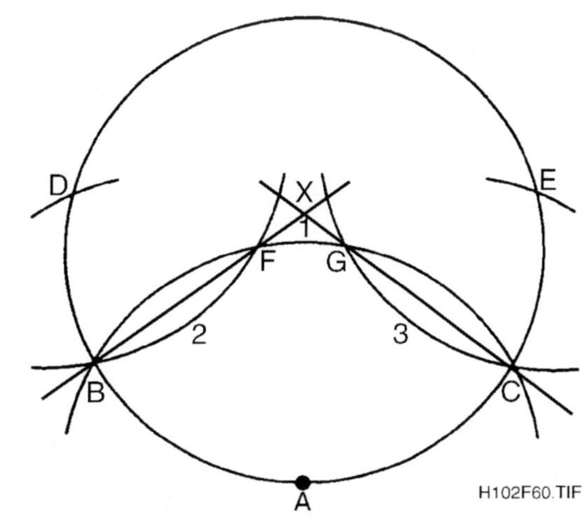

Figure 60. Locating Centers (Method 2 - Step 4)

SUMMARY

This module built on the knowledge gained in the *Basic Math* module. It covered mathematics important to the HVAC trade, with particular emphasis on volume, weight, pressure, vacuum, and temperature calculations. It also introduced basic algebra and geometry, which apprentices must master in order to work successfully in the HVAC trade.

References

For advanced study of topics covered in this task module the following books are suggested:

Modern Refrigeration and Air Conditioning, The Goodheart-Willcox Company, Inc., South Holland, Illinois.

Refrigeration & Air Conditioning Technology, Second Edition, Delmar Publishers, Inc., Albany, New York.

SELF CHECK REVIEW / PRACTICE QUESTIONS

For questions 1 through 3, use the following information and equation.

Air conditioning units are often expressed in terms of "tons of cooling." One ton of cooling represents the amount of cooling produced when one ton (2000 lbs.) of ice is melted over a 24-hour period. The equation for this is as follows:

$$T = \frac{HA}{288,000}$$

T = Tons of Refrigeration Effect
HA = Heat Absorbing Ability
288,000 = Btu equivalent of 1 ton of refrigeration per 24 hours

1. Given a refrigeration unit with a 4-ton capacity, find the heat absorbing ability in Btu's per 24 hours.
    a. 1,152,000 Btu's
    b. 48,000 Btu's
    c. 98,000 Btu's
    d. 12,000 Btu's

2. Using the same information as for question 1, what is the Btu equivalent per hour?
    a. 1,152,000 Btu's
    b. 48,000 Btu's
    c. 98,000 Btu's
    d. 12,000 Btu's

3. Given a refrigeration unit with a capacity of 432,000 Btu's per day, what is its rate in tons?
    a. 15 tons
    b. 4.32 tons
    c. 1.5 tons
    d. 18 tons

4. Given the equation E = IR, find the current (I) if E = 240 and R = 10.
    a. 2.4 amps
    b. 24 amps
    c. 2400 amps
    d. None of the above

5. How much air is contained in a duct that is 15 feet long, 4 feet wide, and $2\frac{1}{2}$ feet high?
    a. 150 cubic feet
    b. 60 cubic feet
    c. 10 cubic feet
    d. 2.0 cubic yards

6. What is the volume of an air conditioning plenum measuring 35 inches by 20 inches by $12\frac{3}{4}$ inches?
    a. 4 cubic feet
    b. 11.1 cubic feet
    c. 8925 cubic inches
    d. 2.4 cubic yards

7. A system requires a charge of 372 ounces. You have a digital refrigerant scale that reads in pounds. How many pounds of refrigerant are required?
    a. 3.72 lbs.
    b. 186 lbs.
    c. 23.25 lbs.
    d. None of the above

8. A large drum has an inside pressure of 361,600 lbs. How many tons is this equal to?
    a. 1808 tons
    b. 18.08 tons
    c. 1.808 tons
    d. 180.8 tons

9. Convert 600 Pa. to in. w.g.
    a. 8820 in. w.g.
    b. 614.7 in. w.g.
    c. 2.412 in. w.g.
    d. .201 in. w.g.

10. Find the absolute pressure if gauge pressure is 164 psig.
    a. 178.7 psia
    b. 14.7 psia
    c. 164 psia
    d. 149.3 psia

11. A manufacturing plant requires a water pressure of 140 psig. What head is required to develop this pressure?
    a. 72 feet
    b. 72 inches
    c. .78 nautical miles
    d. 323 feet

12. Convert 68°F to Celsius.
    a. 100°C
    b. 36°C
    c. 41°C
    d. 20°C

13. Convert -40°C to Fahrenheit.
    a. -72°F
    b. -40°F
    c. +41°F
    d. -11°F

14. You have been given a scrap of sheet metal in the shape of a trapezoid with the dimensions below and need to cut it to cover an opening that is 8 x 14 inches. How do you go about making sure that the piece you cut is a perfect rectangle?

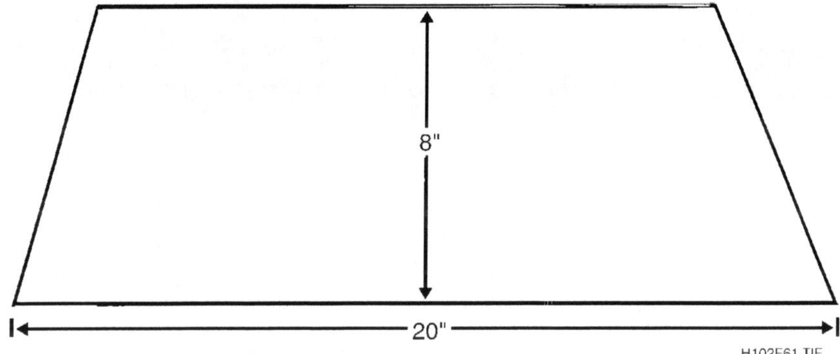

15. You need to construct a sheet metal offset at a 135° angle. Given the piece of sheet metal below on which we have already established a perpendicular, how will you construct the angle?

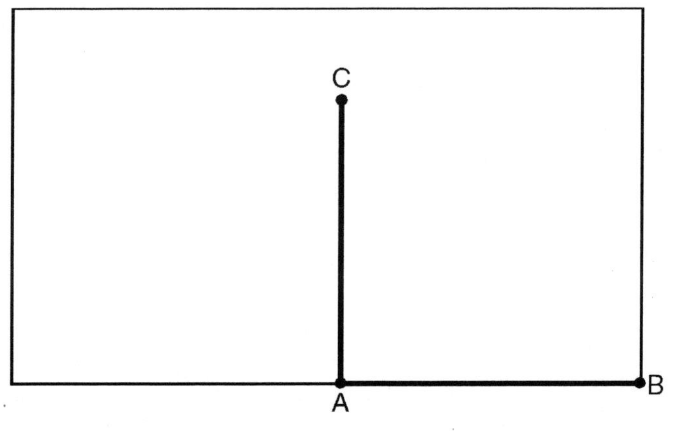

Answer   **Section Reference**

1. a   2.3.1
2. b   2.3.1
3. c   2.3.1
4. b   2.3.1
5. a   3.1.0
6. c   3.1.0
7. c   3.2.0
8. d   3.2.0
9. c   3.3.0
10. a   3.3.1
11. d   3.3.1
12. d   3.5.0
13. b   3.5.0

14. Begin by measuring a 14-inch line (e.g., AB) along the bottom of the trapezoid, then erect two perpendiculars at points A and B. The two 90° angles will create a perfect rectangle. (Section 4.6.3)

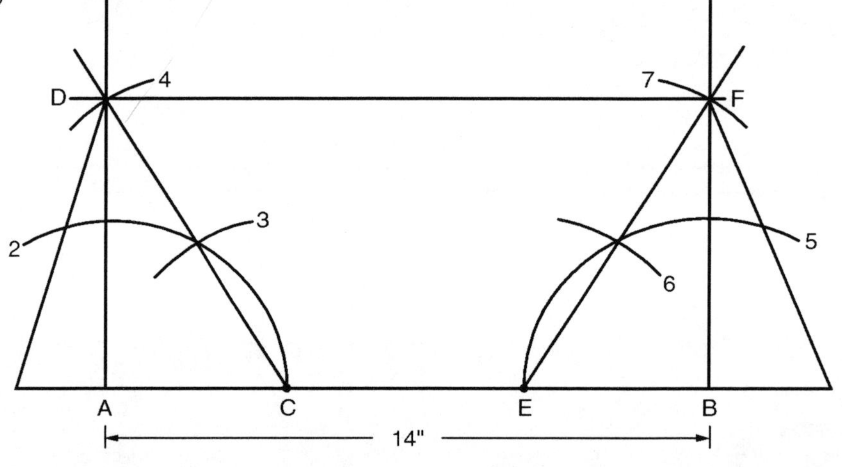

15. Bisect the second 90° angle created by the perpendicular. The 135° angle will be created by adding the first 90° angle to the first 45° angle (90° + 45° = 135°). (Section 4.6.5)

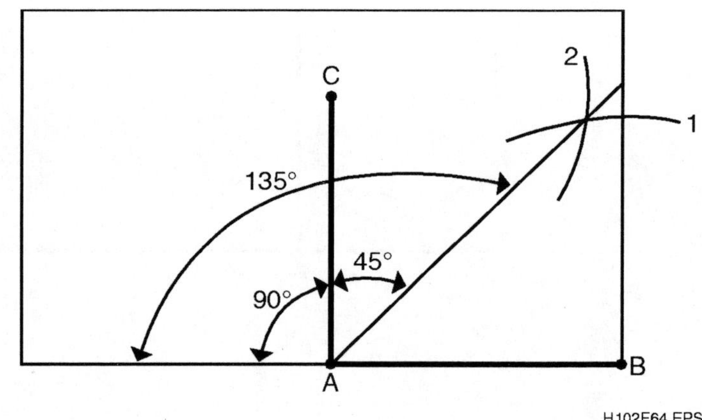

     HVAC TRAINEE TASK MODULE 03102

The NCCER makes every effort to keep these manuals up-to-date and free of technical errors. We appreciate your help in this process. If you have an idea for improving this manual, or if you find an error, a typographical mistake, or an inaccuracy in the NCCER's Craft Training Manuals, please write us, using this form or a photocopy. Be sure to include the exact module number, page number, a description of the problem, and the correction, if possible. Your input will be brought to the attention of the Technical Review Committee. Thank you for your assistance.

Instructors – If you found that additional materials were necessary in order to teach this module effectively, please let us know so that we may include them in the Equipment/Materials list in the Instructor's Guide.

Write: Curriculum and Revision Department
National Center for Construction Education and Research
P.O. Box 141104
Gainesville, FL 32614-1104
Fax: 352-334-0932

Craft ______________________   Module Name ______________________

Module Number ______________________   Page Number(s) ______________________

Description of Problem ______________________

____

____

____

(Optional) Correction of Problem ______________________

____

____

(Optional) Your Name and Address ______________________

____

Tools of the Trade

Module 03103

TOOLS OF THE TRADE

Objectives

Upon completion of this module, the trainee will be able to:

1. Identify and demonstrate the ability to use the following tools:
   - Pipe wrenches
   - Torque wrenches
   - Tinner's and soft face hammers
   - Hand cutting snips
   - Hand and power hacksaws
   - Drill press
   - Measuring tools
2. Explain the procedures for the proper maintenance of the tools listed above.
3. Explain and demonstrate the safety precautions that must be followed when using the hand and power tools listed above.

Prerequisites

Successful completion of the following Task Modules is required before beginning study of this Task Module: Common Core Curricula, HVAC Module 03101.

Required Student Material

1. Student Module
2. Appropriate Personal Protective Equipment

Course Map Information

This course map shows all of the *Wheels of Learning* task modules in the first level of the HVAC curricula. The suggested training order begins at the bottom and proceeds up. Skill levels increase as a trainee advances on the course map. The training order may be adjusted by the local Training Program Sponsor.

Course Map: HVAC, Level 1

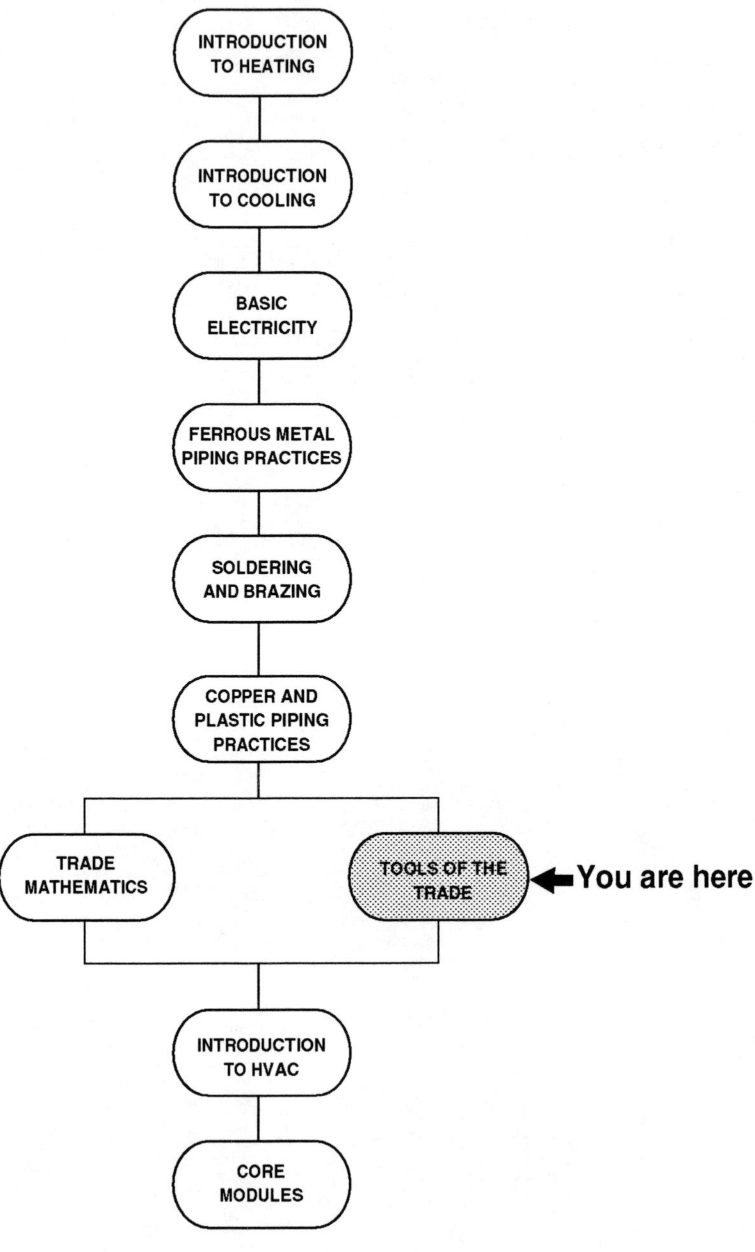

Trade Terms Introduced In This Module

Break-away torque: The torque required to loosen a fastener. This is usually less than the torque required to tighten the fastener.

Run-down resistance: The torque required to overcome the resistance of associated hardware, such as locknuts and lockwashers, when tightening a fastener.

Seizure: In the last stages of rotation in reaching a final torque, seizing or set of the fastener may occur. This is usually accompanied by a noticeable popping effect.

Torque: The resistance to turning or twisting.

1.0.0 INTRODUCTION

For every task there is a tool that will perform the job quickly and effectively. The ability to select the right tool for the job is a skill which every HVAC technician needs to develop.

This module describes hand-operated tools and power tools not previously covered in the core modules, *Introduction To Hand Tools*, Module 00103, and *Introduction To Power Tools*, Module 00104. Some added information about a few of the tools previously studied in the core modules is also given because of its relevance to the HVAC trade.

2.0.0 PIPE WRENCHES

Many types and sizes of pipe wrenches are used in the HVAC trade. Pipe wrenches are heavy-duty wrenches made of durable cast iron or aluminum that are used to assemble and disassemble pipe. It is critical to select the right size pipe wrench for a given job. A pipe wrench that is too small will not hold the pipe firmly, and a handle that is too short will not provide enough leverage. A pipe wrench that is too large can strip the threads, break the pipe or fitting, or cause excessive marring or scratching of the pipe. Several types of pipe wrenches are used by HVAC technicians, including the following:

- Straight pipe wrenches
- Offset pipe wrenches
- Chain wrenches
- Strap wrenches
- Compound leverage wrenches

Figure 1 shows different types of pipe wrenches.

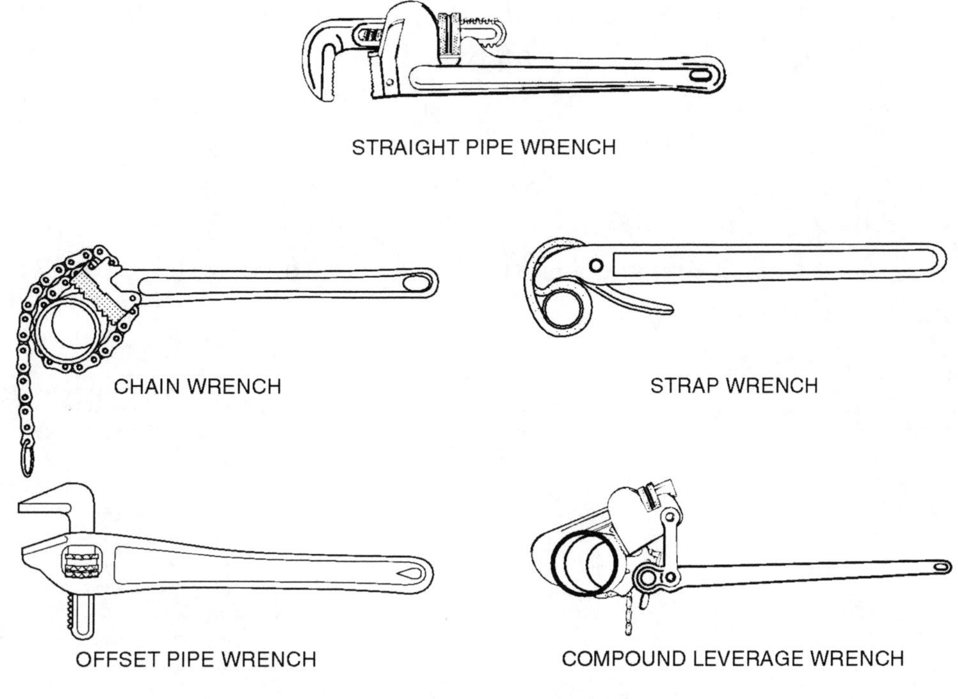

Figure 1.   Pipe Wrenches

2.1.0   STRAIGHT PIPE WRENCHES

Straight pipe wrenches have two jaws: a stationary jaw known as the *heel* and a moveable jaw known as the *hook jaw*. Both jaws have teeth cut into them to provide a firm grip on the pipe. The teeth on the hook jaw face inward, and the teeth on the heel face outward. *Figure 2* shows the components of a straight pipe wrench.

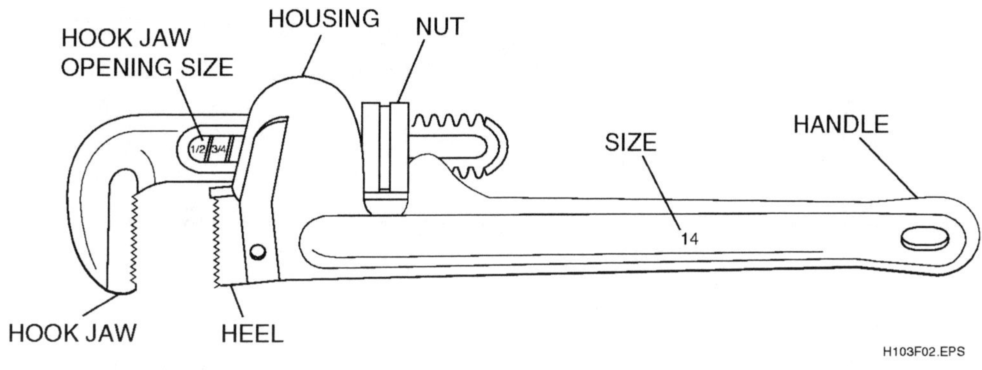

Figure 2.   Components Of A Straight Pipe Wrench

The hook jaw of a straight pipe wrench is adjusted by turning the knurled nut located below the housing. A spring inside the housing allows the hook jaw to float in the housing. This causes the hook jaw to open slightly as you apply the wrench to the pipe. Straight pipe wrenches are sized by the length measured from the inside top of the hook jaw to the end of the handle with the jaws fully open. They range in size from 6 to 48 inches and larger. The size does not refer directly to the pipe size it fits, but each size wrench has recommended pipe capacities. The size of the wrench is stamped into the wrench handle. *Table 1* shows pipe wrench sizes and capacities.

WRENCH SIZE	CAPACITY (JAW OPENING)	RECOMMENDED FOR PIPE SIZES
6	1/8 in. - 1/2 in.	1/8 in.
8	1/8 in. - 3/4 in.	1/4 in. - 3/8 in.
10	1/8 in. - 1 in.	1/2 in. - 3/4 in.
14	1/4 in. - 1-1/2 in.	1 in.
18	1/4 in. - 2 in.	1-1/4 in. - 1-1/2 in.
24	1/2 in. - 2-1/2 in.	2 in. - 2-1/2 in.
36	1/2 in. - 3-1/2 in.	3 in.
48	1 in. - 5 in.	4 in. - 5 in.

Table 1. Pipe Wrench Sizes And Capacities

When using a pipe wrench, either place the wrench on the pipe or fitting with the jaw opening facing you and pull on the handle, or place the wrench on the pipe or fitting with the jaw opening facing away from you and push on the handle. Never place the jaws of a pipe wrench onto the threaded end of a pipe because this will ruin the threads.

WARNING! Be careful when pulling the handle toward you to keep the jaw from slipping off the pipe, causing the handle to hit you. Be sure to brace yourself when pushing on the handle so that you do not fall if the wrench slips off the pipe.

2.2.0 OFFSET PIPE WRENCHES

The offset pipe wrench operates the same way as the straight pipe wrench. It is also made the same way and has similar components. The only difference is that the hook jaw and the heel are offset from the handle at either 45 or 90 degrees. This wrench is useful when working in tight spaces, such as on pipes located close to a ceiling or wall or on other pipes where you cannot grip and turn the pipe with a straight pipe wrench. *Figure 3* shows offset pipe wrenches.

HVAC TRAINEE TASK MODULE 03103

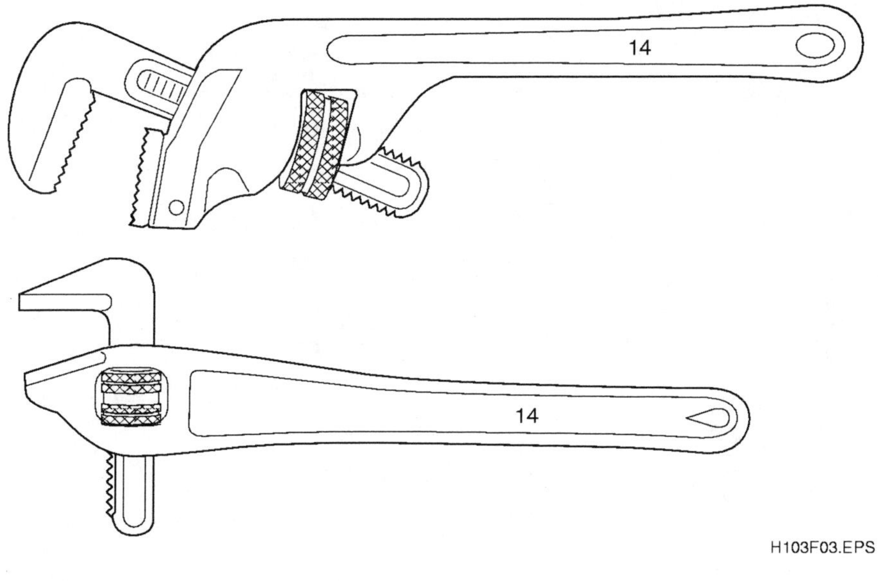

Figure 3. Offset Pipe Wrenches

2.3.0   CHAIN WRENCHES

Chain wrenches, also known as chain tongs, are used in the same manner as pipe wrenches. A chain wrench is a handled pipe wrench with a fixed, serrated jaw. A length of steel chain attached to the pipe wrench wraps around a pipe and connects to the other side of the wrench frame. This wrench holds pipe firmly and distributes the bite of the jaw evenly. A ratchet action allows you to turn the pipe either way and to tighten or loosen the pipe without removing the chain from the pipe. The chain wrench is especially useful in close quarters. Be careful not to use a chain wrench on alloy piping, such as stainless steel, because cross-contamination of the metals will occur. *Figure 4* shows a chain wrench.

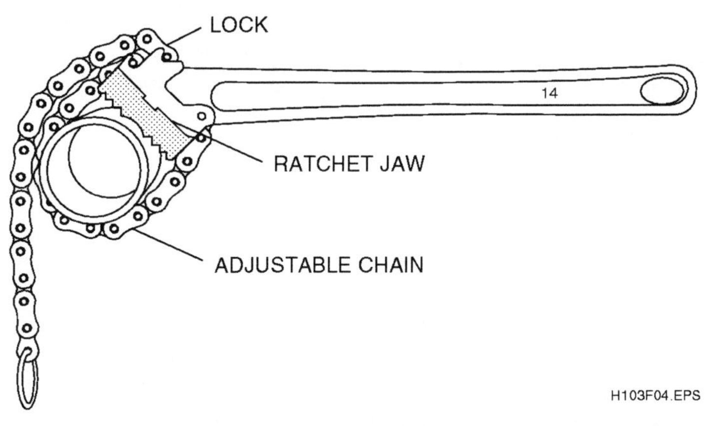

Figure 4. Chain Wrench

2.4.0   STRAP WRENCHES

A strap wrench is similar to a chain wrench except that a heavy-duty cloth or leather strap replaces the chain. A strap wrench is used to remove or install polished or plated pipe because the strap will not scar the pipe as a chain or pipe wrench will. The strap serves as the jaw of the wrench and is adjustable for tight, secure grips. *Figure 5* shows a strap wrench.

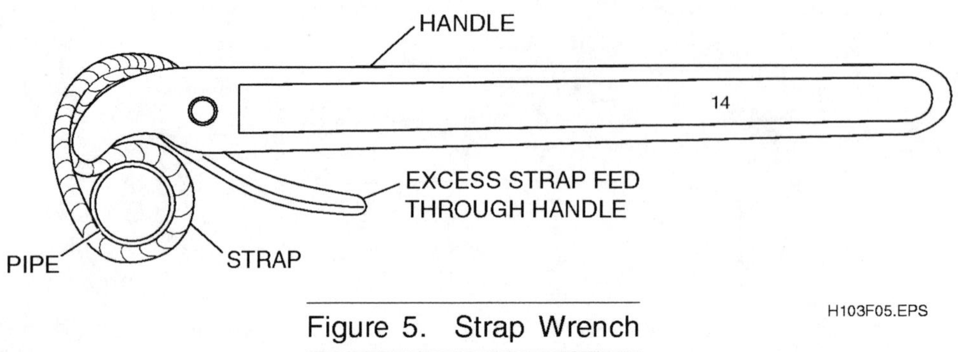

Figure 5.   Strap Wrench

2.5.0   COMPOUND LEVERAGE WRENCHES

A compound leverage wrench is a combination between a hook jaw pipe wrench and a chain wrench. A compound leverage wrench is actually two wrenches on one handle. It is used to provide extra leverage to loosen seized pipe joints or to make up joints without using a vise. The compound leverage wrench is positioned so that the hook jaw grips the pipe fitting and the chain wrench grips the mating pipe. When force is applied to the handle, the hook jaw turns in one direction and the chain wrench turns in the other direction. *Figure 6* shows a compound leverage wrench.

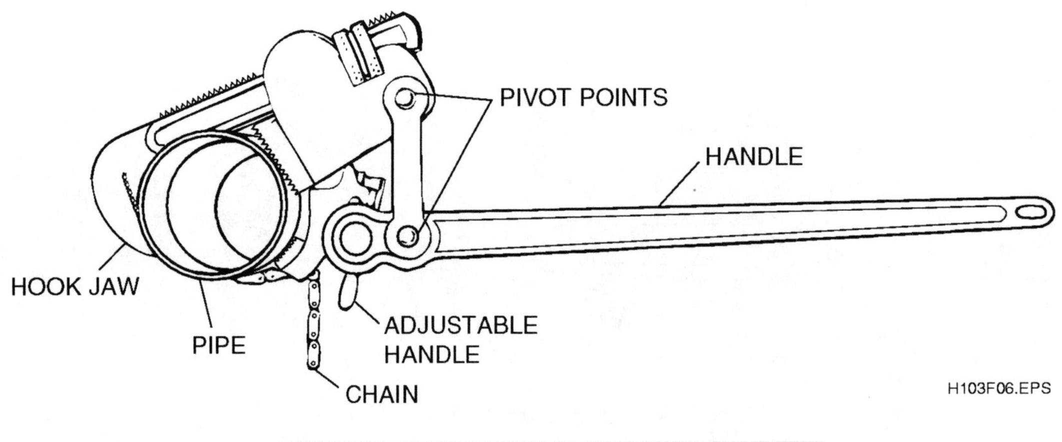

Figure 6.   Compound Leverage Wrench

2.6.0   USING AND CARING FOR PIPE WRENCHES

Follow these guidelines to use and care for pipe wrenches:

- Use pipe wrenches only to turn pipe and fittings. Do not use a pipe wrench to bend, raise, or lift a pipe.

HVAC TRAINEE TASK MODULE 03103

- Do not use a pipe wrench as a hammer. It is not designed for any sort of pounding.
- Do not drop or throw a pipe wrench. If you break or crack the wrench, you may cause injury to yourself or others.
- Check the teeth of the wrench often. They should be kept clean and sharp to prevent the wrench from slipping.
- Apply penetrating oil to the wrench threads to prevent them from sticking.
- Never use a cheater or extender bar on a pipe wrench handle to increase leverage. This may cause the handle or the pipe to bend or break.

3.0.0 TORQUE WRENCHES

The torque wrench is a combination wrench and measuring tool. It measures **torque** - the resistance to turning or twisting. Torque wrenches are used when installing nuts, bolts, and similar fasteners which must be tightened to specific torque values in order to avoid warpage or other damage to parts. The torque wrench consists of a handle and an attached indicator calibrated to measure torque in either foot-pounds (for large fasteners) or inch-pounds (for small fasteners). Most also have calibrated metric scales. A correct socket size is usually attached to the torque wrench handle to tighten the nut or bolt to be fastened.

There are numerous types of torque wrenches. Simple ones, called deflecting beam-type wrenches (*Figure 7*), have a graduated scale and indicator tip that gives a line-on-line reading when deflected. Some give a direct reading on a calibrated dial or digital readout indicator. Others have a micrometer-like mechanism that is set to the desired torque, and then sounds a signal or momentarily releases when the preset torque is reached.

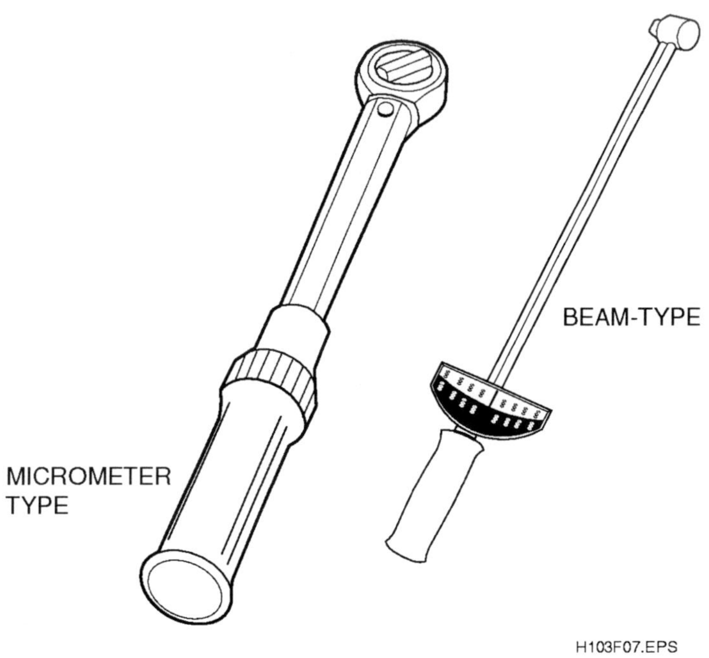

Figure 7. Torque Wrench

The correct size wrench for a job is one that will read between 25% and 75% of the scale when the required torque is applied. This allows for adequate capacity and provides the satisfactory accuracy. Avoid using an oversized torque wrench because the scale divisions are too coarse, making it difficult to get an accurate reading. Too small a wrench will not allow for extra capacity in the event of seizure or run-down resistance. It is important that all threaded fasteners be clean and undamaged in order to get accurate settings. Force should be applied to a torque wrench by pulling whenever possible. This is mainly because there is a greater hazard to the fingers or knuckles when pushing. While pulling is the preferred method, either way will produce accurate results. The calibration of torque wrenches must be checked periodically to be sure of accuracy. The following terms and procedures need to be understood when using a torque wrench:

- **Break-Away Torque** – The torque required to loosen a fastener. This is generally lower than the torque to which it has been tightened. For a given size fastener, there is a direct relationship between tightening torque and break-away torque. This relationship is determined by actual test. Once known, the tightening torque can be checked by loosening and checking break-away torque.
- **Set or Seizure** – In the last stages of rotation in reaching a final torque, seizing or set of the fastener may occur. When this happens there is usually a noticeable popping effect. To break the set, back off and then again apply the tightening torque. Accurate torque settings cannot be made if the fastener is seized.
- **Run-Down Resistance** – The torque required to overcome the resistance of associated hardware, such as locknuts and lockwashers, when tightening a fastener. To obtain the proper torque value where tight threads on locknuts produce a run-down resistance, add the resistance to the required torque value. Run-down resistance must be measured on the last rotation or as close to the makeup point as possible.

4.0.0　HAMMERS

Most of the hammers you use have been previously described in the core module entitled *Introduction To Hand Tools*, Module 00103. Described here are the "Tinner's" and "soft face" hammers used mainly when fabricating HVAC ductwork.

4.1.0 TINNER'S HAMMER

A Tinner's or metalworker's hammer is used to straighten and form sheet metal for ductwork. *Figure 8* shows a Tinner's hammer.

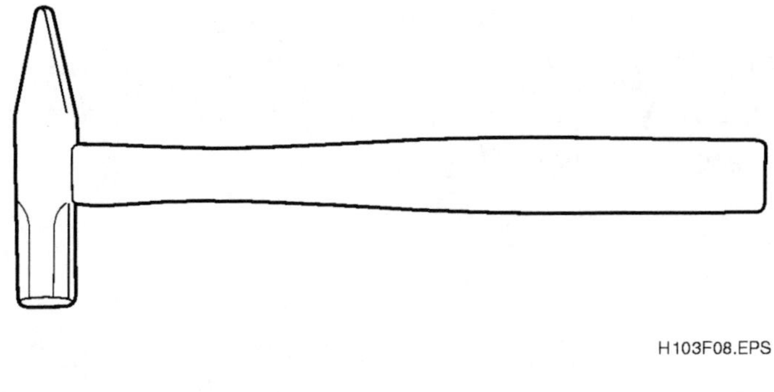

Figure 8. Tinner's Hammer

4.2.0 SOFT FACE HAMMER

Soft face hammers have faces (tips) that yield at the striking point. They are generally used when working on surfaces or equipment that cannot be marred. In most cases the hammer tips are immune to oil, gasoline, and other common chemicals and will not pick up chips and transfer them to the work. Non-sparking soft face hammers are available that have no exposed metal parts. *Figure 9* shows two types of soft face hammers.

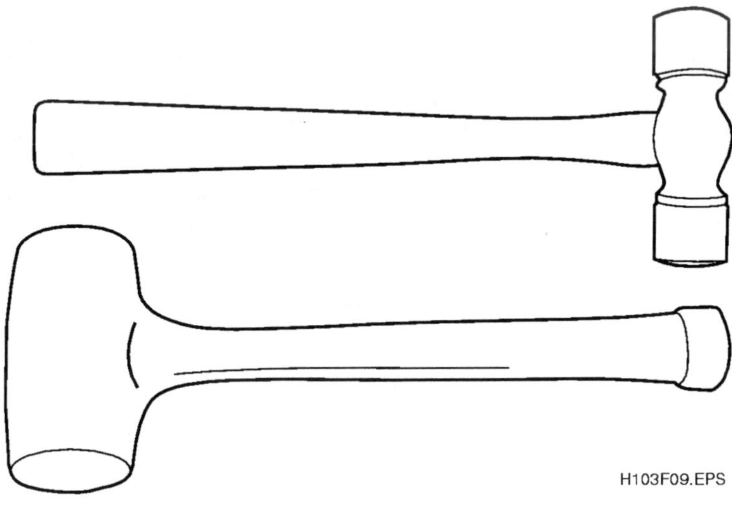

Figure 9. Soft Face Hammers

5.0.0   HAND CUTTING TOOLS

Snips and shears of various types are indispensable for cutting thin sheets of brass, aluminum, copper, black and galvanized iron, and stainless steel (*Figure 10*). Hand snips cannot be used for cutting tempered steel. Snips and shears are of two general classifications: the regular straight cut and the circular cut. They come in various types and sizes. These snips, with the exception of aviation snips, can only cut metal up to 20 gauge thickness. Aviation snips can be used to cut up to 12 gauge thickness. Heavier gauge metals must be cut with either a hacksaw, a chisel, or power tools.

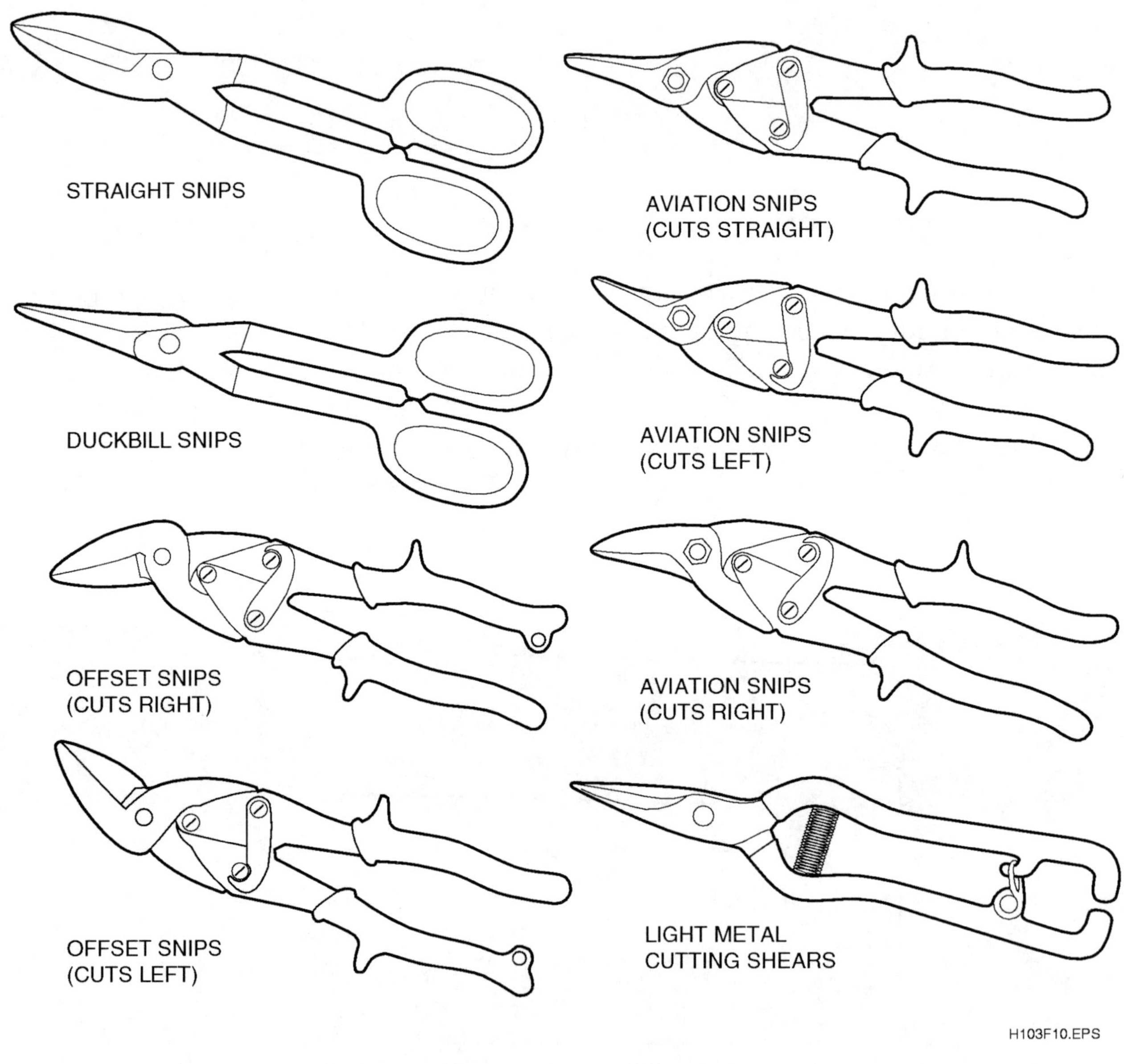

Figure 10.   Hand Snips

HVAC TRAINEE TASK MODULE 03103

5.1.0   TIN SNIPS

Tin snips are used like scissors to cut thin, soft metal. Whatever the purpose of the snips may be, the blades are of two basic types—straight blade or combination blade. The difference between the two is that the straight blade has the face of the blade running up from the cutting edges, whereas the combination blade is curved back from the cutting edge. This curve allows the metal to slip over the top blade when cutting curves. Straight blade snips have greater strength and can therefore be constructed with longer blades.

5.2.0   GENERAL PURPOSE SNIPS

General purpose straight and duckbill snips are used for most general-purpose cutting of 26 gauge or thinner metals. Duckbill snips can also be used for cutting tight curves.

5.3.0   AVIATION AND OFFSET SNIPS

Aviation and offset snips have a compound leverage design, allowing them to cut thicker metal than the general purpose snips. The blade design allows them to be used for cutting small, irregular curves and even inside 90 degree corners. Aviation snips are available in either straight, left-hand or right-hand cutting models. Offset snips are normally available in left-hand or right-hand models. The blades are serrated, which allows for better gripping of the metal.

The color of the handle denotes whether they are straight-line (yellow), left-hand (red) or right-hand (green) cutting models. If the handle grip colors are not identifiable, the right-hand and left-hand snips can be distinguished by the positioning of the upper blade. If the upper blade is on the right, it is a right-hand cut snips; if the upper blade is on the left, it is a left-hand cut snips.

6.0.0   SAWING TOOLS

6.1.0   HAND HACKSAW

A hacksaw is a tool designed for cutting metals, plastics, and other synthetics. It has a pistol grip handle and uses a variety of thin metal blades. Locking wingnuts on the frame secure the blade and allow you to vary the blade tightness. Blades are selected by the number of teeth per inch, called pitch, and by the material composition of the blade. Blades with finer and more numerous teeth per inch are used for quicker cutting of harder, stronger materials. *Figure 11* shows a hacksaw. *Table 2* shows which blades to use for various materials.

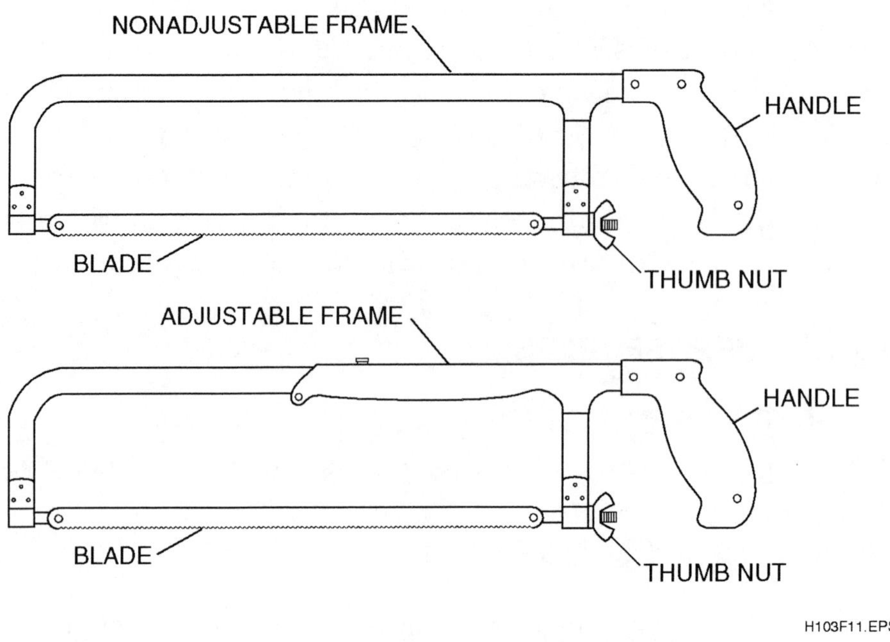

Figure 11.   Hacksaw

STOCK TO BE CUT	PITCH OF BLADE (TEETH PER INCH)	EXPLANATION
Machine steel Cold rolled steel Structural steel	14	The coarse pitch makes the saw free and fast cutting.
Aluminum Babbitt Tool steel High-speed steel Cast iron	18	Recommended for general use.
Tubing Tin Brass Copper Channel iron Steel metal (over 18 gauge)	24	Thin stock will tear and strip teeth on a blade of coarser pitch.
Small tubing Conduit Sheet metal (less than 18 gauge)	32	Recommended to avoid tearing.

Table 2.   Hacksaw Blades For Various Materials

The blade should always be installed in the saw with the teeth pointing away from the handle because the forward stroke is the cutting stroke. Because of their short length and brittleness, hacksaw blades require oiling to prevent rust and excessive wear.

Blade materials include carbon steel, molybdenum alloy steel, tungsten steel alloy and high-speed steel with either tungsten or molybdenum alloy contents. Tooth set refers to the way teeth are "set" from one side or the other to provide clearance (kerf) for the blade to pass through the metal and prevent binding. Two kinds of set are used, a wavy set and a raker set. A raker set allows for one tooth to be offset or bent left and the next one right. A wavy set has several teeth in a row bent one way, then several teeth in a row bent the other way. Coarse blades of 18 or lower pitch are usually raker set, while finer toothed blades are generally wavy set.

6.2.0   POWER HACKSAWS

The power hacksaw is a machine used for sawing all kinds of metal except hardened steel. The frame and blade move back and forth on the workpiece. Power is supplied by an electric motor which operates an eccentric gear mechanism. The power hacksaw can be adjusted so that it feeds and stops itself.

Blades are inserted in the frame in a manner similar to the hand hacksaw, with the teeth pointing toward the workpiece. The metal to be cut is held in a self-contained vise. The metal is measured from the workpiece side of the blade to the end length of the metal. The vise is tightened against the workpiece and the saw is started after the feed has been set and the blade manually moved toward the starting place on the stock. Power hacksaw blades are selected with five factors in mind:

- Length
- Thickness
- Width
- Tooth coarseness (pitch)
- Kind of material

Blades are available in many lengths, thicknesses, widths, and pitches. The most common size is 1" to 1¼" wide with a pitch of 10 to 14 teeth per inch. Almost all power hacksaw blades are raker tooth set. A coolant should be directed onto the blade for efficient cutting.

When using power hacksaws, the following safety and maintenance considerations should be observed:

WARNING!   Always wear proper eye protection and safety gloves around power tools. Do not operate power tools without proper ground fault protection. Before connecting to the power source, make sure the "POWER ON/OFF" switch is in the "OFF" position.

- Make sure to use the right saw blade for the job.
- Always use a sharp saw blade.

- Make sure the saw blade is securely tightened in the saw before starting.
- Let the saw blade do the work. Never force the saw while working.
- Power saws require relatively little maintenance. Follow the manufacturer's instructions.
- Keep the saw and saw motor ventilation passages clean.

7.0.0 DRILL PRESS

The two types of drill presses are the bench type (*Figure 12*) and the floor model. The bench drill press can either be fastened to a standard work bench or may be held in place solely by the weight or shape of the machine. The floor drill press is usually heavy enough so that it need not be secured to the floor, but can be set where desired. The table upon which the workpiece is held is usually adjustable over the entire range of the support column.

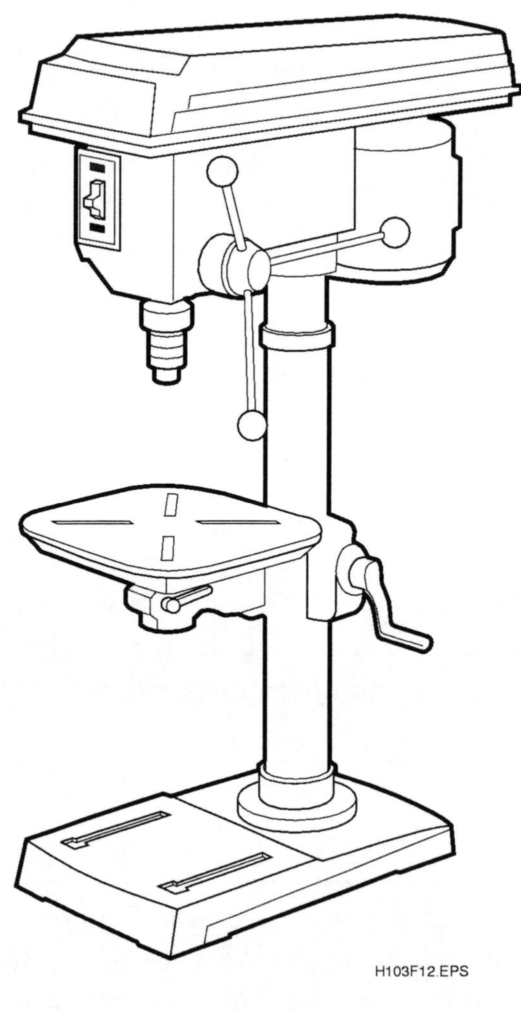

Figure 12. Bench Drill Press

The speed of the drill press can be varied by the placement of drive belts or through the use of gearing. Speed can be regulated for the metal being drilled. The feed of the drill is measured in fractions of an inch per revolution and should vary with the type of workpiece material and size of the drill. The correct speed and feed are dependent upon many conditions but can generally be determined by judgment and experience.

When using electric drill presses, the following safety and maintenance considerations should be observed:

WARNING!   Always wear proper eye protection and safety gloves. Do not operate a drill press without proper ground fault protection. Before connecting to the power source, make sure the "POWER ON/OFF" switch is in the "OFF" position.

- Make sure to use the right drill bit for the job.
- Always use a sharp drill bit.
- Make sure the drill bit is securely tightened in the chuck before starting.
- Let the drill do the work. Never force the drill while drilling. This dulls the drill bit and can damage the drill press bearings.
- Drill presses require relatively little maintenance. Follow the manufacturer's instructions.
- Keep the drill press and ventilation passages clean.

8.0.0   MEASURING DEVICES

MEASURING DEVICES

Accurate measurements are vital in the HVAC trade. Below are some measuring devices that will provide precise measurements in a variety of applications.

8.1.0   CIRCUMFERENCE RULE

The upper edge of the circumference rule (*Figure 13*) can be used for the same general-purpose measuring as the steel rule described in Core Module 00103. The lower edge, however, indicates at a glance the circumference of any given cylinder. For example, a cylinder with a diameter of 6 inches would be 19 inches in circumference.

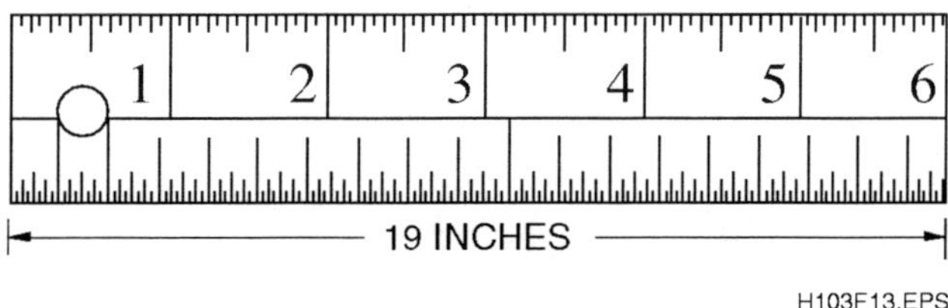

Figure 13.   Circumference Rule

The reverse side of the rule contains various tables indicating container sizes, dry and liquid measures, and container capacity computing rules. The circumference rule is available in 36-inch or 48-inch lengths.

8.2.0   CALIPERS AND DIVIDERS

Calipers (*Figure 14*) are used to measure the thickness or diameter of a piece of work, to measure distances between or on surfaces, and to compare and transfer work dimensions.

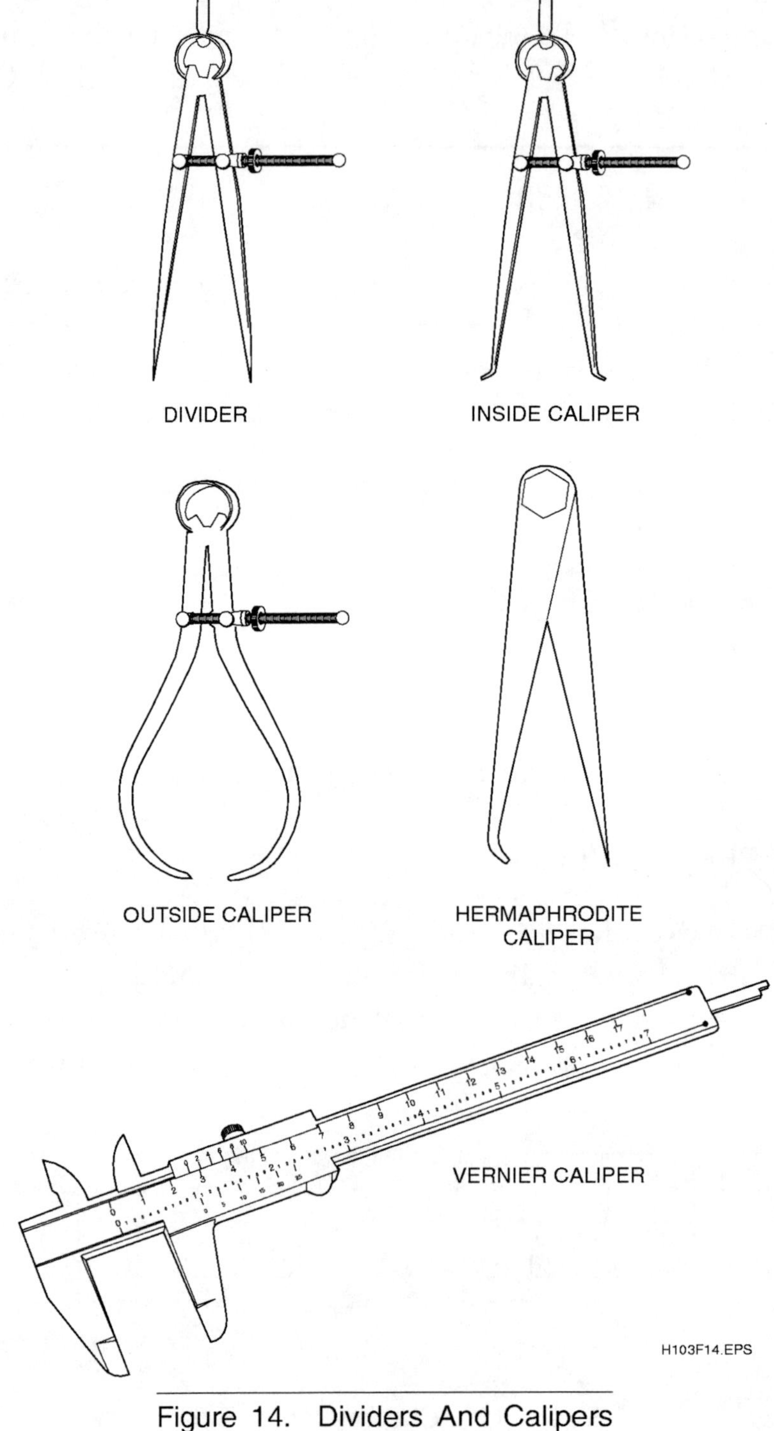

Figure 14.   Dividers And Calipers

Inside calipers are for taking inside measurements of work such as pipes, screw holes and similar positions. Outside calipers are for measuring outside diameters of work such as pipes, tubes and the like.

Dividers are used to measure the distance between lines or points, transfer dimensions from measurements to the work and to scribe lines, arcs and circles.

Keyhole and hermaphrodite calipers are variations of inside and outside calipers. Their principle use is in scribing parallel lines from an edge or for locating the center of cylindrical work. The keyhold caliper differs from the hermaphrodite caliper in that the "leg" does not taper and has a blunt end.

8.3.0 MICROMETER

A micrometer is an instrument that measures in thousands of an inch (*Figure 15*). Metric micrometers are also available that measure in $^1/_{100}$ of a millimeter. An outside micrometer is most commonly used. It is used to measure outside diameters of round objects and thicknesses of flat stock and materials.

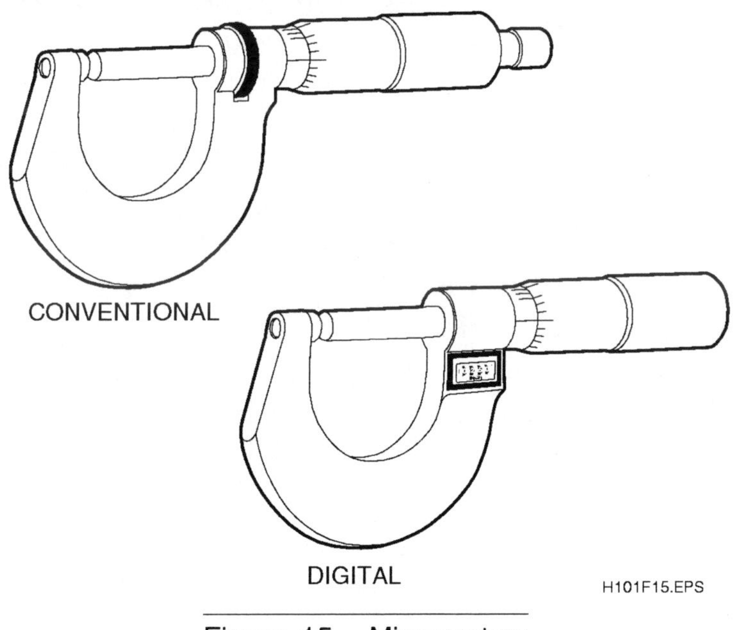

Figure 15.   Micrometers

The micrometer is a fine precision tool and should be handled with extreme care, as if it were a watch. Dust, dirt and grease may damage its internal mechanisms and render it useless.

All micrometers are calibrated alike and operate on the principle that a screw with a pitch of 40 threads per inch will advance $^1/_{40}$ of an inch or .025" (twenty-five one-thousands of an inch), with each complete turn. There are also 40 lines to the inch on the sleeve, the same number of threads as on the screw. Each turn, as indicated above, equals .025". If one turn equals .025", then $^1/_{25}$ of a turn will equal .001", ($^1/_{25}$ of .025") making it possible to measure in thousandths of an inch. One-half and lesser parts of a thousandth can be judged by sight as nearly as possible or a ten-thousandth (vernier) micrometer may be used. Micrometers are available that have direct reading digital readouts that eliminate complicated conversions.

8.4.0 VERNIER CALIPER

The vernier caliper (shown previously in *Figure 14*) is made up of a graduated steel rule or beam with a fixed jaw, a movable jaw which carries a graduated vernier scale, and a mechanism for making fine adjustments. One side of the scale is graduated for reading outside measurements, and the other side is for reading inside measurements.

Making accurate measurements with the vernier caliper requires the same care and sensitive touch as when measuring with a micrometer. Also, the reading of a vernier caliper is very similar to reading a micrometer. Each inch of the beam, like each inch on the sleeve of a micrometer, is graduated into 40 equal parts with the decimal equivalent of $^1/_{40}$ of an inch equal to .025". Again, as with the micrometer, every fourth division line is a little longer and is marked 1, 2, 3, etc., which denotes tenths of an inch. The vernier plate correlates with the micrometer thimble and is graduated into twenty-five equal divisions, representing thousands of an inch. Vernier calipers are available with dial and digital readouts.

8.5.0 THICKNESS (FEELER) GAUGE

Feeler gauges (*Figure 16*) are used to measure small distances or clearances between objects that cannot be measured in any other way. The feeler gauge is made up of a number of thin steel blades which fold into a handle. Feeler gauges of various thicknesses are tried until one fits snugly. The thickness of each blade is marked by a number. The number on each blade indicates the thickness of that blade in thousands of an inch and/or in millimeters. A modification of the feeler gauge is a stepped feeler gauge which is often called a go/no-go gauge.

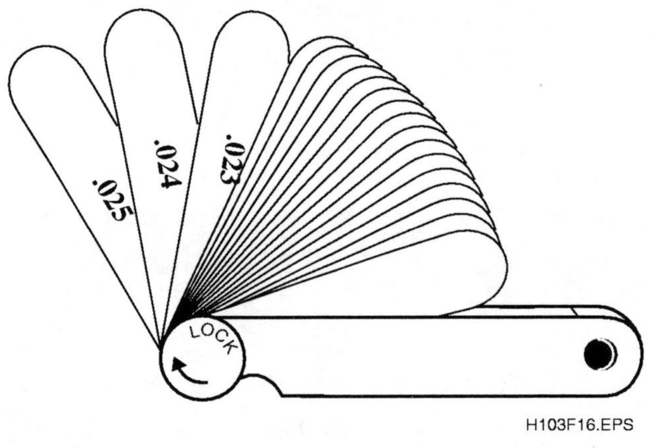

H103F16.EPS

Figure 16. Thickness (Feeler) Gauge

8.6.0   THREAD PITCH GAUGE

Nuts, bolts, and screws come in many sizes from very small to very large. These fasteners can be measured in terms of pitch. Not only can a rule be used to count the number of threads per inch, but one can also use a thread pitch gauge (*Figure 17*). A thread gauge has many leaves. Each leaf measures 1" across and has a specific pitch marked on the side. Proper use of the gauge involves finding the leaf that has the proper number of teeth to fit the number of threads per inch (pitch) on the fastener and then reading the number imprinted on the leaf.

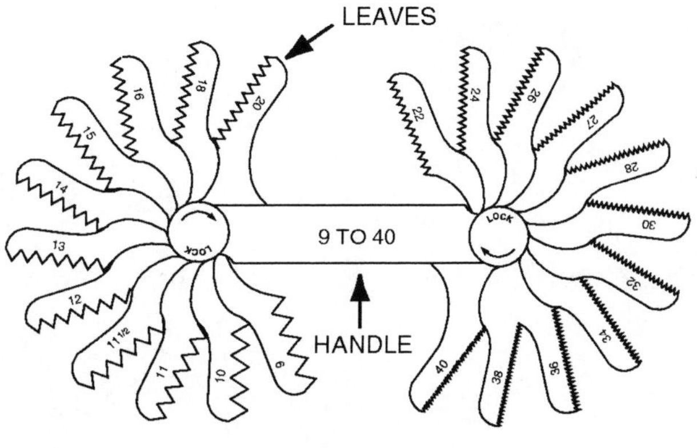

H103F17.EPS

Figure 17.   Thread Pitch Gauge

SUMMARY

The best work is performed by using the right tool for the job. Proper use of the right tool helps increase production. It also takes into consideration the safety of the worker. Poor workmanship and injury can often be traced to the improper use of tools.

References

For advanced study of topics covered in this task module the following books are suggested:

Modern Plumbing, The Goodheart-Willcox Company, Inc., South Holland, Illinois.
Modern Refrigeration and Air Conditioning, The Goodheart-Willcox Company, Inc., South Holland, Illinois.
Refrigeration & Air Conditioning Technology, Second Edition, Delmar Publishers, Inc., Albany, New York.

1. The size of a straight pipe wrench is found on its:
   a. hook jaw.
   b. housing.
   c. handle.
   d. nut.

2. When using a torque wrench to tighten a bolt, a noticeable popping sound occurs. This means that:
   a. the break-away torque has been reached.
   b. the bolt is seized.
   c. the run-down resistance has been reached.
   d. none of the above.

3. Most soft face hammers:
   a. have faces that yield at the striking point.
   b. are used on surfaces that cannot be marred.
   c. will not pick up chips and transfer them to the work.
   d. all of the above.

4. Aviation snips with red handles cut:
   a. straight.
   b. to the left.
   c. to the right.
   d. circles.

5. When using a hacksaw to cut channel iron, a blade with _____ teeth per inch should be used.
   a. 14
   b. 18
   c. 24
   d. 32

6. To measure the circumference of a cylinder, the best tool to use is the:
   a. micrometer.
   b. outside caliper.
   c. inside caliper.
   d. circumference rule.

7. The sleeve of a micrometer is graduated into 40 equal parts, each with a decimal value equal to:

   a. .010 inch.
   b. .025 inch.
   c. .015 inch.
   d. .250 inch.

8. A single tool commonly used for making both inside and outside measurements is called a:

   a. micrometer.
   b. divider.
   c. vernier caliper.
   d. hermaphrodite caliper.

9. A tool used to measure clearances between objects is called a:

   a. thickness gauge.
   b. vernier caliper.
   c. feeler gauge.
   d. both a and c.

10. A thread pitch gauge is used to measure:

   a. the threads per inch of fasteners.
   b. the taper per inch of fasteners.
   c. the outside dimensions of fasteners.
   d. the inside dimensions of fasteners.

ANSWERS TO SELF CHECK REVIEW / PRACTICE QUESTIONS

<u>Answer</u>	<u>Section Reference</u>
1. c	2.1.0
2. b	3.0.0
3. d	4.2.0
4. b	5.3.0
5. c	6.1.0
6. d	8.1.0
7. b	8.3.0
8. c	8.4.0
8. d	8.5.0
10. a	8.6.0

Copper and Plastic Piping Practices

Module 03104

COPPER AND PLASTIC PIPING PRACTICES

Objectives

Upon completion of this module, the trainee will be able to:

1. State the precautions that must be taken when installing refrigerant piping.
2. Select the right tubing for a job.
3. Cut and bend tubing.
4. Join tubing by using flare and compression fittings.
5. Determine the kinds of hangers and supports needed for refrigerant piping.
6. Insulate refrigerant piping.
7. State the basic requirements for pressure-testing a system once it has been installed.
8. Follow basic safety precautions for the installation, operation and maintenance of refrigerating and air conditioning equipment.

Prerequisites

Successful completion of the following Task Modules is required before beginning study of this Task Module: Common Core Curricula, HVAC Modules 03101 through 03103.

Required Student Material

1. Trainee Task Module
2. Appropriate Personal Protective Equipment

Course Map Information

This course map shows all of the *Wheels of Learning* task modules in the first level of the HVAC curricula. The suggested training order begins at the bottom and proceeds up. Skill levels increase as a trainee advances on the course map. The training order may be adjusted by the local Training Program Sponsor.

Course Map: HVAC, Level 1

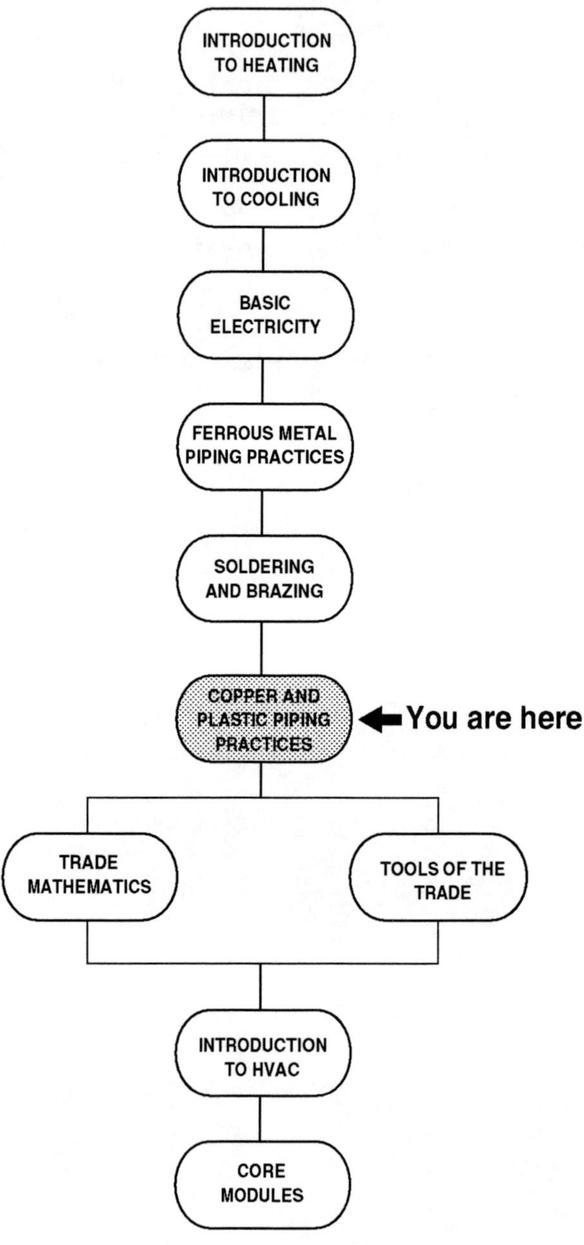

TABLE OF CONTENTS

Trade Terms Introduced In This Module

ACR tubing: Copper tubing made especially for refrigeration and HVAC work. It is especially clean and is usually charged with dry nitrogen. The ends are sealed to prevent contamination.

Annealing: Heat treating to soften metal. Soft copper tubing is made by annealing hard copper.

Brazing: A means of joining metal using an alloy with a melting point higher than that of solder, but lower than that of the metals being joined.

Compression joint: A method of connection in which tightening a threaded nut compresses a compression ring to seal the joint.

Flare fitting: A fitting in which one end of each tube to be joined is flared outward using a special tool. The flared tube ends mate with the threaded flare fitting and are secured to the fitting with flare nuts.

Halocarbon refrigerants: Short for halogenated hydrocarbons. A class of refrigerants that includes most of the refrigerants used in residential and small commercial air conditioning systems.

Inside diameter (I.D.): The distance between the inner walls of a pipe. Used as the standard measure for tubing used in heating and plumbing applications.

Insulation: A substance that retards the flow of heat.

Outside diameter (O.D.): The distance between the outer walls of a pipe. Used as the standard measure for ACR tubing.

Piping: A generic term used to designate thick wall pipe which can be threaded and joined with threaded fittings.

Specification: A document that describes the quality of the materials and work required. Specifications are the source for quality of tubing, fixtures, hangers, etc. to be used on a project.

Swaged joint: A method of creating a pipe joint in which the diameter of one of the pipes to be joined is expanded using a special tool. The other pipe then fits inside the swaged pipe.

Sweating: A method of joining pipe in which solder is applied to the joint and heated until it flows into the joint.

Tubing: Thin wall pipe; generally, pipe that can be easily bent.

1.0.0 INTRODUCTION

An HVAC technician must be able to work with several kinds of piping. Copper pipe is used to transport refrigerant in residential and smaller commercial air conditioning systems. In large commercial and industrial systems, welded steel piping is more common. Steel is also used with special refrigerants, such as ammonia. Steel and plastic piping are used to transport water in HVAC systems that use water as a heat exchange medium. Black iron pipe is used to transport natural gas for heating systems. This module focuses on handling, cutting, bending, and joining copper and plastic piping.

2.0.0 INSTALLATION PRECAUTIONS

There are six basic piping precautions that an HVAC technician must learn in order to avoid costly mistakes:

1. Protect refrigerant piping from exposure to dirt and other contaminants and don't leave open ends of piping stock exposed. Even tiny amounts of contaminants can cause damage in an HVAC system.
2. Remove the charge from a pressurized system before opening any solder joints. It is dangerous to work on piping when the system is under pressure and it is illegal to vent refrigerant to the atmosphere.
3. Only ACR copper piping and fittings should be used in refrigeration work. The copper piping and fittings (type M or DWV) used in plumbing are usually unsuitable for HVAC work.
4. Use as few fittings as possible. Fewer fittings mean less chance for leaks and pressure drops.
5. Exercise caution in making every solder connection. Use the correct solder and recommended techniques.
6. Pitch horizontal lines in the direction of refrigerant flow; otherwise, oil may cling to the inner walls of the tubing. Tubing pitch will allow oil to flow in the right direction. It also avoids backward flow during shutdown. The recommended pitch is $1/2$" per 10 feet of run.

3.0.0 MATERIALS

Residential and small commercial refrigeration systems using **halocarbon** (halogenated hydrocarbon) **refrigerants** normally use copper tubing. However, welded steel pipe and fittings may be specified where diameters above three inches are required. Aluminum, stainless steel and plastic tubing are used in some systems.

Tubing pre-charged with refrigerant is available in a variety of sizes to simplify installation. The tubing is charged with refrigerant and sealed at both ends with quick-connect fittings that are used to attach the tubing to the system. No soldering is required; therefore, the connections are less likely to leak. Pre-charged tube sets often come with insulation already installed. Because they are pre-charged with refrigerant, there is less likelihood of charging error. Refrigerant line sizes, applications and charges are fairly predictable, enabling the manufacturer to build pre-charged line sets that will fit many systems.

Large commercial refrigeration systems are not as simple to design and operate. Line sizes are larger and evaporator refrigerant temperatures vary widely, from +35°F down to -40°F and below. Structural location distances are more variable and some systems have refrigerant pressures at or near vacuum conditions. Oil return problems at low temperatures are more critical; therefore, the use of pre-charged line sets is almost nonexistent in these systems.

3.1.0 SELECTING COPPER TUBING AND FITTINGS

The size (diameter) of the tubing to be used depends on the heating or cooling capacity of the HVAC system. Soft copper tubing is suitable for small systems. It ranges in size from $\frac{1}{8}$" to $1\frac{3}{8}$" in diameter. It is easily bent with hand tools and comes in rolls of 25 and 50 feet; larger rolls may be special-ordered. Because of its length, soft copper needs fewer connections. This saves time on installation and decreases the chances of a refrigerant leak. On small-capacity jobs, soft copper may be joined with flare fittings. This eliminates the need for soldering.

Hard copper tubing has a thicker wall and can therefore withstand higher system pressures. It comes in 20-foot lengths and should not be bent. Sections must be joined by soldering or brazing and bends must be made using fittings such as tees and elbows.

Hard and soft copper may be used on the same job because not all refrigerant lines are under high pressure.

The term **tubing** generally applies to thin-wall materials that are joined together by methods other than threads cut into the tube wall. **Piping** is the term applied to thick pipe-wall material (iron and steel) into which threads can be cut and to which fittings can be joined by threading onto the pipe.

Another distinction between tubing and piping is the method of size measurement. Piping uses inside and outside diameter measurements, depending on the pipe size. Tubing uses inside and outside diameter measurements depending on the pipe *use*. Tubing sizes used for plumbing and heating are expressed in terms of **inside diameter (I.D.)** (*Figure 1*). Air conditioning and refrigeration tubing is expressed in terms of **outside diameter (O.D.)**.

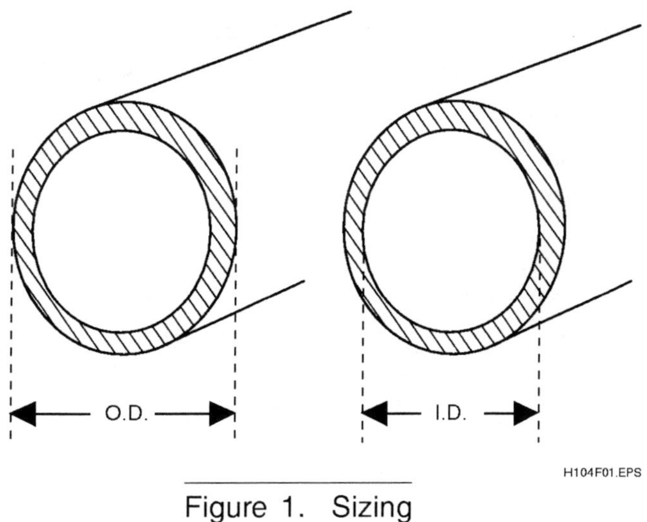

Figure 1. Sizing

The tubing used in all domestic refrigeration systems is special **annealed** (softened by heat treatment) copper. It is known as **ACR tubing**; that is, it has been manufactured specifically for use in air conditioning and refrigeration work.

ACR tubing is pressurized with nitrogen gas to keep out air, moisture and impurities and to protect against oxides that are normally formed during **brazing** (hard soldering). Nitrogen should also be fed through the tubing during the brazing process to prevent harmful oxidation. The ends of new stock are plugged to keep out moisture and dirt. These plugs should be replaced after cutting a length of tubing for use.

4.1.0 TYPES OF COPPER TUBING

Copper is divided into three classifications based on the wall thickness. Type K (heavy wall) and type L (medium wall) tubing are ACR approved. Type M, thin wall, is not used on pressurized refrigerant lines since it does not have the wall thickness to meet safety standards. It is, however, used on water lines and condensate drains.

Heavy-wall, type K, is meant for special use where heavy corrosion might be expected. Type L is most frequently used for normal refrigeration applications. Both K and L copper tubing are available in soft or hard drawn types.

Soft copper tubing is available in sizes from $^1/_8$" O.D. to $1^3/_8$" O.D. Line sizes from $^1/_4$" to $^3/_4$" O.D. are common. This tubing may be soldered or used with flared or other mechanical fittings. It is easily bent or shaped, but it must be held in place by clamps or other hardware as it cannot support its own weight.

Hard-drawn tubing is also widely used in commercial refrigeration and air conditioning systems. It comes in straight lengths of 20 feet and in sizes from $^3/_8$" O.D. to 6" O.D. The lengths are charged with nitrogen and plugged at each end to maintain a clean, moisture-free, internal condition. It is intended for use with formed fittings to make the necessary bends or changes in direction. It is more self-supporting and therefore needs fewer supports than soft copper tubing.

4.2.0 CUTTING TUBING

Cutting copper tubing is accomplished with one of two tools:

1. Hand-held tube cutter
2. Hacksaw and sawing fixture

The tube cutter is the preferred method because it makes a cleaner joint and leaves no metal particles.

4.2.1 Hand-Held Tube Cutter

To use the tube cutter (*Figure 2*), place it on the tube at the desired cutting point. Tighten the knob, forcing the cutting wheel against the tube. The cut is made by rotating the cutter around the tube under constant pressure. Afterward, the built-in deburring blade is used to remove any burrs from inside the tube. There are a variety of models available and the cutting ranges vary from $^1/_8$" O.D. to as much as $4^1/_8$" O.D.

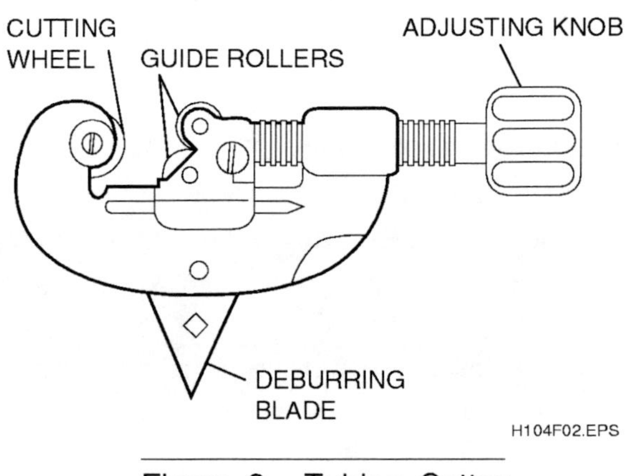

Figure 2. Tubing Cutter

4.2.2 Hack Saw And Sawing Fixture

For larger size hard-drawn tubing, a hack saw and sawing fixture may be used. The fixture helps square the ends and allows more accurate cuts. The blade of the hack saw should have at least 32 teeth per inch. Avoid getting saw cuttings inside the tubing. File the end to produce a smooth surface and carefully clean the inside of the tubing with a cloth.

Portable power tools with abrasive wheels to cut, clean and buff the tubing are available. They may be used for production pipe cutting.

4.3.0 BENDING

It is best to bend rather than cut soft-drawn tubing. Smaller-diameter tubing can easily be bent by hand. Care must be taken to avoid flattening the tube with a bend that is too sharp. For large tubing, the minimum bending radius to which a tube may be curved is up to 10 times the tube's diameter. For smaller tubing, it may be up to 5 times the tube's diameter.

Tube-bending springs (*Figure 3*) are available to be placed on the outside of the tube to prevent it from collapsing during hand bending. Tube bending equipment (*Figure 4*) is also available for making accurate and reliable bends in both soft and hard tubing. Various sizes of forming attachments are available to bend tubing of diameters up to ⁷⁄₈" at any angle up to 180 degrees.

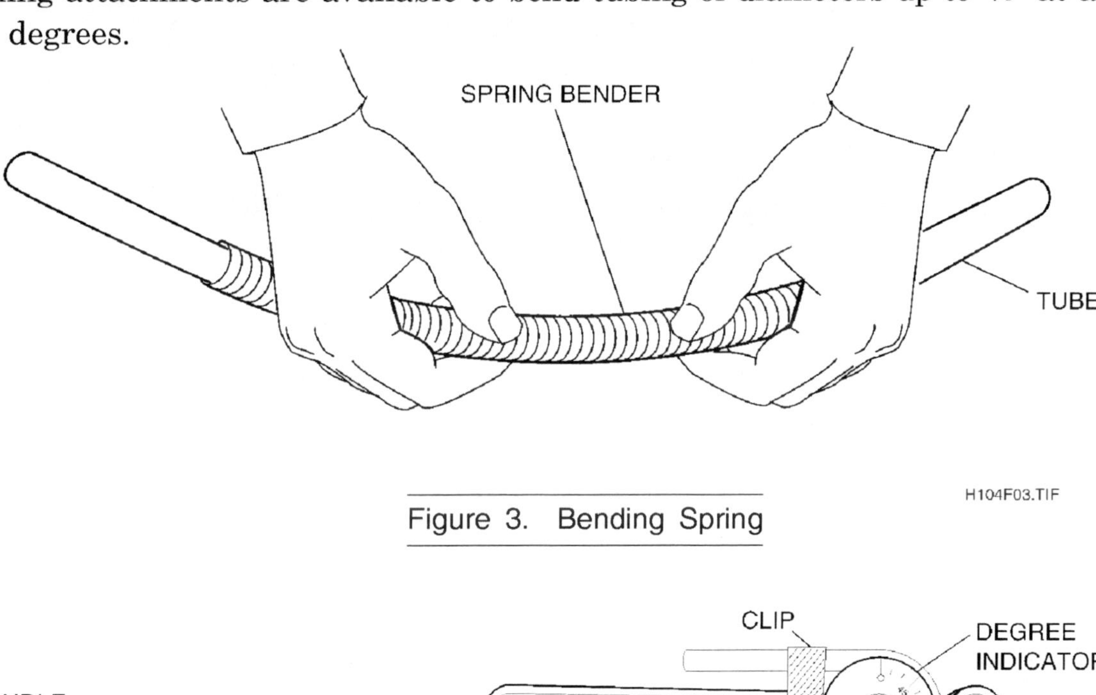

Figure 3.   Bending Spring

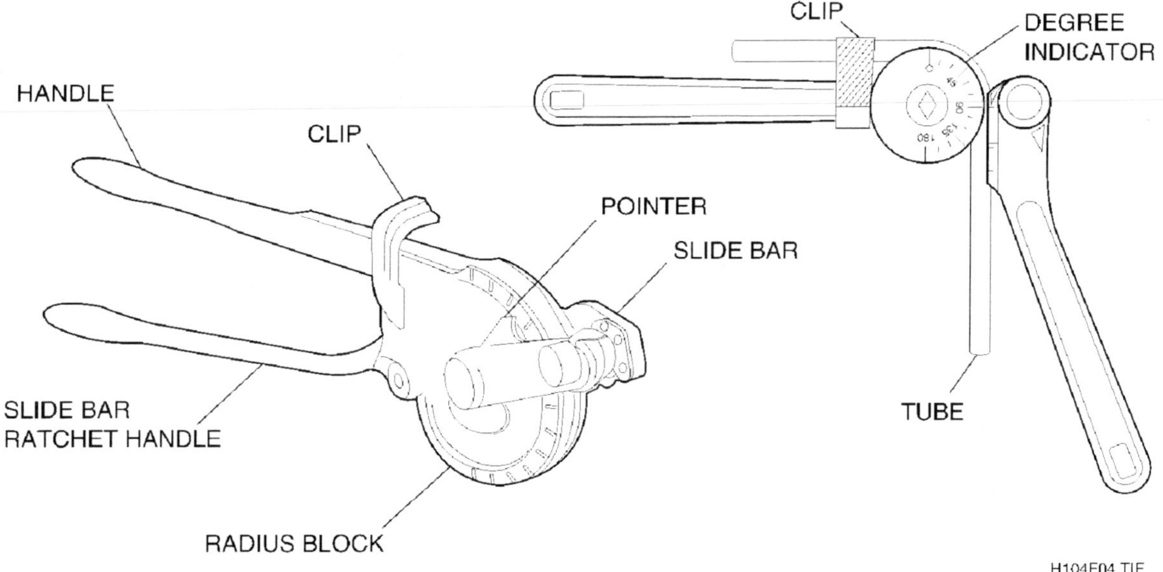

Figure 4.   Tube Bending Equipment

4.4.0 JOINING

There are two accepted methods of joining tubing. Because copper tubing used for refrigerant piping is too thin to thread, mechanical coupling or heat bonding methods are used. Mechanical coupling includes flared and compression fittings that are semipermanent (can be taken apart). Heat bonding is accomplished by soldering or brazing. They are considered to be permanent, but can be separated by reheating the joint. Soft solder should not be used on the high pressure side of a refrigeration system, as the soft solder is vulnerable to vibration.

4.4.1 Flared Connections

Soft-drawn copper fittings should be leakproof and easily dismantled with the right tools. The flared connection (*Figure 5*) is a popular method of joining soft copper tubing.

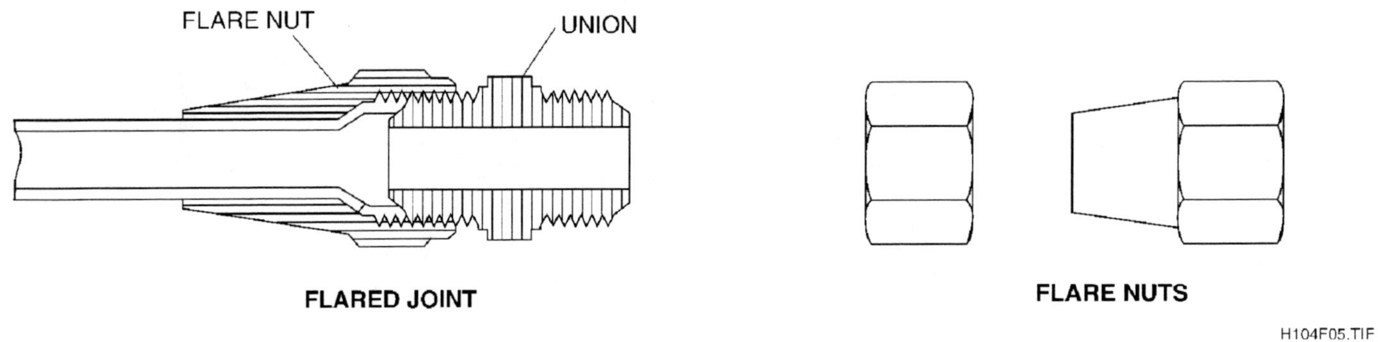

Figure 5. Flared Connection And Flare Nuts

A special flaring tool, shown in *Figure 6*, expands the tube's end into the shape of a cone, or flare.

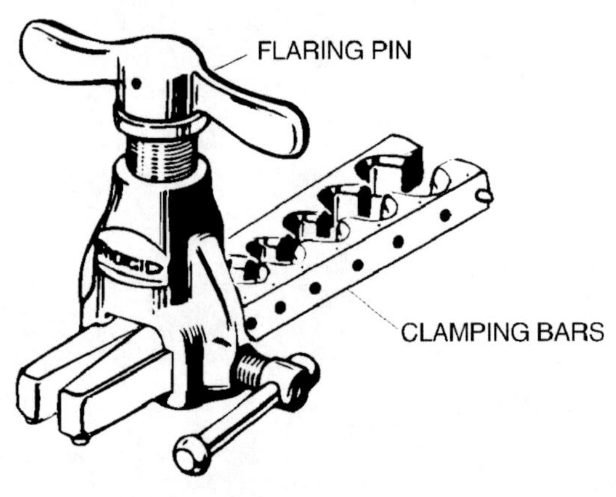

Figure 6. Flaring Tool

Two kinds of **flare fittings** are popular:

1. The single thickness flare forms a 45-degree cone that fits up against the face of a flare fitting (*Figure 7*).

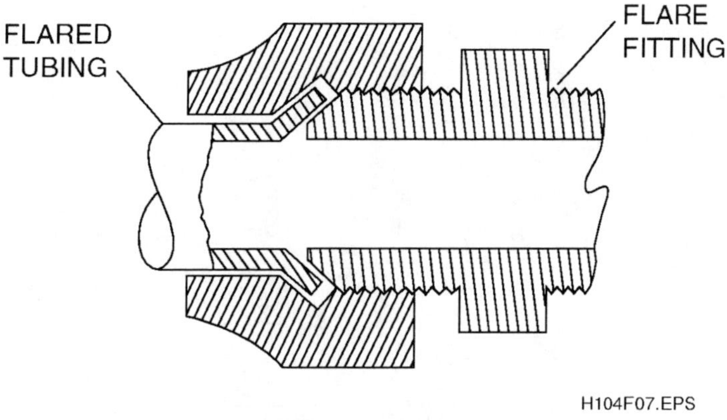

Figure 7.   Single Thickness (Angle) Flare

Under excessive pressure or expansion, larger size tubing with a single flare may be weak. In these applications a double thickness flare is preferable. Double thickness flare connections can also be taken apart and reassembled more easily without damage.

2. The double thickness flare and the method of forming it are shown in *Figure 8*. In one operation the single thickness flare is formed, then the lip is folded back onto itself and compressed.

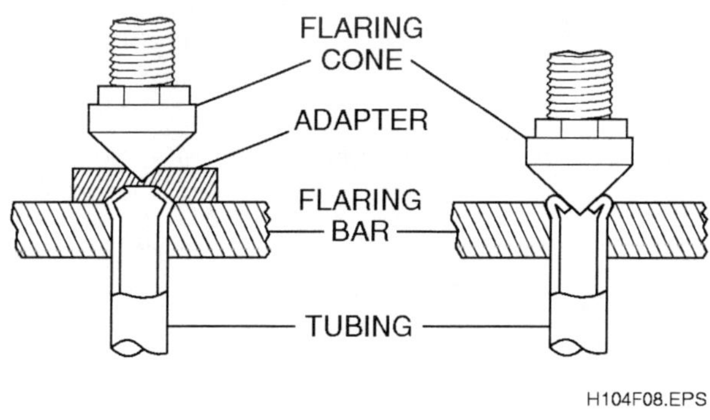

Figure 8.   Double Thickness Flare

Flare fittings used in refrigeration and air conditioning consist of a variety of elbows, tees, unions, etc. (*Figure 9*). They are drop-forged brass and are accurately machined to form the 45-degree flare face. Fittings are based on the size of tubing to be used. Flare nuts are hexagon-shaped. A flare nut wrench is used with these fittings. The fitting body usually has a flat surface which will accommodate an open-end **swaged joint**.

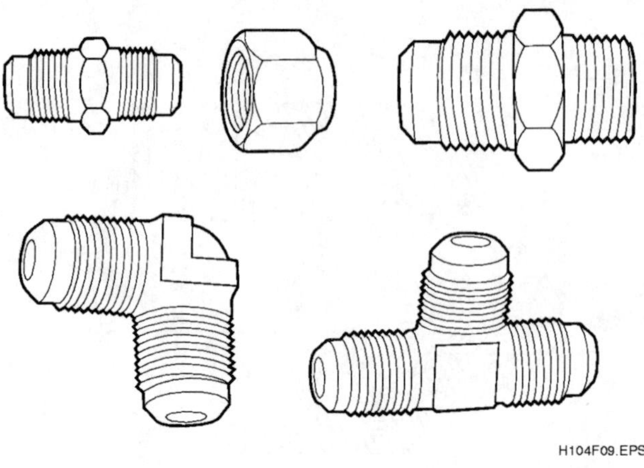

Figure 9. Types Of Flare Fittings

When joining copper tubing of the same diameter, some workers feel it is more reliable to join two pieces by making a swaged connection (*Figure 10*). This type of joint is created with a swaging tool, which is similar to the flaring tool. The block holds the tube and the punch is forced into the end of the tubing. When properly done and soldered, swaging produces very secure soldered joints, thereby reducing leak hazards; however, it does take more time.

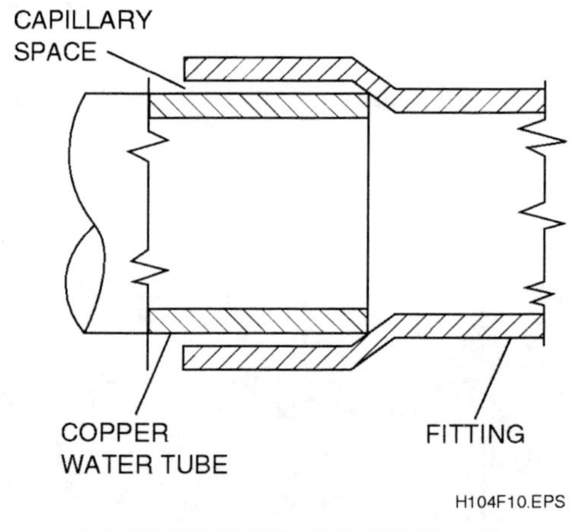

Figure 10. Swaged Joint

4.4.2 Compression Joints

Compression joints (*Figure 11*) are sometimes used for joining refrigerant tubing. This method is popular because it takes less time than making flared or heat-bonded connections. A seal is obtained by connecting the tubing to the threaded fitting with a coupling nut and compression ring. The joint is formed by compressing the ring between the nut and the threaded fitting, forming a gas-tight seal. This particular type of coupling comes in sizes from 1/4" to 1 1/8" O.D. Adapters are available to join these couplings to other fittings.

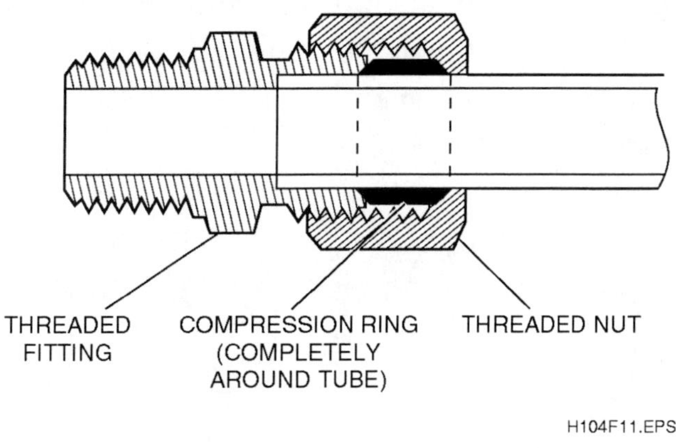

Figure 11. Compression Fitting

4.4.3 Elbow Fittings

Elbows produce a large pressure drop in a piping system. With equal velocities, the amount of this pressure drop depends upon the sharpness of the turn. Long radius elbows, rather than short radius elbows, should be used whenever possible. When laying out offsets, 45-degree ells are recommended over 90-degree ells (*Figure 12*).

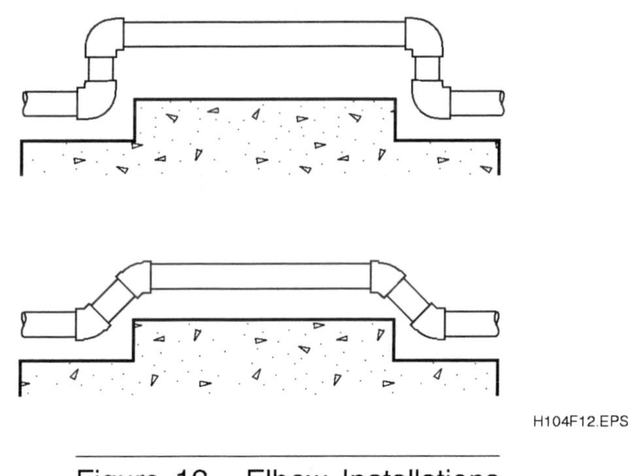

Figure 12. Elbow Installations

4.4.4 Tee Fittings

If not properly installed, tee fittings can cause a condition known as "bullheading," which results in turbulence, greatly adding to the pressure drop. Bullheading may also cause hammering in the line. If more than one tee is installed in the line, a straight piece of pipe 10 diameters in length between tees is recommended to reduce turbulence.

4.4.5 Sweat Fittings

As stated earlier, copper tubing is often joined by soldering or brazing techniques in which a soft metal is melted in the joint between two tubes. This type of heat bonding is also known as **sweating**. There are special fittings known as sweat fittings made for soldering or brazing copper tubes (*Figure 13*). Sweat fittings may be made of copper or brass. They are made slightly larger than the tubes to be joined, leaving only enough room for solder to flow into the joint. Some sweat fittings, like the transition fitting shown in *Figure 13*, allow copper tubing to be joined with threaded pipe.

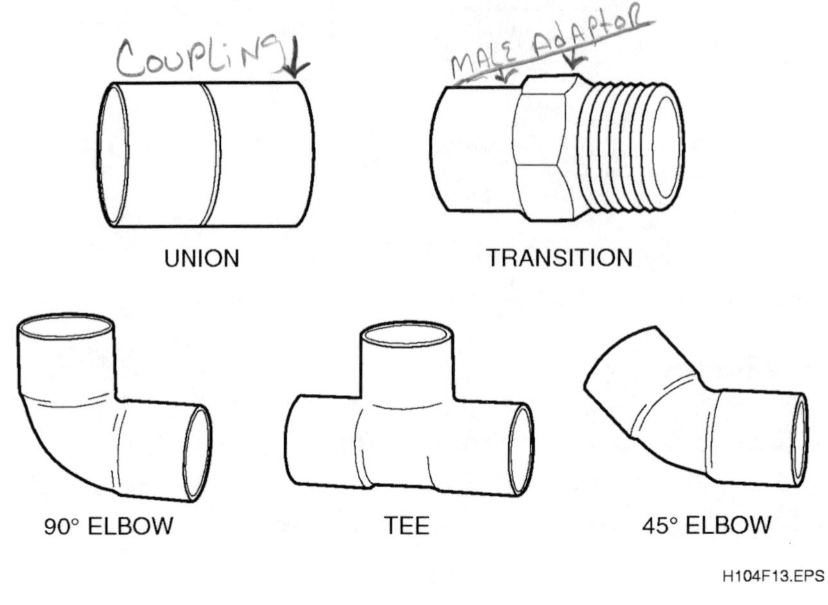

Figure 13.   Sweat Fittings

5.0.0   PLASTIC PIPE

Plastic pipe is used to vent furnaces, drain condensate from air conditioners, and transport water in water-cooled HVAC systems. There are several types of plastic pipe, each with a different application.

5.1.0   ABS (AERYLONITRILEBUTADIENE STYRENE) PIPE

ABS pipe is rigid and has good impact strength at low temperatures. It is used for water, vents, and drains. In an unpressurized system, it can withstand heat up to 180°.

5.2.0   PE (POLYETHYLENE) PIPE

PE pipe is used in cold water systems such as water source heat pumps. It is flexible and, like ABS, has good impact strength at low temperatures. It is joined with clamps.

5.3.0   PVC (POLYVINYL CHLORIDE) PIPE

PVC is a rigid pipe with high impact strength. It can be used in high-pressure systems at low temperatures. PVC pipe is usually joined with cement. It can also be threaded and joined to steel pipe with a transition fitting.

5.4.0 CPVC (CHLORINATED POLYVINYL CHLORIDE) PIPE

The applications and joining methods for CPVC are very similar to those of PVC. However, CPVC can be used for hot water (up to 180 degrees) in a pressurized system (up to 100 psi).

5.5.0 JOINING PLASTIC PIPE

Plastic pipe may be cut with a tubing cutter, a hacksaw, or a special cutter known as a plastic tubing shear. The cut must be square to be sure of a good joint. Once cut, the pipe is deburred inside and out, using a knife or file.

To join PVC or CPVC pipe, the pipe end is cleaned, then primer and cement are applied to both the pipe and fitting. The pipe is then inserted into the fitting and turned one quarter turn to spread the cement. Some PVC and CPVC pipe (Schedule 80) can be threaded; however, the same die should not be used for both plastic and metal because a die used for metal will not be sharp enough to cut plastic.

The procedure for joining ABS pipe with cement is the same as above, except that no primer is required.

6.0.0 HANGERS AND SUPPORTS

Air conditioning and heating installations planned by architects and engineering consultants include plans and **specifications** which completely describe the proposed system. Specifications are based upon codes or ordinances and must be adhered to. A specification for pipe hangers, for example, may read as follows: "All piping shall be supported with hangers spaced not more than 10 feet apart (on center). Hangers shall be the malleable iron split-ring type and shall be as manufactured by XYZ Hangers, Inc. or other approved vendor."

Hangers (*Figure 14*) are used for horizontal or vertical support of pipes and piping. The principle purpose of hangers and brackets is to keep the piping in alignment and to prevent it from bending or distorting. Horizontal hangers can be attached to wooden structures with lag screws or large nails.

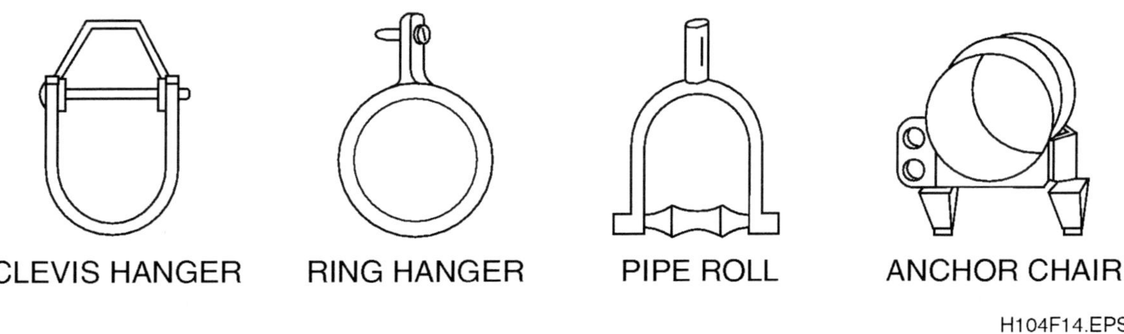

Figure 14. Hangers

Pipe hangers are used primarily to support pipe, but they may also be used as a vibration isolator. If vibration problems are not anticipated, ordinary plumbing practices may be used.

For fastening to beams and other metal structures, beam clamps or C-clamps (*Figure 15*) are used. Other horizontal support clamps and brackets are shown in *Figure 16*. Vertical hangers (*Figure 17*) or pipe risers consist of a friction clamp which can be attached to structural site components to support the vertical load of the pipe. Special fasteners are used to attach hangers to masonry, concrete or steel.

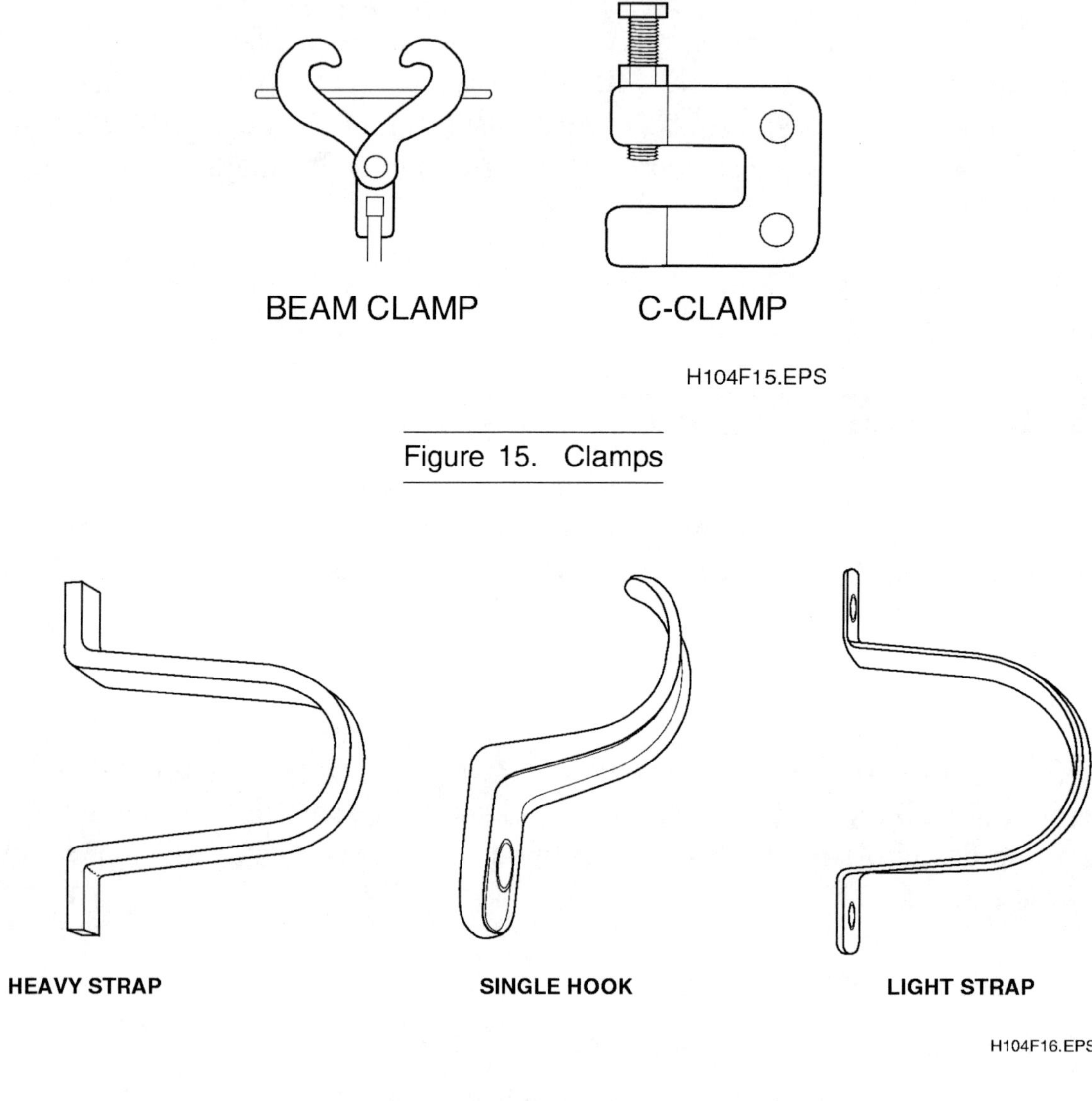

Figure 15.   Clamps

Figure 16.   Horizontal Clamps

HVAC TRAINEE TASK MODULE 03104

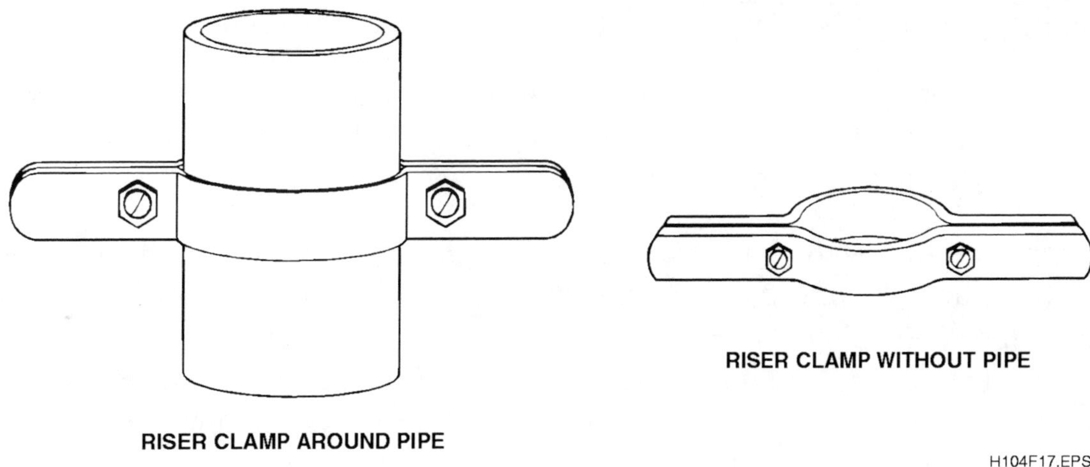

Figure 17. Vertical Hangers

System installers will be required to make the final connection between the evaporator of the air handling unit and the ductwork. This is usually a canvas connection. Canvas will eliminate any vibration carry-over from the air handling unit to the ductwork. Good weatherproof canvas is a necessity and must be correctly installed.

7.0.0 INSULATING

Under certain temperature and humidity conditions, condensation will form on refrigerant and cold water piping and may drip into equipment or occupied areas. Refrigeration piping can pick up heat from the air, causing an air conditioning system to lose efficiency. Similarly, heat can escape from hot water piping. To prevent these conditions, some piping is insulated.

Insulation is a material that prevents the transfer of heat. Cork, glass fibers, mineral wool and polyurethane foams are examples of insulating materials. Insulation should be fire resistant, moisture resistant and vermin proof.

ACR tubing can be purchased with factory-installed insulation. If it is necessary to install the insulation at the job site, it should be added before the tubing is connected. That way, the insulation can be slid onto the tubing. The inside of the insulation is usually powdered to allow it to slip on easily.

If the insulation cannot be installed before the tubing is connected, it must be slit lengthwise to fit onto the pipe. Slit seams and connecting seams must then be sealed with adhesive. Do not use tape. Insulation should not be stretched, because its effectiveness will be reduced.

Some pipes are always insulated, while others are insulated only under certain conditions. Local building codes and job specifications will usually describe insulation requirements. Which pipes to insulate and under what conditions will be covered in detail in a later module.

8.0.0 PRESSURE TESTING

When the installation is complete, it must be inspected to make sure that the work has been done in accordance with the job specifications and applicable codes. Then, the system must be leak-tested to be sure that all connections are secure.

Testing under pressure is the best method of leak testing. This test is done using a mix of 5 percent refrigerant gas and 95 percent nitrogen. Nitrogen is inexpensive and, if there is a leak, you do not risk losing expensive refrigerant and violating laws regarding the release of refrigerants to the atmosphere. The 5 percent mix specified above is within EPA regulations. The test pressure will vary with the size of the system and the refrigerant used; it should never exceed the maximum system test pressure as stated on the system nameplate. A safe maximum is 170 psig.

8.1.0 TECHNIQUE

When nitrogen is used to build up pressure for testing at code specifications, the technician must be very careful. First, the refrigerant cylinder should be disconnected to prevent nitrogen backing into it. Second, there must be a hand shutoff valve, a pressure regulator, a pressure gauge and a pressure relief valve in the charging line. The relief valve should be adjusted to open one or two psi above the test pressure.

WARNING!   Oxygen, air, acetylene, or other gases should never be used to pressure test a refrigerant system. Oxygen will cause an explosion if it comes into contact with refriger-ant oil. Acetylene is highly flammable. The only gas other than refrigerant that should be introduced into an HVAC system is nitrogen.

After pressure has built up in the piping, it is a good practice to rap each joint and connection with a rubber mallet to be sure that all connections are leakproof. Soap bubbles or a commercial test liquid can be used to check for leaks when using nitrogen. If no leaks are found, the nitrogen is usually left in the system for 24 hours. Again, if no leaks are found, the inspector sometimes checks further by producing a vacuum in the system. If the vacuum is maintained over a specified period of time, the installation is approved. Electronic leak detectors are available to test for refrigerant leaks. They will detect tiny amounts of refrigerant and sound an audible alarm when a leak is detected.

9.0.0 PIPING CODES

Most cities, counties and states have adopted codes which may be based on suggested national standards, but which may be subject to local interpretation and reflect local conditions. In earthquake-prone areas, for example, there may be special requirements for installing equipment and piping. In flood-prone areas, other special requirements will apply. The job specifications should detail these requirements. These codes are usually divided into electrical, plumbing and refrigeration, and other building codes. Code violation corrections are expensive and are the responsibility of the installer. Standards serve as guidelines to improve the performance or reliability of a system or component, but codes and ordinances are specific and mandatory rules to be complied with.

10.0.0 SAFETY

HVAC systems operate under high pressure and are powered by electricity. Therefore, anyone working with HVAC systems must know and practice applicable safety precautions in order to protect themselves and their co-workers from injury and to prevent equipment damage.

Specific safety precautions will be covered as you progress through the training program. Here are a few general precautions that you should follow whenever your job or your training place you near the equipment.

- Do not attempt any work that you have not been specifically trained to do, and then, do it only under the direct supervision of your instructor or foreman. For example, do not operate system valves unless you know exactly what the result will be. In general, before you work independently on an HVAC system, you must know the temperature and pressure conditions that exist at every point in the system, and how those conditions can be affected by malfunctions or by changes in valve positions.
- Disconnecting piping in a system that is under pressure can cause serious injury, as well as release of refrigerant to the atmosphere. The only proper way to reduce the pressure and avoid release of refrigerant is to remove the refrigerant charge using approved recovery methods and equipment.
- Wear goggles and gloves when working with refrigerants, and don't work with refrigerants in poorly ventilated areas. Also, be aware that some refrigerants become toxic when exposed to an open flame.
- Unless it is absolutely necessary to work with electrical power applied, always shut off power and use approved lockout/tagout procedures to avoid electrical shock.
- Only nitrogen should be used to pressure test a system and recommended pressures should be used. If testing pressures are not given on the name plate, never exceed national or local codes or manufacturer recommendations. Oxygen or acetylene must never be used when pressure testing for leaks. Oxygen will explode when exposed to oil and acetylene is highly flammable.

SUMMARY

An HVAC service technician must be able to work with several kinds of piping, including copper, steel, and plastic. Copper is the most common type used in the HVAC industry. It provides the passageways for refrigerant in most systems. Only ACR copper tubing is used for refrigerant because it is designed to withstand the pressures in HVAC systems.

Soft copper is available in rolls of 25 and 50 feet; diameters range from $1/8$" to $1^3/8$". Soft copper tubing can easily be cut, bent, and joined. Hard copper tubing typically comes in 20-foot lengths with diameters up to 6".

Refrigerant tubing and piping must be protected from exposure to dirt and moisture. Pipe ends must remain capped, and a tubing cutter should be used, rather than a hacksaw, because it leaves no metal particles. If a hacksaw is used, the pipe must be thoroughly cleaned inside and out before it is joined.

Copper pipe can be joined with flare or compression fittings. However, it is more common to solder or braze the connections. Piping must be properly supported and insulated. The requirements for pipe hangers and insulation are usually specified in the job specifications or by local building codes.

Once the piping installation is complete, it must be inspected and tested. The most effective method for leak testing is to pressurize the system with a charge containing 5% refrigerant and 95% nitrogen.

Anyone working with pressurized systems and refrigerants must follow special safety practices to avoid injury to personnel or damage to equipment.

References

For advanced study of topics covered in this task module the following books are suggested:

Modern Plumbing, The Goodheart-Willcox Company, Inc., South Holland, Illinois.

Modern Refrigeration and Air Conditioning, The Goodheart-Willcox Company, Inc., South Holland, Illinois.

Pipefitter's Handbook, Third Edition, Industrial Press, Inc., 200 Madison Avenue, New York, New York.

Refrigeration & Air Conditioning Technology, Second Edition, Delmar Publishers, Inc., Albany, New York.

SELF CHECK REVIEW / PRACTICE QUESTIONS

1. Horizontal refrigerant lines are pitched in the direction of refrigerant flow because:
    a. refrigerant won't flow uphill.
    b. it makes sure that oil flows in the right direction.
    c. it makes less work for the compressor.
    d. it takes advantage of the law of gravity.

2. The term "tubing" applies to:
    a. thin wall piping that is not threaded.
    b. soft copper pipe only.
    c. hard copper pipe only.
    d. small-diameter plastic piping.

3. In an HVAC system, copper tubing with an outside diameter up to _____ is used to circulate refrigerant through the system.
    a. 1"
    b. 6"
    c. $^3/_4$"
    d. $1^3/_8$"

4. The preferred tool for cutting copper tubing is a:
    a. pipe threader.
    b. hatchet.
    c. hacksaw.
    d. tubing cutter.

5. A non-heat joining method for copper tubing is a:
    a. swage joint.
    b. flare fitting.
    c. compression fitting.
    d. both b and c.

6. Which type(s) of plastic piping can be threaded?
    a. PVC
    b. ABS
    c. PE
    d. Type K

7. Which of the following is used to secure pipe to a beam?
   a. ring hanger
   b. pipe roll
   c. trapeze hanger
   d. C-clamp

8. If in doubt about which pipes to insulate:
   a. insulate all of them.
   b. follow the job specifications and/or local codes.
   c. insulate none of them.
   d. insulate the refrigerant and hot water lines.

9. When nitrogen is used to pressure-test a system, the charging line must include a hand shutoff valve, a pressure regulator, a pressure gauge, and a:
   a. stethoscope.
   b. rubber mallet.
   c. refrigerant bottle.
   d. pressure relief valve.

10. Oxygen and acetylene should never be used for pressure testing because:
    a. they do not give accurate test results.
    b. they are highly explosive under some circumstances.
    c. they are not readily available.
    d. they are too costly.

notes

ANSWERS TO SELF CHECK REVIEW / PRACTICE QUESTIONS

Answers	**Section Reference**
1. b	2.0.0
2. a	3.1.0
3. b	4.1.0
4. d	4.2.0
5. d	4.4.1, 4.4.2
6. a	5.5.0
7. d	6.0.0
8. b	7.0.0
9. d	8.1.0
10. b	8.1.0

The NCCER makes every effort to keep these manuals up-to-date and free of technical errors. We appreciate your help in this process. If you have an idea for improving this manual, or if you find an error, a typographical mistake, or an inaccuracy in the NCCER's Craft Training Manuals, please write us, using this form or a photocopy. Be sure to include the exact module number, page number, a description of the problem, and the correction, if possible. Your input will be brought to the attention of the Technical Review Committee. Thank you for your assistance.

Instructors – If you found that additional materials were necessary in order to teach this module effectively, please let us know so that we may include them in the Equipment/Materials list in the Instructor's Guide.

Write: Curriculum and Revision Department
National Center for Construction Education and Research
P.O. Box 141104
Gainesville, FL 32614-1104
Fax: 352-334-0932

Craft _______________________ Module Name _______________________

Module Number _______________________ Page Number(s) _______________________

Description of Problem _______________________

____

____

____

(Optional) Correction of Problem _______________________

____

____

____

(Optional) Your Name and Address _______________________

____

____

Soldering and Brazing

Module 03105

SOLDERING AND BRAZING

Objectives

Upon completion of this module, the trainee will be able to:

1. Assemble and operate the tools used for soldering.
2. Prepare tubing and fittings for soldering.
3. Identify the purposes and use of solder and solder fluxes.
4. Solder copper tubing and fittings.
5. Assemble and operate the tools used for brazing.
6. Prepare tubing and fittings for brazing.
7. Identify the purposes and use of filler metals and fluxes used for brazing.
8. Braze copper tubing and fittings.
9. Identify the inert gases that can safely be used to purge tubing when brazing.

Prerequisites

Successful completion of the following Task Modules is required before beginning study of this Task Module: Common Core Curricula, HVAC Modules 03101 through 03104.

Required Student Material

1. Trainee Task Module
2. Appropriate Personal Protective Equipment

Course Map Information

This course map shows all of the *Wheels of Learning* task modules in the first level of the HVAC curricula. The suggested training order begins at the bottom and proceeds up. Skill levels increase as a trainee advances on the course map. The training order may be adjusted by the local Training Program Sponsor.

Course Map: HVAC, Level 1

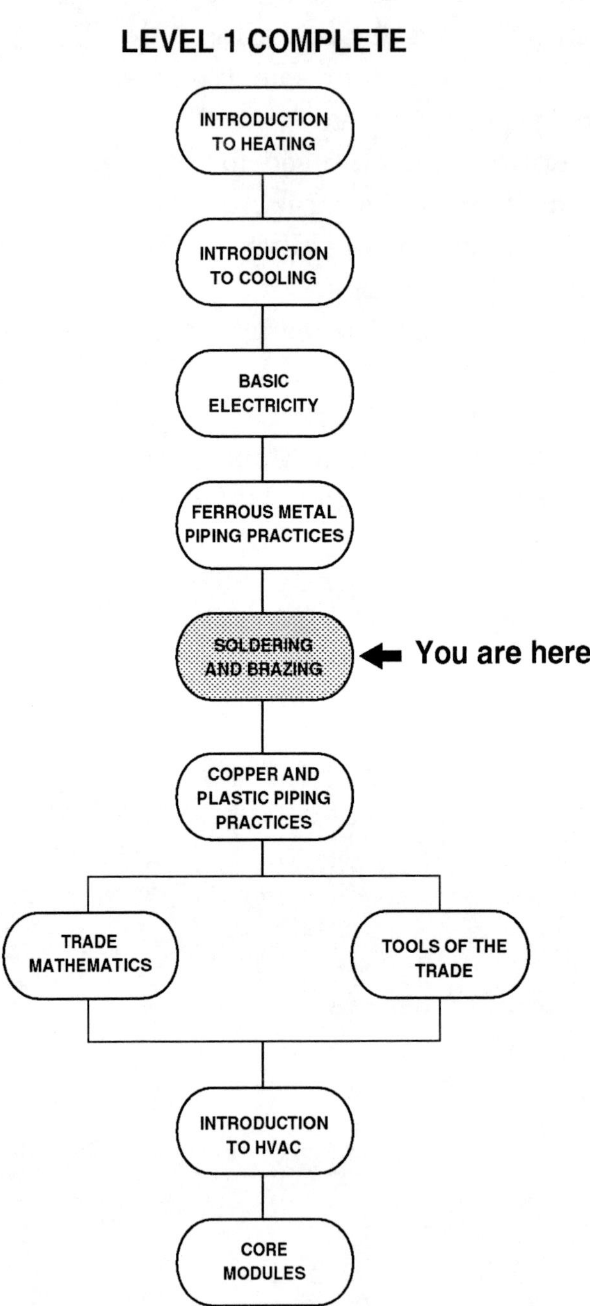

TABLE OF CONTENTS

Trade Terms Introduced In This Module

Alloy: Any substance made up of two or more metals.

Brazing: Method for joining metals with a nonferrous filler metal using heat above 800°F but below the melting point of the base metals being joined.

Capillary action: Movement of a liquid, in this case, nonferrous filler metal, along the surface of a solid in a kind of spreading action.

Flux: A chemical substance that prevents oxides from forming on the surface of metals as they are heated for soldering, brazing or welding.

Hard soldering: See brazing.

Nonferrous: Group of metals and metal alloys which contain no iron.

Oxidation: The process by which the oxygen in the air combines with metal to produce tarnish and rust.

Purging: Releasing compressed gas to the atmosphere through some part or parts, such as a hose or pipeline, for the purpose of removing contaminants from that part or parts.

Solder: A fusible alloy used to join metals.

Soft soldering: See soldering.

Soldering: Method for joining metals with a nonferrous filler metal using heat below 800°F and below the melting point of the base metals being joined.

Sweat soldering: See soldering.

Wetting: Process that reduces the surface tension so that molten (liquid) solder flows evenly throughout the joint.

1.0.0 INTRODUCTION

Soldering and brazing are two methods used for joining copper tubing and fittings. Both methods fasten the metals together using a nonferrous filler metal that adheres to the surfaces being joined. The filler metal is distributed between the closely fitted surfaces by capillary action. The difference between soldering and brazing is the temperature needed to melt the filler metal in order to make the joint. Soldering uses filler metals that melt at temperatures below 800°F, usually in the 375° to 500°F range; filler metals used for brazing melt at temperatures above 800°F.

2.0.0 SOLDERING

Soldering, also commonly called **sweat soldering** and **soft soldering**, is the most common method of joining copper tubing and fittings. It is of little value where strength is required and is usually used as a sealing process. Soldered joints are used in piping systems that carry liquids at temperatures of 250°F or below. Soldered joints are typically used for:

- Domestic water lines
- Sanitary drain lines
- Hot water heating systems

Soldering involves joining two metal surfaces by using heat and a nonferrous filler metal. A **nonferrous** filler metal is a metal that contains no iron and is therefore nonmagnetic. The melting point of the filler metal must be lower than that of the two metals that are being joined.

Soldered joints depend on capillary action to pull and distribute the melted solder into the small gap between the fitting and the tubing. When the joint is filled, the solder will form a tiny bead around the joint called a fillet. **Capillary action** is the flow of liquid, in this case solder, into a small space between two surfaces. Capillary action is most effective when the space between the tubing and the fitting is between 0.002 and 0.005 inch. To maintain the proper spacing, it is important to check the joint for correct alignment before soldering. *Figure 1* shows capillary action.

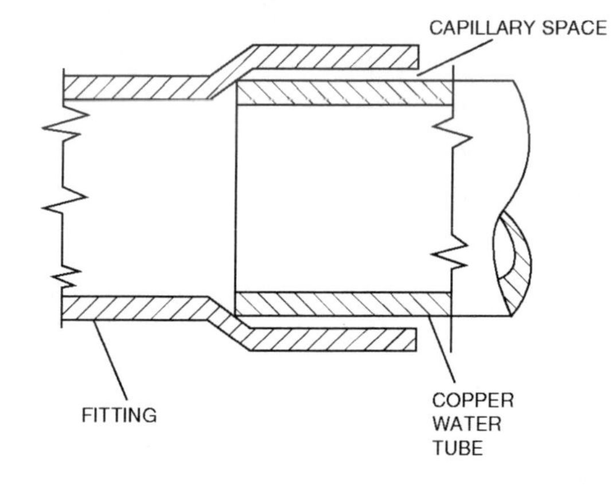

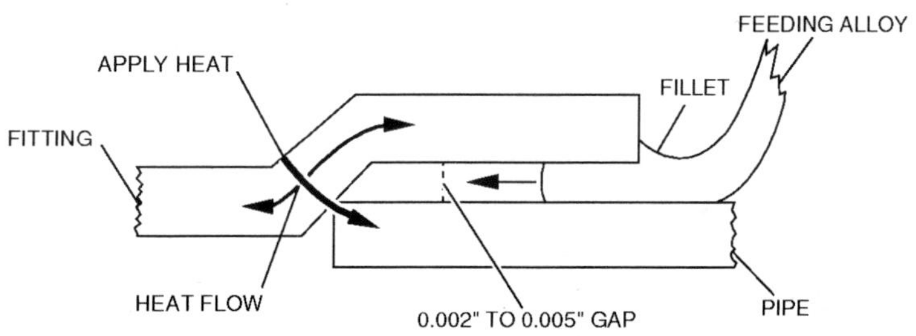

Figure 1. Capillary Action

To properly solder copper tubing and fittings, you must understand the following materials and procedures:

- Solders and soldering fluxes
- Preparing tubing and fittings for soldering
- Soldering joints

2.1.0 SOLDERS AND SOLDERING FLUXES

Solder is a nonferrous metal or metal alloy with a melting point below 800°F. An **alloy** is any substance made up of two or more metals. The Federal Safe Drinking Water Act Amendments of 1986 mandates the use of lead-free solder for drinking water supply piping. Therefore, use of soft solder composed of 50 percent tin and 50 percent lead is not permitted for joining copper pipe. The most common solder used on copper tubing is an alloy made of 95 percent tin and 5 percent antimony. This type of solder is used for potable water connections because it contains no lead. It is usually recommended for applications requiring greater joint strength. The tin-antimony alloy solder melts between 430 and 480°F and solidifies rapidly. Generally, 95-5 percent solder in the form of wire is supplied on spools for easier use.

Choosing proper flux is very important. Soldering **flux** performs many functions, and the wrong flux can ruin the soldered joint. Flux performs the following functions:

- It chemically cleans and protects the surfaces of the tubing and its fitting from oxidation. **Oxidation** occurs when the oxygen in the air combines with the recently cleaned metal. Oxidation produces tarnish or rust in metal and prevents solder from adhering.
- It allows the soldering alloy or filler metal to flow easily into the joint.
- It "floats out" remaining oxides ahead of the molten filler metal.
- It promotes wetting of the metals. **Wetting** is the process that reduces the surface tension so that the molten solder flows evenly throughout the joint.

Fluxes can be classified into three general groups: highly corrosive, less corrosive, and noncorrosive. The fluxing process must render the flux inert, that is, lacking any chemical action. If not rendered inert, the flux gradually destroys the soldered joint.

The fluxes used for joining copper tubing and copper fittings are noncorrosive fluxes. One type is composed of water and white rosin dissolved in an organic or benzoic acid base. The best noncorrosive fluxes for joining copper pipe and fittings are compounds of mild concentrations of zinc and ammonium chloride with petroleum bases.

An oxide film begins forming on copper immediately after it has been mechanically cleaned. Therefore, it is important to apply flux immediately to all recently cleaned copper fittings and tubing. Flux should be applied to the clean metal with a brush or swab, never with fingers. Not only is there a chance of causing infection in a cut, but there is also a chance that flux could be carried to the eyes or mouth. In addition, body contact with the cleaned fittings and tubing adds to unwanted contamination of the metal.

Flux must be stirred before each use. If a can of flux is not closed immediately after use, or if a can is not used for a considerable length of time, the chlorides separate from the petroleum base.

CAUTION   Brazing flux and soldering flux are not the same. Do not allow these fluxes to become mixed or interchanged. Carelessness can ruin work.

2.2.0   PREPARING TUBING AND FITTINGS FOR SOLDER

To prepare tubing and fittings for soldering, the tubing must be measured, cut, and reamed, and the tubing and fittings must be cleaned. It is critical that proper cleaning techniques be used in order to produce a solid, leakproof joint. Follow the procedure below to prepare the tubing and fittings for soldering.

Step 1   Measure the distance between the faces of the two fittings (face-to-face method). *Figure 2* shows the face-to-face method.

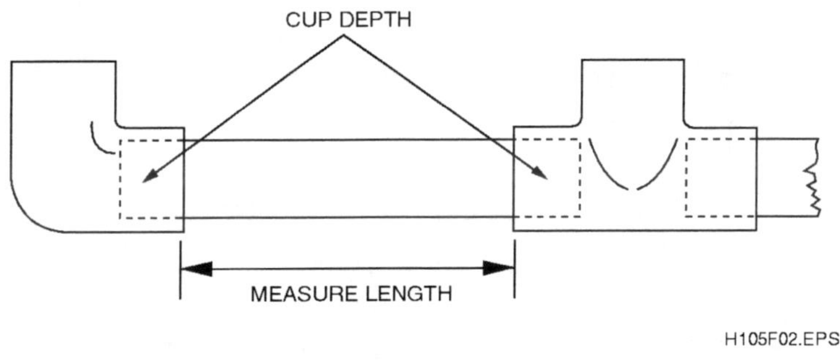

Figure 2.   Face-To-Face Method

Step 2   Determine the cup depth engagement of each of the fittings. The cup depth engagement is the distance that the tubing penetrates the fitting. This distance can be found by measuring the fitting or by using a manufacturer's makeup chart. *Table 1* shows a manufacturer's makeup chart.

Step 3   Add the cup depth engagement of both fittings to the measurement found in step 1 to find the length of tubing needed.

Step 4   Cut the copper tube to the correct length using a tubing cutter.

Step 5   Ream the inside and outside of both ends of the copper tube using a reamer.

Pipe Size	Depth of Cup	Pipe Size	Depth of Cup
1/4	5/16	2	1-11/32
3/8	3/8	2-1/2	1-15/32
1/2	1/2	3	1-21/32
5/8	5/8	3-1/2	1-29/32
3/4	3/4	4	2-5/32
1	29/32	5	2-21/32
1-1/4	31/32	6	3-3/32
1-1/2	1-3/32	--	--

Table 1.   Manufacturer's Makeup Chart

CAUTION   Care must be taken when cleaning copper tubing and fittings to remove all of the abrasions on the copper without removing a large amount of metal. Abrasions can weaken or ruin a copper joint. Do not touch or brush away filings from the tube or fitting with your fingers because your fingers will also contaminate the freshly cleaned metal.

Step 6   Clean the tubing and fitting to a bright finish using No. 00 steel wool, emery cloth, or a special copper-cleaning tool.

CAUTION   No more than two hours should be allowed to elapse between cleaning and soldering the joint(s).

Step 7   Using a brush or swab, apply flux to the copper tubing and to the inside of the copper fitting socket immediately after cleaning them.

Step 8   Insert the tube into the fitting socket, and push and turn the tube into the socket until the tube touches the inside shoulder of the fitting.

Step 9   Wipe away any excess flux from the joint.

Step 10 Check the tube and fitting for proper alignment before soldering.

2.3.0   SOLDERING JOINTS

Because soldering requires relatively low heat, heating equipment that mixes acetylene, butane, or propane directly with air is all that is needed. This means that you need only one tank of gas and a torch. Follow the procedure below to solder a joint.

WARNING!   Always solder in a well-ventilated area because fumes from the flux can irritate your eyes, nose, throat, and lungs.

Step 1   Obtain either an acetylene tank and related equipment or a propane bottle and torch. Obtain the required solder.

Step 2   Set up the equipment according to the manufacturer's instructions.

Step 3   Light the heating equipment according to the type of equipment you are using and the manufacturer's instructions.

WARNING!   When lighting the torch be sure to wear gloves and goggles. Point the torch away from your body when lighting it. Use only a spark lighter to light the torch. Do not use a match, cigarette lighter, or cigarette to light the torch.

Step 4   Heat the tubing first, and then move the flame onto the fitting (*Figure 3*).

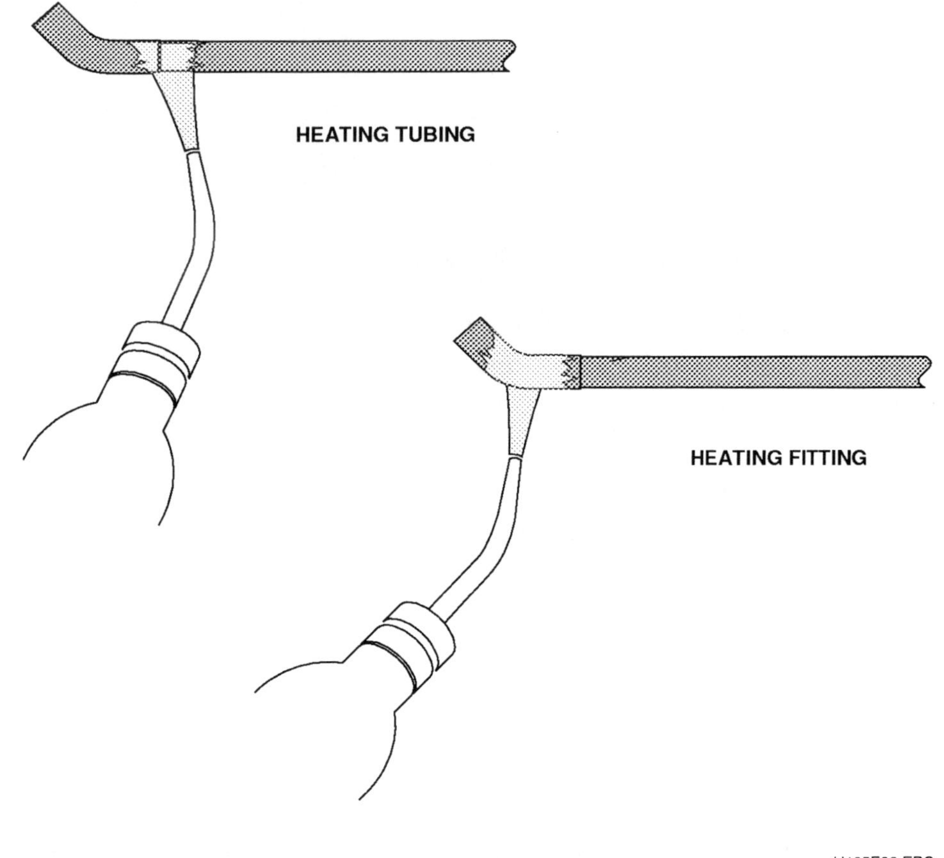

Figure 3.   Heating Tubing And Fitting

CAUTION   The inner cone of the flame should barely touch the metal being heated. Do not direct the flame into the socket because this will burn the flux. Be sure to keep the flame moving on the metals instead of holding the flame in one place.

Step 5   Move the flame away from the joint.

Step 6   Touch the end of the solder to the area between the fitting and the tube. The solder will be drawn into the joint by capillary action. The solder can be fed upward or downward into the joint.

Note   If the solder does not melt on contact with the joint, remove the solder and heat the joint again. Do not melt the solder with the flame.

Step 7   Continue to feed the solder into the joint until a ring of solder appears around the joint, indicating that the joint is filled. On $^3/_4$" diameter tubing and smaller, the solder can be fed into the joint from one point. On larger tubing, the solder should be applied from the 6-o'clock position to the 12-o'clock position on the tubing. Generally, the amount of solder used is equal to the diameter of the tubing. For example, with $^3/_4$" tubing, $^3/_4$" of solder will fill the joint.

Step 8   Allow the joint to cool.

Note   Water can be applied to the joint to speed the cooling process.

Step 9   Wipe the joint clean with a soft cloth after the joint has cooled.

3.0.0   BRAZING COPPER FITTINGS AND TUBING

Brazing, like soldering, uses nonferrous filler metals to join base metals that have a melting point above that of the filler metals. Brazing is performed above 800°F. Brazed tubing and fittings are used in:

- Low-pressure steam lines
- Refrigeration lines
- Medical gas lines
- Compressed air lines
- Vacuum lines
- Fuel lines
- Other chemical lines that need extra corrosion resistance in the piping joints

Brazing, also known as **hard soldering**, produces mechanically strong, pressure-resistant joints. The strength of a brazed joint results from the ability of the filler metal to penetrate the base metal. However, penetration can only occur if the base metals are properly cleaned, the proper flux and filler metal are selected, and the clearance gap between the outside of the tubing and the inside of the fitting is 0.003 to 0.004 inch.

To properly braze copper tubing and fittings, you must understand the following:

- Filler metals and fluxes
- Preparing tubing and fittings for brazing
- Setting up heating equipment
- Lighting an oxyacetylene torch
- Brazing joints

3.1.0 FILLER METALS AND FLUXES

Filler metals used to join copper tubing are of two groups: alloys that contain 30 percent to 60 percent silver (the BAg series) and copper alloys that contain phosphorous (the BCuP series). *Table 2* lists brazing filler materials according to their American Welding Society (AWS) classification and principal elements.

AWS Classification	Percent of Principal Element					
	Silver	Phosphorous	Zinc	Cadmium	Tin	Copper
BCuP-2	--	7 - 7.5	--	--	--	Balance
BCuP-3	4.75 - 5.25	5.75 - 6.25	--	--	--	Balance
BCuP-4	5.75 - 6.25	7 - 7.5	--	--	--	Balance
BCuP-5	14.5 - 15.5	4.75 - 5.25	--	--	--	Balance
BAg-1	44 - 46	--	14 - 18	23 - 25	--	14 - 16
BAg-2	34 - 36	--	19 - 23	17 - 19	--	25 - 27
BAg-5	44 - 46	--	23 - 27	--	--	29 - 31
BAg-7	55 - 57	--	15 - 19	--	4.5 - 5.5	21 - 23

Table 2. Brazing Filler Materials

WARNING!   BAg-1 and BAg-2 contain cadmium. Heating when brazing can produce highly toxic fumes. Use adequate ventilation and avoid breathing the fumes.

The two groups of filler metal differ in their melting, fluxing, and flowing characteristics. These characteristics should be considered when selecting a filler metal. When joining copper tubing, any of these filler metals can be used; however, the most often used filler metals for close tolerances are BCuP-3 and BCuP-4. BCuP-5 is used where close tolerances cannot be held, and BAg-1 is used as a general purpose filler metal.

Fluxes, so important in the soldering process, are even more necessary in brazing. In addition to protecting the surface from oxidation and aiding the flow of filler material, brazing fluxes serve to indicate the temperature of the metal. Without flux it is almost impossible to know when the base metal reaches the correct temperature. Brazing fluxes are applied using the same methods and rules as soldering fluxes. Brazing fluxes are more corrosive than soldering fluxes, so care must be taken never to mix a soldering flux with a brazing flux. For best results, use the flux recommended by the manufacturer of the brazing filler metals.

When copper tubing is joined to wrought copper fittings with copper-phosphorus alloys (BCuP series), flux can be omitted because the copper-phosphorus alloys are self-fluxing on copper. However, fluxes are required for joining all cast fittings.

3.2.0 PREPARING TUBING AND FITTINGS FOR BRAZING

To prepare tubing and fittings for brazing, you must follow the same procedures as you would to prepare tubing and fittings for soldering. It is critical that proper cleaning techniques be used in order to produce a solid, leakproof joint. Follow the procedure below to prepare the tubing and fittings for brazing.

Step 1 Measure the distance between the faces of the two fittings.

Step 2 Determine the cup depth engagement of each of the fittings.

Note Cup depths of fittings used with brazing are shorter than the cup depths of fittings used with soldering. The reason is that less penetration is needed for brazing than with soldering. This distance can be found by measuring the fitting or by using a manufacturer's makeup chart. *Figure 4* shows a manufacturer's brazing fitting makeup chart.

Step 3 Add the cup depth engagement of both fittings to the measurement found in step 1 to find the length of tubing needed.

Step 4 Cut the copper tubing to the correct length using a tubing cutter.

Step 5 Ream the inside and outside of both ends of the copper tubing using a reamer.

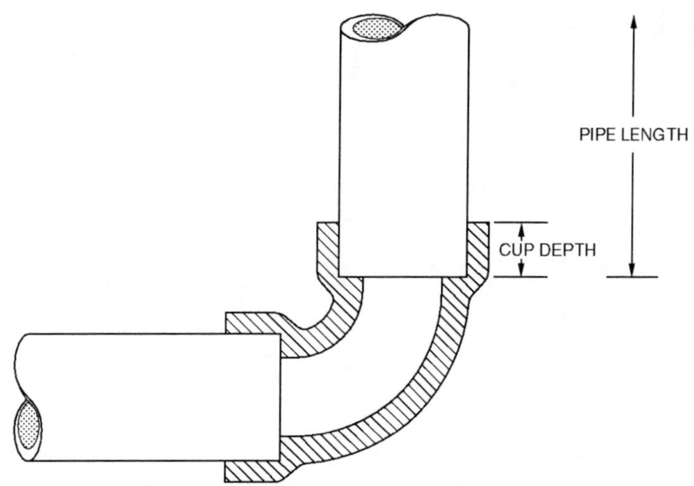

PIPE SIZE (in.)	CUP DEPTH (in.)
1/4	17/64
3/8	5/16
1/2	3/8
3/4	13/32
1	7/16
1-1/4	1/2
1-1/2	5/8
2	21/32

PIPE SIZE (in.)	CUP DEPTH (in.)
2-1/2	25/32
3	53/64
3-1/2	7/8
4	29/32
5	1
6	1-7/64
7	1-7/32
8	1-5/16

H105F04.EPS

Figure 4. Manufacturer's Brazing Fitting Makeup Chart

CAUTION   Care must be taken when cleaning the tubing and fittings to remove all the abrasions on the copper without removing a large amount of metal. Abrasions can weaken or ruin a copper joint. Do not touch or brush away filings from the tube or fitting with your fingers because your fingers will also contaminate the freshly cleaned metal.

Step 6   Clean the tubing and the fitting using No. 00 steel wool, emery cloth, or a special copper-cleaning tool.

CAUTION   No more than two hours should be allowed to elapse between cleaning and soldering the joint(s).

Step 7   Apply flux to the copper tubing and to the inside of the copper fitting socket immediately after cleaning them.

Step 8   Insert the tube into the fitting socket, and push and turn the tube into the socket until the tube touches the inside shoulder of the fitting.

Step 9   Wipe away any excess flux from the joint.

Step 10   Check the tube and fitting for proper alignment before brazing.

3.3.0 SETUP OF BRAZING HEATING EQUIPMENT

The brazing heating procedure differs from soldering in that different equipment is required to raise the temperature of the metals to be joined above 800°F. Actually, most brazed joints are made at temperatures between 1200°F and 1550°F. Because of the higher temperatures needed, oxygen-acetylene (oxyacetylene) brazing equipment is used for brazing. The flame is produced by burning acetylene, which is a fuel gas, mixed with pure oxygen. *Figure 5* shows oxyacetylene brazing equipment.

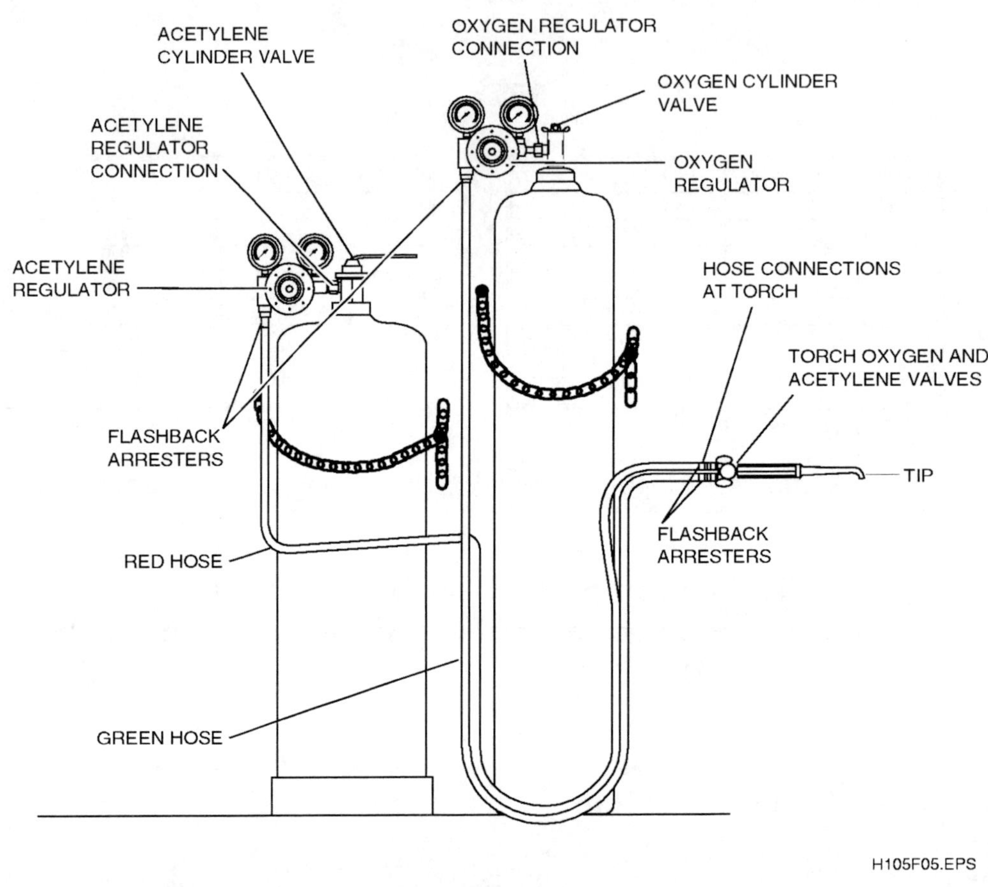

Figure 5. Oxyacetylene Brazing Equipment

3.3.1 Handling Oxygen And Acetylene Cylinders

Working with oxyacetylene brazing equipment requires that basic safety precautions be followed. Oxygen and acetylene are compressed and shipped under medium to high pressures in cylinders. Because their use is so common, technicians often get careless about handling them. Oxygen is supplied in cylinders at pressures of about 2000 psi. Acetylene cylinders are pressurized at about 250 psi. These cylinders should not be moved unless the protective caps are in place. Dropping a cylinder without the cap installed may result in breaking the valve off the cylinder. This allows the pressure inside to escape, causing the cylinder to propel like a rocket.

During use, transportation and/or storage, oxygen and acetylene cylinders must be secured with a stout cable or chain in the upright position to prevent them from falling and injuring people or damaging equipment. When stored at the job site, oxygen and acetylene cylinders must be stored separately with at least 20 feet between them, or with a five foot high, $1/2$ hour minimum fire wall separating them. Store empty cylinders away from partially full or full cylinders and make sure they are properly marked to clearly show that they are empty.

Oxygen can cause ignition even when no flame or spark is around to set it off, especially when it comes in contact with oil or grease. Never handle oxygen cylinders with oily hands or gloves. Keep grease away from the cylinders and do not use oil or grease on cylinder attachments or valves. Never use an oxygen regulator for any other gas or try to use a regulator with oxygen that has been used for other service.

A pressure-reducing regulator set for not more than 15 psig must be used with acetylene. Acetylene becomes unstable and volatile above 15 psig. The valve wrench should be left in position on open acetylene valves. This enables quick closing in an emergency. It is good practice to open the acetylene valve as little as possible, but never more than $1^{1}/_{4}$ turns.

3.3.2   Initial Setup Of Oxyacetylene Equipment

Follow the procedure below to set up oxyacetylene brazing equipment.

WARNING!   Do not handle acetylene and oxygen cylinders with oily hands or gloves. Keep grease away from the cylinders and do not use oil or grease on cylinder attachments or valves. The mixture of oil and oxygen will cause an explosion.

WARNING!   Make sure that the protective caps are in place on the cylinders before transporting or storing the cylinders.

Step 1   Install and securely fasten the oxygen and acetylene cylinders in a bottle cart or in an upright position.

WARNING!   Do not allow anyone to stand in front of the oxygen cylinder valve when opening because the oxygen is under high pressure (about 2000 psig) and could cause severe injury when released.

Step 2   Install the oxygen regulator (*Figure 5*) on the oxygen cylinder.

- Remove the cylinder protective cap.
- Open (crack) the oxygen cylinder valve just long enough to allow a small amount of oxygen to pass through the valve, then close it.

- Turn the adjusting screw on the oxygen regulator (*Figure 6*) counterclockwise to decrease the potential of overpressurizing the hose and torch during hookup.
- Using a suitable wrench, install the oxygen regulator on the cylinder. Oxygen cylinders and regulators have right-hand threads. Tighten the nut snugly. Be careful not to overtighten the nut because this may strip the threads.

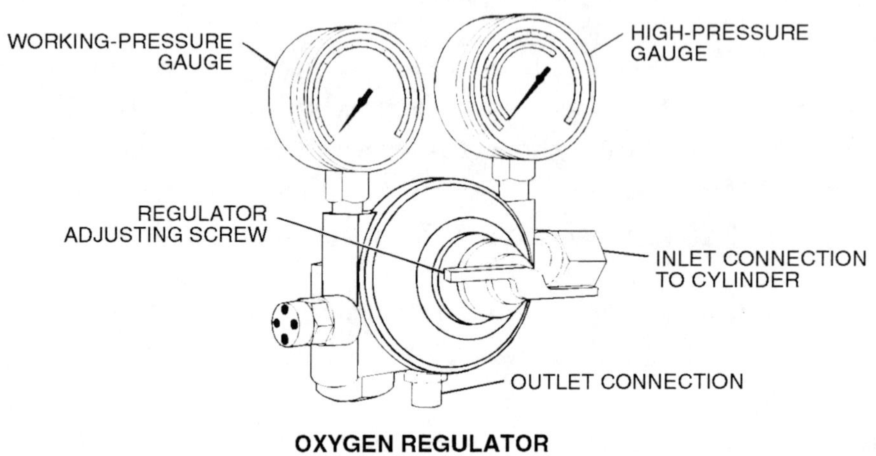

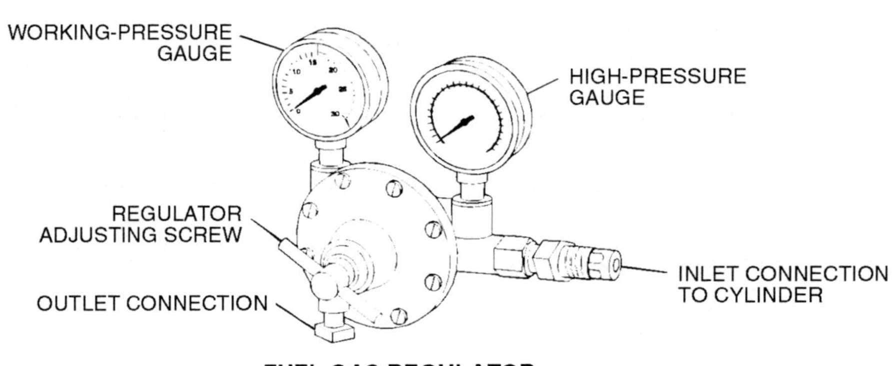

H105F06.EPS

Figure 6.   Oxygen And Acetylene Regulators

WARNING!   Acetylene gas is flammable. Do not allow open flames near it.

Step 3   Install the acetylene regulator (*Figure 5*) on the acetylene cylinder.

Note   Acetylene is stored in the cylinder at a pressure of about 250 psig.

- Remove the cylinder protective cap.
- Open (crack) the acetylene cylinder valve, using the cylinder key, just long enough to allow a small amount of acetylene to pass through the valve, then close it.
- Turn the adjusting screw on the acetylene regulator (*Figure 6*) counterclockwise to decrease the potential of overpressurizing the hose and torch during hookup.

HVAC TRAINEE TASK MODULE 03105

- Using a suitable wrench, install the acetylene regulator on the cylinder. Acetylene cylinders and regulators have left-hand threads. Tighten the nut snugly. Be careful not to overtighten the nut because this may strip the threads.

Step 4 Install the hoses and brazing torch.

- Install flashback arresters on the oxygen and acetylene regulators (*Figure 5*).
- Connect the green hose to the oxygen gauge and the red hose to the acetylene gauge. Tighten the hoses snugly. Be careful not to overtighten the fittings because this may strip the threads.

WARNING!   Do not stand in front of the oxygen gauge because the pressure may blow the face of the gauge outward, causing personal injury.

CAUTION   Open the oxygen cylinder valve slowly because a sudden release of pressure could damage the gauges.

Step 5 Purge (clean) the oxygen hose.

- Open the oxygen cylinder valve slowly until a small amount of pressure registers on the oxygen high pressure gauge (*Figure 6*); then open the valve completely.
- Turn the oxygen regulator adjusting screw clockwise until a small amount of pressure shows on the oxygen working pressure gauge (*Figure 6*). Allow a small amount of pressure to build up and purge the oxygen hose, cleaning it.
- Turn the oxygen regulator adjusting screw counterclockwise to decrease the pressure.

Step 6 Purge (clean) the acetylene hose.

- Open the acetylene cylinder valve slowly until a small amount of pressure registers on the acetylene high pressure gauge (*Figure 6*). Close the acetylene cylinder valve, then open it again about ³/₄ turn.
- Turn the acetylene regulator adjusting screw clockwise until a small amount of pressure shows on the acetylene working pressure gauge (*Figure 6*). Allow a small amount of pressure to build up and purge the acetylene hose, cleaning it.
- Turn the acetylene regulator adjusting screw counterclockwise to decrease the pressure.

Step 7 Install flashback arresters on the torch (*Figure 5*).

Step 8 Install the brazing torch on the ends of the hoses and close the valves on the torch.

WARNING!   Never adjust the acetylene regulator higher than 15 psig because acetylene becomes unstable and volatile at this pressure.

Step 9   Check the oxyacetylene equipment for leaks.

- Adjust the acetylene regulator adjusting screw for 10 psig on the working pressure gauge.
- Adjust the oxygen regulator adjusting screw for 40 psig on the working pressure gauge.

WARNING!   Do not use a soap with an oil base for leak testing because the mixture of oil and oxygen may cause an explosion.

- Close the oxygen and acetylene cylinder valves and check for leaks. If the working pressure gauges remain at 10 and 40 psig, there are no leaks in the system. If the readings drop, there is a leak. Use soapsuds to check the oxygen or acetylene related connections for leaks.
- Open both valves on the torch to release the pressure in the hoses. Watch the working pressure gauges until they register zero, then close the valves on the torch.
- Turn the oxygen and acetylene regulator valves counterclockwise to release the pressure in the regulators.

Step 10 Coil the hoses and hang them on the hose holder.

3.3.3   Lighting The Oxyacetylene Torch

After the oxyacetylene brazing equipment has been properly set up, the torch can be lit and the flame adjusted for brazing. There are three types of flames: neutral, carburizing, and oxidizing. The neutral flame burns equal amounts of oxygen and acetylene. The inner cone is light blue in color surrounded by a darker blue outer flame envelope that results when the oxygen in the air combines with the superheated gases from the inner cone. A neutral flame is used for almost all brazing applications.

A carburizing flame has a white feather created by excess fuel. The length of the feather depends on the amount of excess fuel in the flame. The outer flame envelope is brighter than that of a neutral flame and is much lighter in color. The excess fuel in the carburizing flame produces large amounts of carbon. The carburizing flame is cooler than the neutral flame and is not recommended for brazing.

An oxidizing flame has an excess amount of oxygen. Its inner cone is shorter, much bluer in color, and more pointed than the cone of a neutral flame. The outer flame envelope is very short and often fans out at the ends. An oxidizing flame is the hottest flame and should not be used for brazing. _Figure 7_ shows the types of flames.

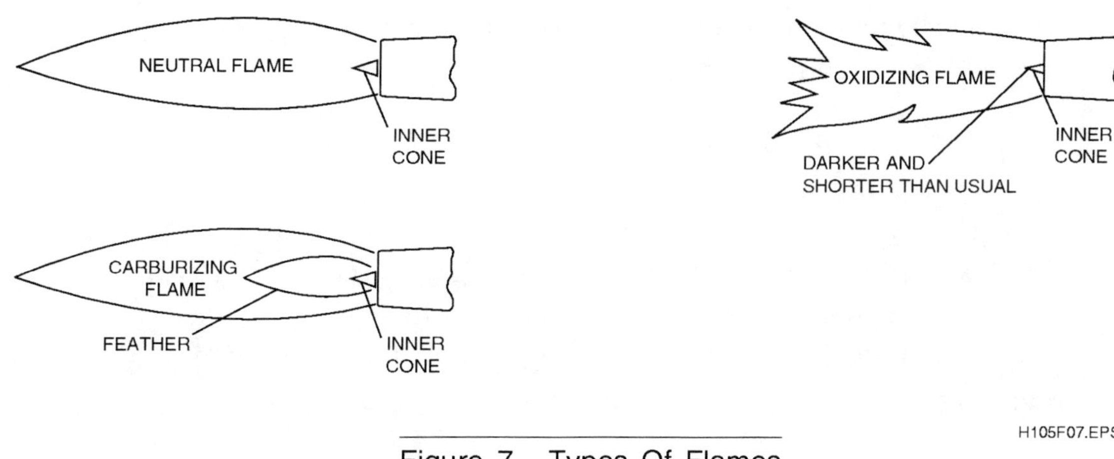

Figure 7.   Types Of Flames

Follow the procedure below to light an oxyacetylene torch.

Step 1   Set up the oxyacetylene torch according to the procedure in section 3.3.2. Make sure the correct tip is installed before lighting the torch. Refer to *Table 3* for recommended tip sizes on the pipe size being brazed. Adjust regulators for pressure settings recommended by the torch manufacturer.

Note   Pressures are not standardized for oxyacetylene torches. Refer to the manufacturer's instructions for recommended gas pressures for the pipe size being brazed.

TIP SIZE (NO.)	ROD SIZE INCHES	PIPE AND FITTING DIA. INCHES
4	3/32	1/4 - 3/8
5	1/8	1/2 - 3/4
6	3/16	1 - 1-1/4
7	1/4	1-1/2 - 2
8	5/16	2 - 2-1/2
9	3/8	3 - 3-1/2
10	7/16	4 - 6

Table 3.   Tip Sizes Used For Common Pipe Sizes

Step 2   Adjust the torch oxygen system.

WARNING!   Do not stand in front of the oxygen gauge because the pressure may blow the face of the gauge outward, causing personal injury.

CAUTION   Open the oxygen cylinder valve slowly because a sudden release
              of pressure could damage the gauges.

- Open the oxygen cylinder valve (*Figure 5*) slightly until pressure registers on the oxygen high pressure gauge (*Figure 6*); then open the valve fully.
- Turn the oxygen regulator adjusting screw clockwise until pressure shows on the oxygen working pressure gauge (*Figure 6*).
- Open the oxygen valve on the torch handle.
- Turn the oxygen regulator adjusting screw clockwise until about 20 to 25 psig registers on the oxygen working pressure gauge.

Note   Always adjust the pressure with the torch valve open. When it is
           closed, the pressure may register higher.

- Close the oxygen valve on the torch handle.

Step 3   Adjust the torch acetylene system.

- Open the acetylene cylinder valve (*Figure 5*) slightly until pressure registers on the acetylene high pressure gauge (*Figure 6*); then open the valve about $1/2$ turn.

Note   Be sure to leave the key on the acetylene cylinder valve so that the
           valve can be closed quickly in case of an emergency.

- Turn the acetylene regulator adjusting screw clockwise until pressure shows on the acetylene working pressure gauge (*Figure 6*).
- Open the acetylene valve on the torch handle.
- Turn the acetylene regulator adjusting screw clockwise until about 5 psig registers on the acetylene working pressure gauge.
- First close, then open the torch acetylene valve about $1/2$ turn.

Step 4   Light the oxyacetylene torch.

WARNING!   When lighting the torch be sure to wear gloves and
               goggles.

               Hold the striker near the end of the torch tip. Do not cover
               the tip with the striker. Always use a striker to light the
               torch. Never use matches, cigarettes or an open flame
               because this could result in severe burns or cause the
               lighter to explode. Also, make sure the torch is not
               pointed toward people or toward any flammable material.

               Any time a flame appears from a leak in a hose, shut off
               the gas immediately.

- Hold the striker in one hand and the torch in your other hand. Strike a spark in front of the escaping acetylene gas.
- Open the oxygen valve on the torch until the flame jumps away from the tip about $1/8$ inch.
- Open the oxygen valve on the torch slowly to add to the burning acetylene.

Note   Observe the luminous cone at the tip of the nozzle and the long, greenish envelope around the flame, which is excess acetylene. As you continue to add oxygen, the envelope of acetylene should disappear. The inner cone will appear soft and luminous, and the torch will make a soft, even blowing sound. This indicates a neutral flame, which is the ideal flame for brazing. If too much oxygen is added, the flame will become more pointed and white in color, and the torch will make a sharp whistling sound.

Step 5   Shut off the torch when finished brazing.

- Shut off both the oxygen and acetylene valves on the torch.
- Shut off both the oxygen and acetylene cylinder valves completely.
- Open both valves on the torch to release the pressure in the hoses. Watch the working pressure gauges until they register zero, then close the valves on the torch.
- Turn the oxygen and acetylene regulator valves counterclockwise to release the pressure in the regulators.

Step 6   Coil the hoses and hang them on the hose holder.

3.4.0   PURGING

Oil inside the tubing or part being brazed can vaporize when the heat of the brazing torch is applied. Oil vapor mixed with air will explode if ignited. Also, when copper is heated during brazing, it reacts with the oxygen in the air to form copper oxide. If air is allowed to flow into the tubing while brazing, copper oxide forms within the tubing. Refrigerants will later wash away the copper oxide particles, which can plug orifices, cause abrasion and pollute the system. Other harmful chemicals will also form in the system. As a precaution, all the air must be removed from the tubing being brazed. This can be done best by **purging** the tubing with an inert gas like nitrogen or carbon dioxide.

The pressure in a nitrogen cylinder is about 2000 psig and in a carbon dioxide cylinder about 800 psig. An accurate pressure regulator and an adjustable pressure relief valve must always be used when purging with either of these two gases. The relief valve should be adjusted to open one or two psi above the purging pressure. Refer to *Figure 8* for a pressure regulator system. For purging, use the lowest pressure (typically 2 psi) that allows just enough gas to keep air out of the tubing being brazed. As a rule of thumb, flow is sufficient when it can be felt with the palm of your hand. The greater the number of connections being brazed, the more important the purging procedure becomes.

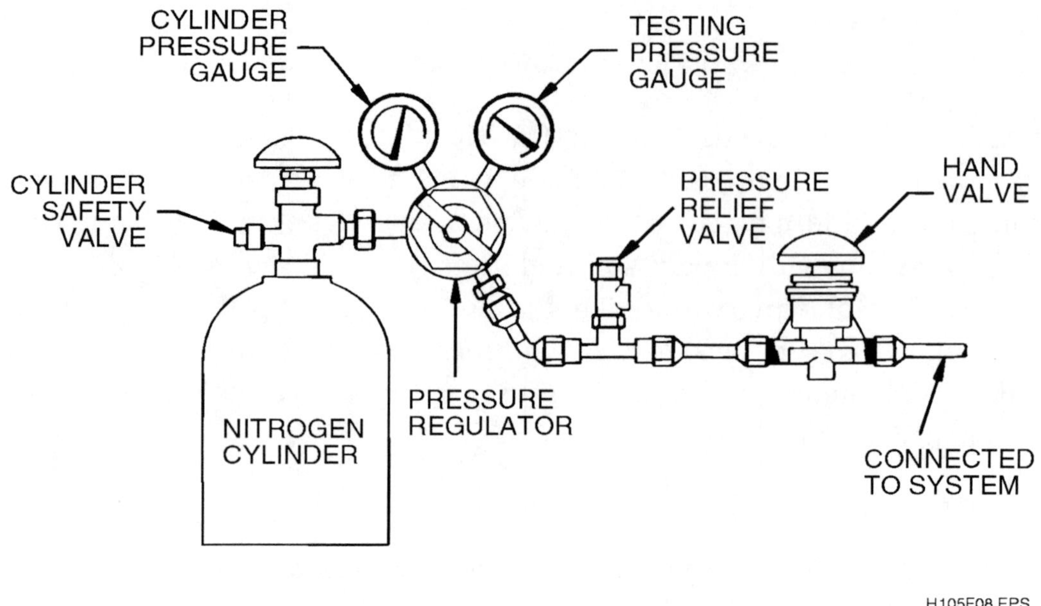

Figure 8.   Pressure Regulator System

WARNING!   Never use oxygen, refrigerant or compressed air to purge tubing. An explosion can result when oil and oxygen are mixed.

3.5.0   BRAZING JOINTS

Follow the procedure below to braze a joint.

Step 1   Set up and light the oxyacetylene brazing equipment as described in section 3.3.
- Refer to *Table 3* for suggested filler metal rod size for use with the size tubing being brazed.
- Adjust the torch to produce a neutral flame.

Step 2   Set up the nitrogen gas to purge the tubing following the guidelines and precautions described in section 3.4.

Step 3   Apply the heat to the tubing first. Watch the flux. It will first bubble and turn white and then melt into a clear liquid. At this time, shift the flame to the fitting and hold it there until the flux on the fitting turns clear.

Step 4   Continue to move the heat back and forth over the tubing and the fitting.

Note Allow the fitting to receive more heat than the tubing by pausing at the fitting as you continue to move the flame back and forth. When heating tubing that is 1¹/₂" in diameter or larger, move the heat around the entire circumference of the joint so that the entire joint is heated to brazing temperature.

Step 5 Touch the filler metal rod to the joint. If the filler metal melts on contact with the joint, the brazing temperature has been met. If the filler metal does not melt on contact, continue to heat and test the joint until the filler metal melts.

Step 6 Hold the filler metal rod to the joint, and allow the filler metal to enter into the joint while holding the torch slightly ahead of the filler metal and directing most of the heat to the shoulder of the fitting.

Step 7 Continue to fill the joint with the filler metal until the filler metal has completely penetrated the joint.

Note If the tubing is 2 inches or more in diameter, 2 torches can be used to evenly distribute the heat. For larger joints, small sections of the joint can be heated and brazed. Be sure to overlap the previously brazed section as you continue around the fitting. *Figure 9* shows working in overlapping sectors.

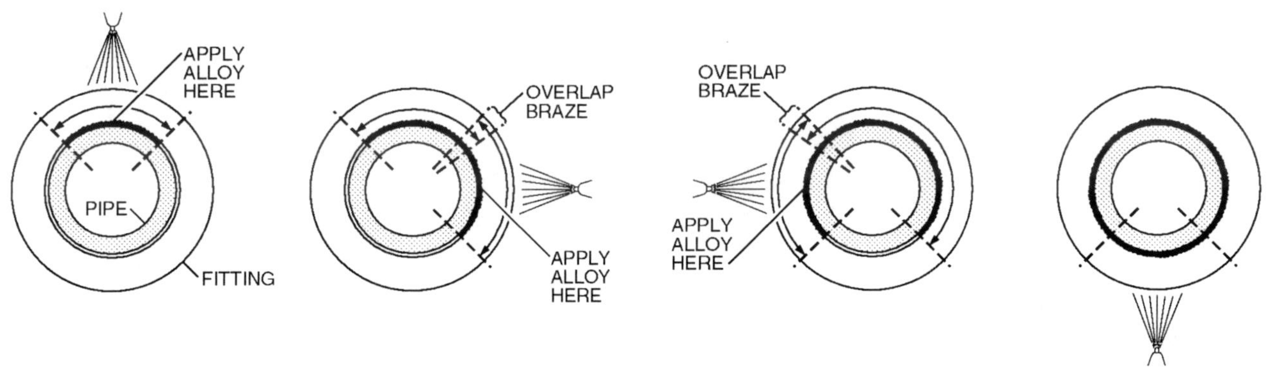

Figure 9. Working In Overlapping Sectors

H105F09.EPS

Step 8 Wash the joint with warm water to clean excess, dried, or hardened brazing flux from the joint.

Note If the flux is too hard to be removed with water, chip the excess flux off, using a small chisel and light peen hammer, and then wash the joint with warm water. Joints clean best when they are still warm.

Step 9 Allow the joints to cool naturally.

SUMMARY

Soldering is one of the most widely used methods for joining metals. Soldering is a procedure which fastens metals together with nonferrous metal of a low melting point that adheres to the surfaces being joined. The filler metal is distributed between the closely fitted surfaces by capillary action. If the solder melts at a temperature above 800°F, the process is called hard soldering or brazing.

Soldering and brazing procedures require a step-by-step approach that includes:

- Cleaning the joints
- Applying flux
- Heating the parts to an exact temperature
- Bringing the nonferrous filler metal into contact with the tubing where it melts and flows into the joint by capillary action
- Cleaning and cooling the joint

Should more training in soldering and brazing procedures be desired, the trainee should contact local vocational and trade schools or colleges for available courses. Also, many distributors of welding and gas supplies often have videos, etc. that they loan to interested individuals or groups.

References

For advanced study of topics covered in this task module the following materials are suggested:

Brazing (VHS Video/Slides/Book), Carrier Corporation, Literature Services, Syracuse, New York.
Modern Plumbing, The Goodheart-Willcox Company, Inc., South Holland, Illinois.
Modern Refrigeration and Air Conditioning, The Goodheart-Willcox Company, Inc., South Holland, Illinois.
Refrigeration & Air Conditioning Technology, Second Edition, Delmar Publishers, Inc., Albany, New York.
Standards and Codes, American Society of Mechanical Engineers (ASME), New York, New York.
Standards and Codes, American Welding Society (AWS), Miami, Florida.

SELF CHECK REVIEW / PRACTICE QUESTIONS

1. Filler metals used for soldering usually melt at a temperature range between:
   a. 1000°F to 1200°F.
   b. 375°F to 500°F.
   c. 200°F to 375°F.
   d. 750°F to 850°F.

2. The melting point of filler metal used for soldering or brazing must be _______ the two metals that are being joined.
   a. higher than
   b. the same as
   c. lower than
   d. different than

3. Which of the following are the characteristics of the filler metal (solder) used for soldering?
   a. nonferrous metal
   b. metal alloy
   c. has a melting point below 800°F
   d. all of the above

4. The most commonly used solder in the HVAC trade consists of:
   a. 95% tin, 5% antimony.
   b. 50% tin, 50% lead.
   c. 96% tin, 4% copper.
   d. 94% tin, 6% silver.

5. Which of the following is not a characteristic of both soldering and brazing fluxes?
   a. promotes wetting of metals
   b. can be used interchangeably
   c. cleans and protects against oxidation
   d. allows filler metal to flow into the joint

6. Because soldering uses relatively low heat, torches that mix air directly with _______ can be used.
   a. acetylene
   b. butane
   c. propane
   d. all of the above

7. Filler metals used for brazing usually melt at a temperature range between:
   a. 850°F to 1200°F.
   b. 375°F to 500°F.
   c. 1200°F to 1550°F.
   d. 750°F to 850°F.

8. The oxygen cylinder used with oxyacetylene heating equipment contains oxygen stored at a pressure of about:
   a. 250 psig.
   b. 1000 psig.
   c. 2000 psig.
   d. none of the above.

9. When using oxyacetylene heating equipment for brazing, the acetylene regulator should never be adjusted to supply a pressure higher than:
   a. 250 psig.
   b. 15 psig.
   c. 10 psig.
   d. 25 psig.

10. The _______ flame is used for almost all brazing applications.
   a. oxidizing
   b. carburizing
   c. feather
   d. neutral

notes

<u>Answers</u>	<u>Section Reference</u>
1. b	1.0.0
2. c	2.0.0
3. d	2.1.0
4. a	2.1.0
5. b	2.1.0
6. d	2.3.0
7. c	3.3.0
8. c	3.3.1, 3.3.2
9. b	3.3.1, 3.3.2
10. d	3.3.3

The NCCER makes every effort to keep these manuals up-to-date and free of technical errors. We appreciate your help in this process. If you have an idea for improving this manual, or if you find an error, a typographical mistake, or an inaccuracy in the NCCER's Craft Training Manuals, please write us, using this form or a photocopy. Be sure to include the exact module number, page number, a description of the problem, and the correction, if possible. Your input will be brought to the attention of the Technical Review Committee. Thank you for your assistance.

Instructors – If you found that additional materials were necessary in order to teach this module effectively, please let us know so that we may include them in the Equipment/Materials list in the Instructor's Guide.

Write: Curriculum and Revision Department
National Center for Construction Education and Research
P.O. Box 141104
Gainesville, FL 32614-1104
Fax: 352-334-0932

Craft _______________________ Module Name _______________________

Module Number _______________________ Page Number(s) _______________________

Description of Problem _______________________

____

____

____

(Optional) Correction of Problem _______________________

____

____

(Optional) Your Name and Address _______________________

____

Ferrous Metal Piping Practices
Module 03106

FERROUS METAL PIPING PRACTICES

Objectives

Upon completion of this module, the trainee will be able to:

1. Identify the types of ferrous metal pipes.
2. Measure the sizes of ferrous metal pipes.
3. Identify the common malleable iron fittings.
4. Cut, ream and thread ferrous metal pipe.
5. Join lengths of threaded pipe together and install fittings.
6. Identify uses for builder's specifications.
7. Identify the main points to consider when installing pipe runs.
8. Explain the method used to join grooved piping.

Prerequisites

Successful completion of the following Task Modules is required before beginning study of this Task Module: Common Core Curricula, HVAC Modules 03101 through 03105.

Required Student Materials

1. Trainee Task Module
2. Appropriate Personal Protective Equipment

Course Map Information

This course map shows all of the *Wheels of Learning* task modules in the first level of the HVAC curricula. The suggested training order begins at the bottom and proceeds up. Skill levels increase as a trainee advances on the course map. The training order may be adjusted by the local Training Program Sponsor.

Course Map: HVAC, Level 1

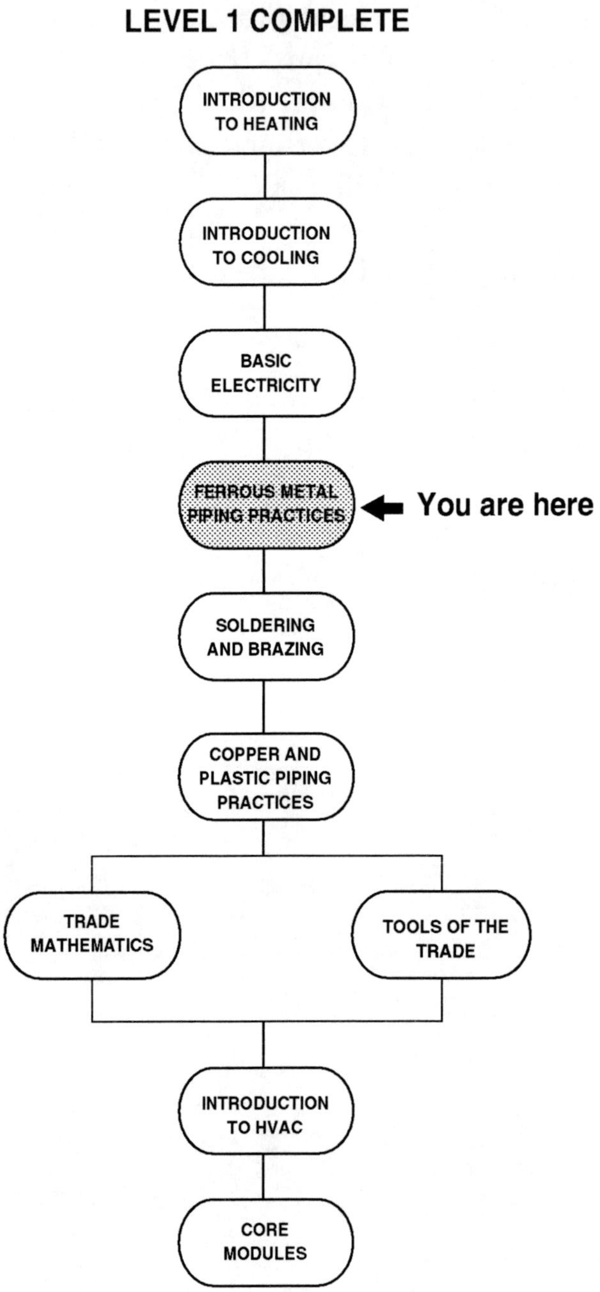

LEVEL 1 COMPLETE

TABLE OF CONTENTS

Trade Terms Introduced In This Module

Black iron pipe: Carbon steel pipe that gets its black coloring from the carbon in the steel.

Bushing: Pipe fitting with male threads on the outside and female threads on the inside. Most often used to connect the male end of a pipe to a fitting of a larger size.

Cap: Female pipe fitting closed at one end; used to close off the end of a piece of pipe.

Chain vise: Vise used to clamp pipe and other round metal objects. It has one stationary metal jaw and a chain that fits over the pipe and is clamped to secure the pipe.

Chain wrench: Adjustable tool for holding and turning large pipe up to 4 inches in diameter. A flexible chain replaces the usual wrench jaws.

Coupling: Pipe fitting containing female threads on both ends. Couplings are used to join two pipes in a straight run or to join a pipe and fixture.

Cross: Pipe fitting with four female openings at right angles to one another.

Die: Tool used to cut external threads by hand or machine.

Die stock: Tool used to hold and turn dies when cutting external threads.

Elbow: Angled pipe fitting having two openings; used to change the direction of a run of pipe.

Galvanized pipe: Carbon steel pipe that has been coated with zinc to prevent rust.

Grooved pipe: Piping method for connecting piping systems. Use of grooved piping eliminates the need for threading, flanging or welding when making connections. Connections are made with gaskets and couplings installed using a wrench and lubricant.

Inside diameter: Measurement made across the inside width (internal opening) of a pipe.

Nipple: Short length of pipe, usually less than 12 inches, with male threads on both ends; used to join fittings.

Nominal size: Approximate dimension(s) of standard material.

Outside diameter: Measurement made across the outside width of a pipe, including wall thickness plus internal opening.

Pipe dope: Putty-like pipe joint material used for sealing threaded pipe joints.

Pipe wrench: Wrench with adjustable, slightly curved, toothed jaws. Grips pipe firmly when pressure is applied to the handle.

Plug: Pipe fitting with external threads and head that is used for closing the opening in another fitting.

Reamer: Tool used to remove the burr from the inside of a pipe that has been cut with a pipe cutter.

Standard yoke (pipe) vise: Holding device used to hold pipe and other round objects. It has one movable jaw that is adjusted with a threaded rod.

Strap wrench: Tool for gripping pipe. The strap is made of nylon web.

Tee: Pipe fitting shaped like the letter T. Each leg of the tee can be joined to a pipe or another fitting.

Union: Pipe fitting used to join two lengths of pipe. It permits disconnecting the two pieces of pipe without cutting.

1.0.0 INTRODUCTION

Ferrous metal piping is piping that contains or is made from iron. Steel pipe is made from a combination of iron ore and carbon. The two most common types of carbon steel pipe are black iron pipe and galvanized pipe. Steel pipe has many uses in the field, including:

- Hot and cold water distribution
- Steam and hot water heating systems
- Gas and air piping systems
- Drainage and vent systems
- Fire protection systems

Some of the many advantages for using steel pipe in a plumbing system include:

- Durability
- Structural strength
- Low material cost
- Low thermal conductivity
- Low expansion properties

2.0.0 STEEL PIPE

2.1.0 BLACK IRON PIPE

Black iron pipe is manufactured exactly like galvanized pipe. The only difference between the two is that black iron pipe is not coated with zinc. Carbon gives the black iron pipe its color. Black iron pipe is most often used in gas, heat, chilled water, steam or air pressure applications. It is used where corrosion will not affect its uncoated surfaces.

2.2.0   GALVANIZED PIPE

Galvanized pipe is steel pipe that has been dipped in molten zinc. This protects the surfaces from abrasive or corrosive materials and gives the pipe a dull, grayish color. Galvanized pipe is most often used when plumbing specifications require steel pipe.

2.3.0   SIZES AND WALL THICKNESS

Pipe is listed in inches by its **nominal size**. For pipe sizes up to and including 12 inches, nominal size is an approximation of the **inside diameter**. From 14 inches on, the nominal size reflects the **outside diameter** of the pipe. There are times when the nominal size of a pipe and its actual inside or outside diameter will differ greatly, but nominal size, not actual size, is always used to describe and select piping. *Figure 1* shows the inside and outside diameters for a 1 inch nominal-sized steel pipe.

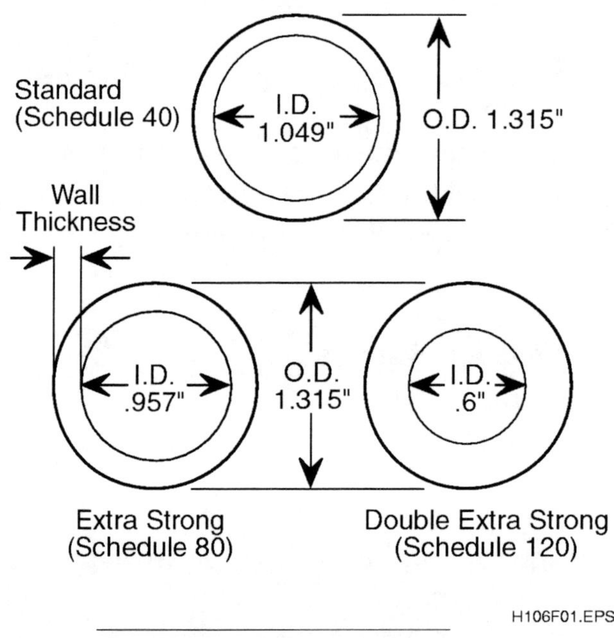

Figure 1.   Pipe Diameters

There are two ways to describe the wall thickness of a pipe. The first is by schedule. As schedule numbers get larger, pipe walls get thicker and stronger. The schedule numbers used for pipe are 5, 10, 20, 30, 40, 60, 80, 100, 120, 140 and 160. It is important to remember that a schedule number only describes the wall thickness of a pipe of a given nominal size. Thus, ³/₄ inch schedule 40 does not have the same wall thickness as 1 inch schedule 40.

The second way to describe pipe wall thickness is by manufacturers' weight. There are three classifications in common use. In ascending order of wall thickness, they are:

- STD - Standard
- XS - Extra Strong
- XXS - Double Extra Strong

HVAC TRAINEE TASK MODULE 03106

The wall thickness and the inside diameter differ with each weight. The thicker the wall, the smaller the inside diameter, but the more pressure the pipe will withstand. Standard weight will prove adequate in most plumbing situations; however, the two stronger weights are available when higher pressures require their use. Appendices A and B provide weights and dimensions of carbon steel pipe schedules and wall thicknesses other than schedules for the various sizes of pipe.

Because all schedules or weights for a specific pipe size have the same outside diameter, the same threading dies will fit all of them.

2.4.0 THREADS

Steel pipe is joined either by welding or by threading the end of the pipe and using threaded fittings. There are two American National Standard Pipe Threads: tapered pipe and straight pipe. Only tapered pipe threads are used for HVAC work because they produce leak-tight and pressure-tight connections. When tight, they also produce a mechanically-rigid piping system. Tapered threads can be cut by hand with a die and stock or with an electric pipe threading machine.

The tapered thread used on pipe (*Figure 2*) is V-shaped with an angle of 60 degrees, very slightly rounded at the top. The taper is $\frac{1}{16}$ of an inch per inch of length ($\frac{1}{32}$ inch per inch each wall). There are about seven perfect threads and two or more imperfect threads for each joint. The actual number of perfect threads used (usable threads) depends on the size of the pipe being threaded. As shown in *Figure 2*, the first group of threads are perfect threads; they are sharp at the top and bottom. The remaining threads are non-perfect because they are not completely cut, resulting in rounded or imperfect edges. They have no sealing power. If the perfect threads are marred or broken, they also lose their sealing power.

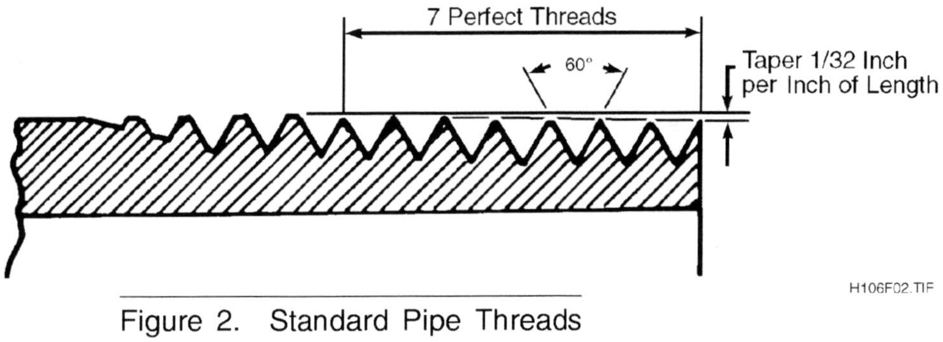

Figure 2. Standard Pipe Threads

Thread diameters refer to the nominal size of steel pipe. Threads are designated by specifying in sequence the nominal size, number of threads per inch, and the thread series symbols. For example, the thread specification $\frac{3}{4}$ - 14 NPT means:

$\frac{3}{4}$ = $\frac{3}{4}$ inch nominal size
14 = 14 threads per inch
NPT = American National Standard Taper Pipe Thread

Taper pipe threads are engaged or made-up in two phases, hand tight engagement and wrench makeup. *Table 1* shows dimensions for hand tight engagement as well as other NPT specifications for commonly used pipe sizes. In practice, about three turns are done by hand, followed by three or four turns with a wrench. When a pipe is threaded properly, about three threads should remain showing after the total makeup of a pipe and fitting.

Nominal Pipe Size	Threads Per Inch	No. Of Usable Threads	Hand Tight Engagement	Thread Makeup	Total Thread Length
1/8	27	7	3/16	1/4	3/8
1/4	18	7	1/4	3/8	9/16
3/8	18	7	1/4	3/8	5/8
1/2	14	7	5/16	1/2	3/4
3/4	14	8	5/16	9/16	13/16
1	11-1/2	8	3/8	11/16	1
1-1/4	11-1/2	8	7/16	11/16	1
1-1/2	11-1/2	8	7/16	3/4	1
2	11-1/2	9	7/16	3/4	1-1/16
2-1/2	8	9	11/16	1-1/8	1-9/16
3	8	10	3/4	1-3/16	1-5/8
3-1/2	8	10	13/16	1-1/4	1-11/16
4	8	10	13/16	1-5/16	1-3/4
5	8	11	15/16	1-3/8	1-13/16
6	8	12	15/16	1-1/2	1-15/16

Table 1. American National Standard Taper Pipe Thread (NPT) Dimensions

2.5.0 PIPE FITTINGS

Pipe fittings for steel pipe are generally made of cast iron, malleable iron, or galvanized (zinc coated) iron. Malleable iron has a slight temper produced by prolonged annealing of ordinary cast iron. This process makes the iron tough. Also, it can be bent or pounded to some extent without breaking. Malleable iron fittings are typically used for gas piping. Cast iron fittings are used for steam and hydronic system piping. Galvanized fittings are used for water piping such as used with exterior cooling towers and drip condensate piping systems.

2.5.1 Tees

Tees can be purchased in a great number of sizes and patterns. They are used to make a branch which is 90 degrees to the main pipe. If all three outlets are the same size, the fitting is called a regular tee (*Figure 3*). If outlet sizes vary, the fitting is called a reducing tee.

HVAC TRAINEE TASK MODULE 03106

Tees are specified by giving the straight-through (run) dimensions first, then the side-opening dimensions. For example, a tee with one run outlet of 2 inches, a second run outlet of 1 inch and a branch outlet of $^3/_4$ inches is known as a 2 x 1 x $^3/_4$ tee (always state the large run size first, the small run size next and the branch size last). Tees are also available with male threads on a run or branch outlet.

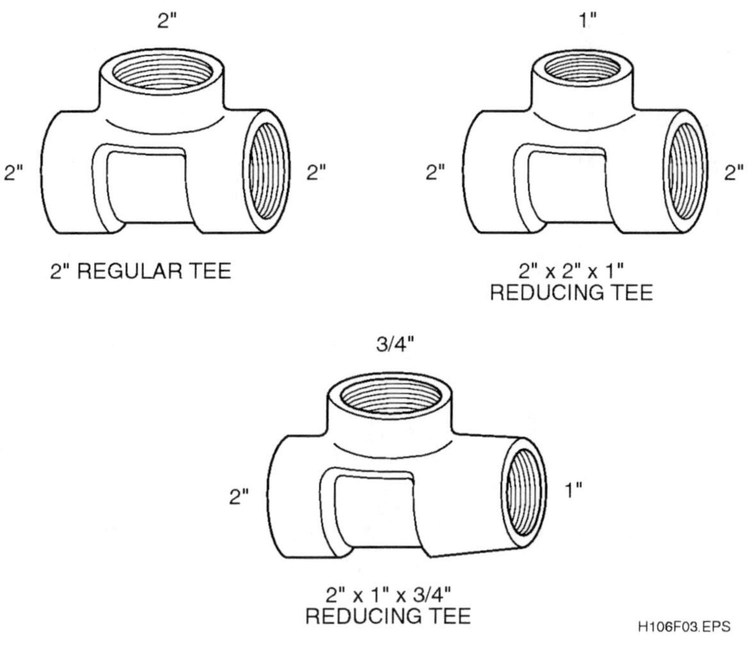

Figure 3.   Tee Fittings

2.5.2   Elbows

Elbows, often called ells, are used to change the direction of pipe. See *Figure 4*. The most common ells are the 90-degree ell, the 45-degree ell, the street ell, which has a male thread on one end, and the reducing ell, which has outlets of different sizes. Ells are also available to make $11^1/_4$, $22^1/_2$ and 60-degree bends.

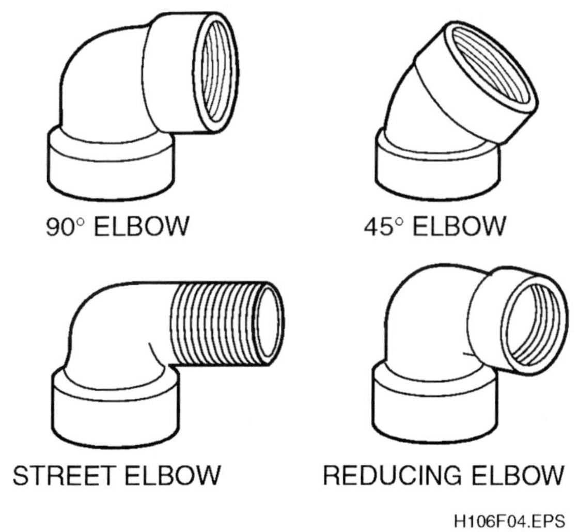

Figure 4.   Elbows

2.5.3   Unions

Unions make it possible to disassemble a threaded piping system. After disconnecting the union, the length of pipe on either end of the union may then be unscrewed. The two most common types of union are the ground joint and the flange (*Figure 5*).

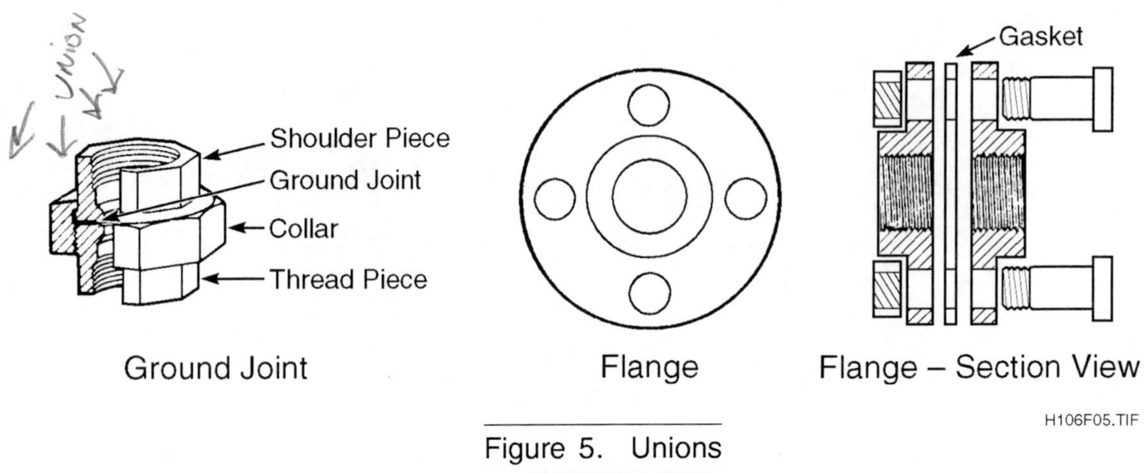

Figure 5.   Unions

The ground joint union connects two pipes by screwing the thread and shoulder pieces onto the pipes. Then, both the shoulder and thread parts are drawn together by the collar. This union creates a gas-tight and water-tight joint.

The flange union also connects two separate pipes. The flanges screw to the pipes to be joined, and are then pulled together with nuts and bolts. A gasket between the flanges makes this connection gas-tight and water-tight.

2.5.4   Couplings

Couplings (*Figure 6*) are short fittings with female threads in both openings. They are used to connect two lengths of pipe when making straight runs. The pipes can be of the same size or different sizes. Couplings cannot be used in place of unions because they cannot be disassembled.

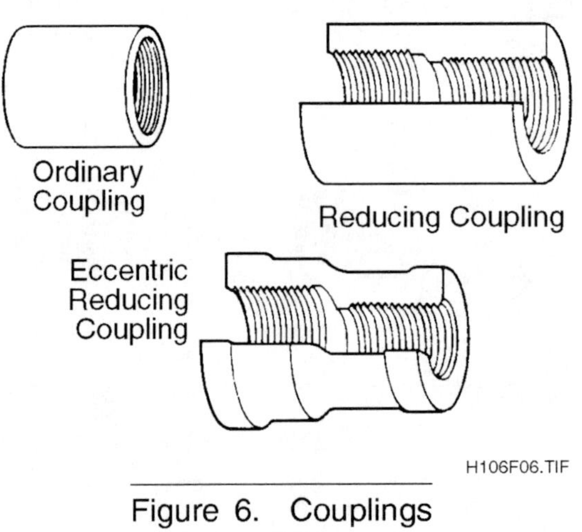

Figure 6.   Couplings

HVAC TRAINEE TASK MODULE 03106

2.5.5 Nipples And Crosses

Nipples (*Figure 7*) are pieces of pipe 12 inches or less in length, threaded on both ends, and used to make extensions from a fitting or to join two fittings. Nipples are manufactured in many sizes beginning with the close or all-thread nipple. **Crosses** are four-way distribution devices.

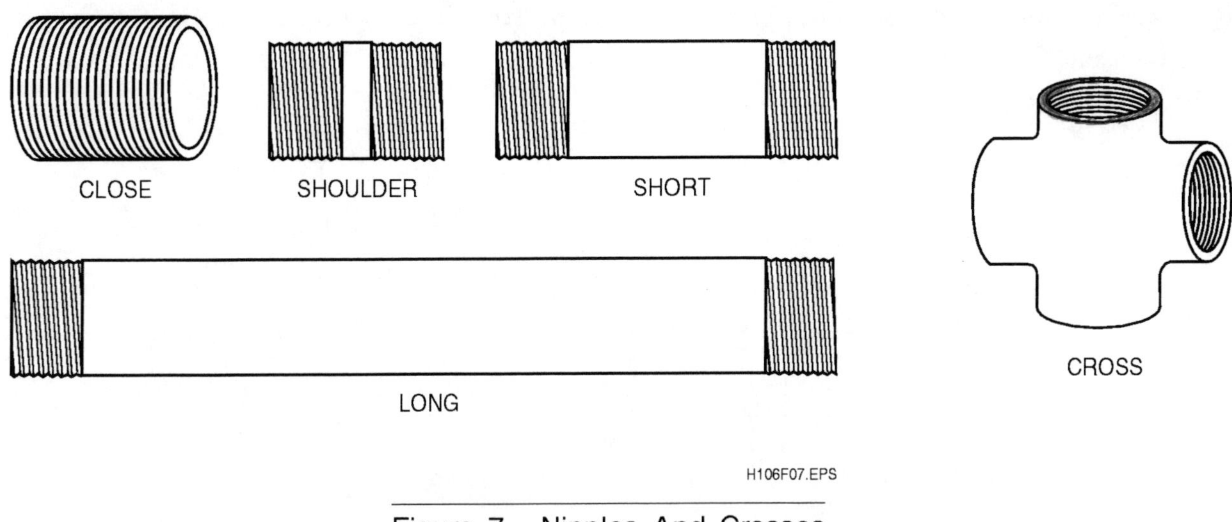

Figure 7. Nipples And Crosses

2.5.6 Plugs, Caps, And Bushings

Plugs are male threaded fittings used to close openings in other fittings. There are a variety of heads (square, slotted and hexagon) found on plugs, as shown in *Figure 8*.

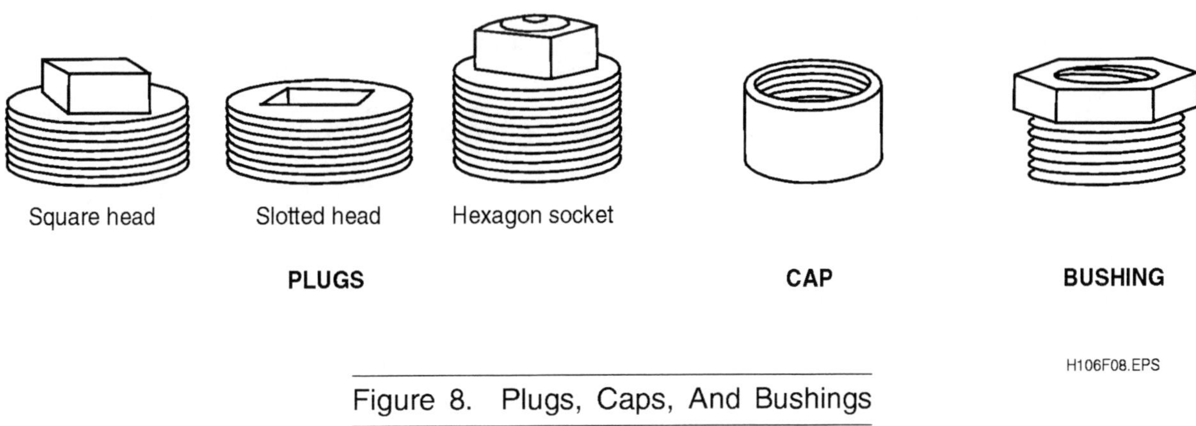

Figure 8. Plugs, Caps, And Bushings

Caps are fittings with a female thread. They are used for the same purpose as a plug except that the cap fits on the male end of a pipe or nipple.

Bushings are fittings with a male thread on the outside and a female thread on the inside. They are usually used to connect the male end of a pipe to a fitting of a larger size. The ordinary bushing has a hexagon nut at the female end.

3.1.0   PIPE CUTTER

Pipe cutters (*Figure 9*) may have from one to four cutting wheels. A single cutting wheel requires enough area so that the cutter may be passed all the way around the pipe. The more cutting wheels a cutter has, the less room it requires to cut the pipe. The pipe cutter is made of a cutting wheel, adjusting screw, and depending on the type, either guide rollers or additional cutting wheels.

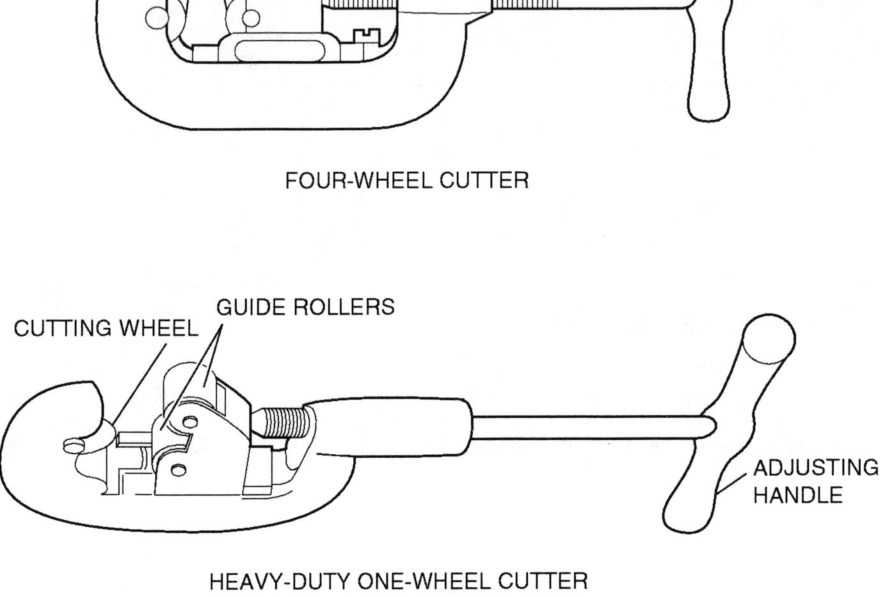

Figure 9.   Pipe Cutters

If the tube or pipe becomes mashed or flattened while being cut, replace the cutting wheel or wheels. Lubricating oil needs to be applied periodically to all movable parts on the pipe cutter to insure smooth operation.

3.2.0   REAMER

After cutting a piece of pipe, a burr (rough edge) will be left on the inside of the pipe. A **reamer** (*Figure 10*) is used to remove the burr. This process is called reaming.

If the burr is left unattended, it will collect deposits and slow the flow of liquid within the pipe. Since reamers are tapered, one reamer can de-burr many pipe sizes.

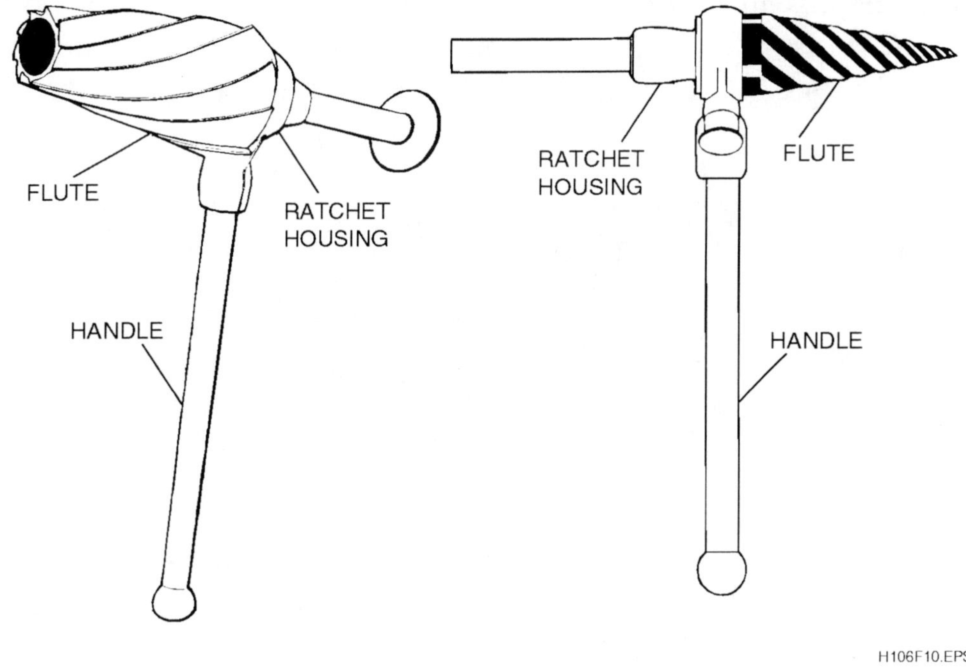

Figure 10. Pipe Reamers

3.3.0 PIPE THREADERS

There are two types of pipe threaders: hand threaders and power threaders. Hand threaders (*Figure 11*) are made up of two parts: the **die** and the **stock** (handle). Dies are used to cut the threads, and the stock is the device that holds the die. The pipe die consists of the holder and cutters. Although they are mostly used manually, a hand threader could be used with a power drive or power vise, which threads automatically.

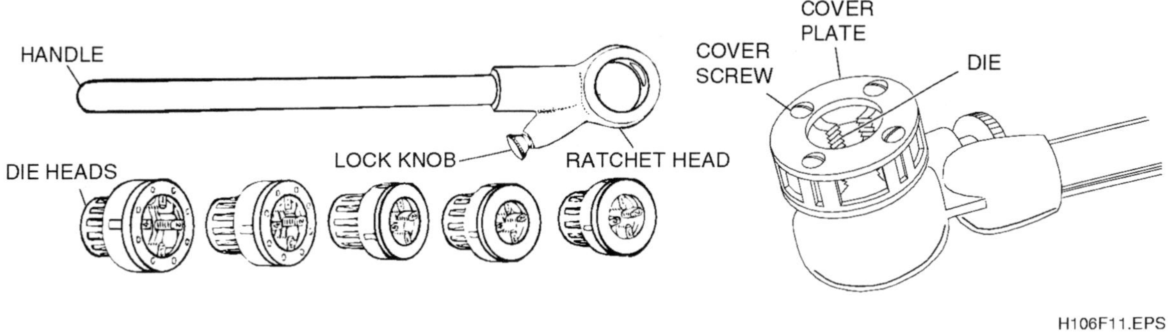

Figure 11. Hand Threader

A pipe threading machine (*Figure 12*) is used when large quantities of pipe require threading. This machine rotates, threads, cuts and reams pipe. The pipe is mounted through the machine speed chuck. After the chuck is tightened, this tool will perform one or more of the above operations.

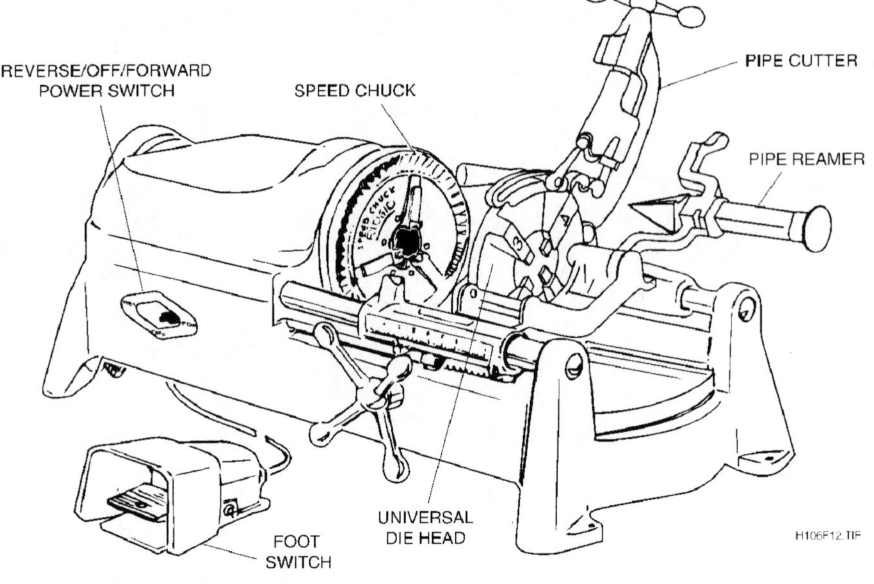

Figure 12. Typical Threading Machine

3.4.0 PIPE VISES AND WRENCHES

3.4.1 Vises

The **standard yoke vise** (*Figure 13*) is the most commonly used vise. Its jaws hold pipe firmly and prevent it from turning. This vise can handle pipe from $1/8$" to $3^{1/2}$" in diameter, depending on vise size.

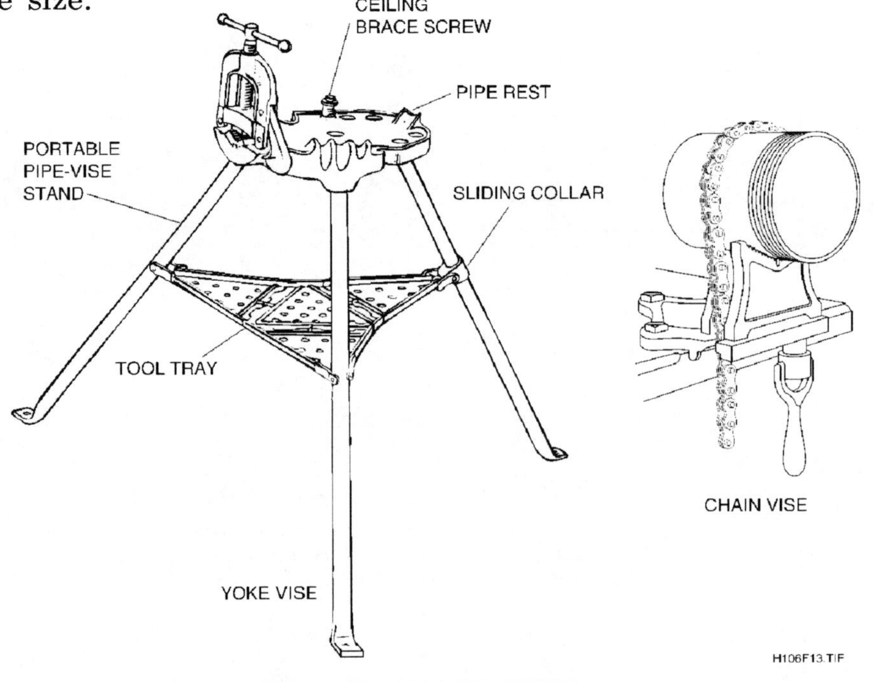

Figure 13. Pipe Vises

The **chain vise** is used in the same way as the standard yoke vise; however, the chain vise can hold much larger pieces of pipe than the standard yoke vise. The chain must be kept oiled or it will become stiff. A stiff chain will make the chain vise operate poorly.

3.4.2   Wrenches

The **pipe wrench** (*Figure 14*) is used to grip and turn round stock. It has teeth that are set at an angle. This angle allows the teeth of the wrench to grip in one direction only.

The **chain wrench** is used on pipe that is over two inches in diameter. The chain must be oiled often to prevent it from becoming stiff or rusty.

The **strap wrench** is used to hold chrome-plated or other types of finished pipe. The strap wrench does not leave jaw marks or scratches on the pipe. Resin applied to the strap adds to the holding power of the wrench by reducing slippage.

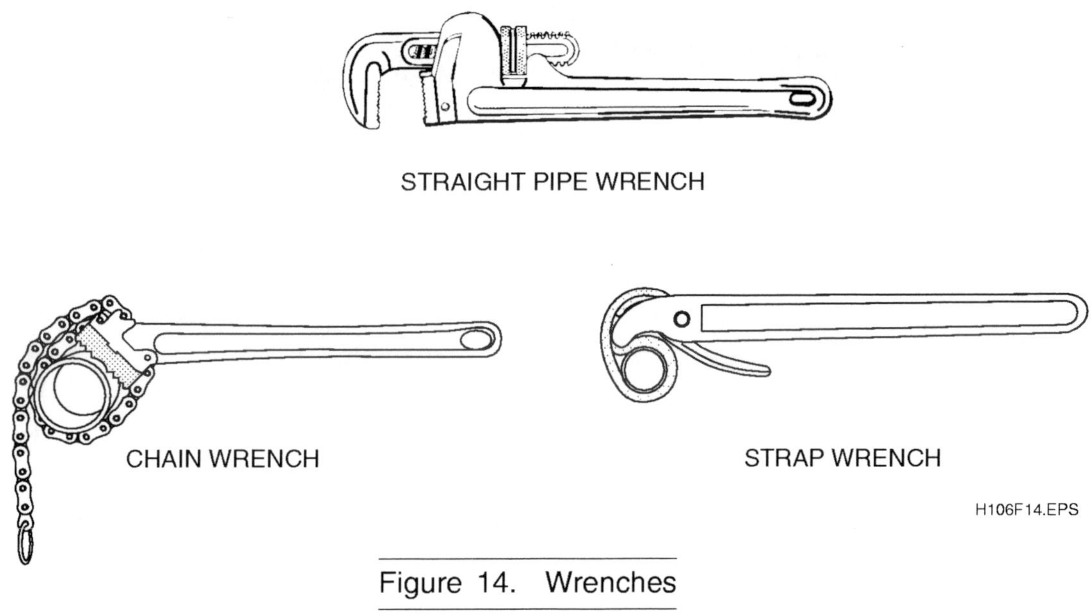

Figure 14.   Wrenches

3.5.0   PIPE DOPE

Joint compound, commonly called **pipe dope**, is a putty-like material applied to seal a joint and provide lubrication for assembly. The dope should be applied only to the male threads. Teflon tape made specifically as a pipe joint sealer may also be used with the proper application. No more than three wraps in the direction of the thread should be made, nor should it overlap the threaded end of the pipe.

4.0.0   JOINING PROCEDURES

The following steps are to be performed when joining threaded steel pipes.

4.1.0   MEASURING

Dimensions on pipe drawings specify the location of center lines and/or points of center lines, they do not specify pipe lengths. This system is also used in the fabrication and installation of pipe assemblies. To determine actual pipe lengths, threaded pipe can be measured by a variety of methods: end-to-end, end-to-center, face-to-end, center-to-center, or face-to-face. *Figure 15* shows the different measuring techniques.

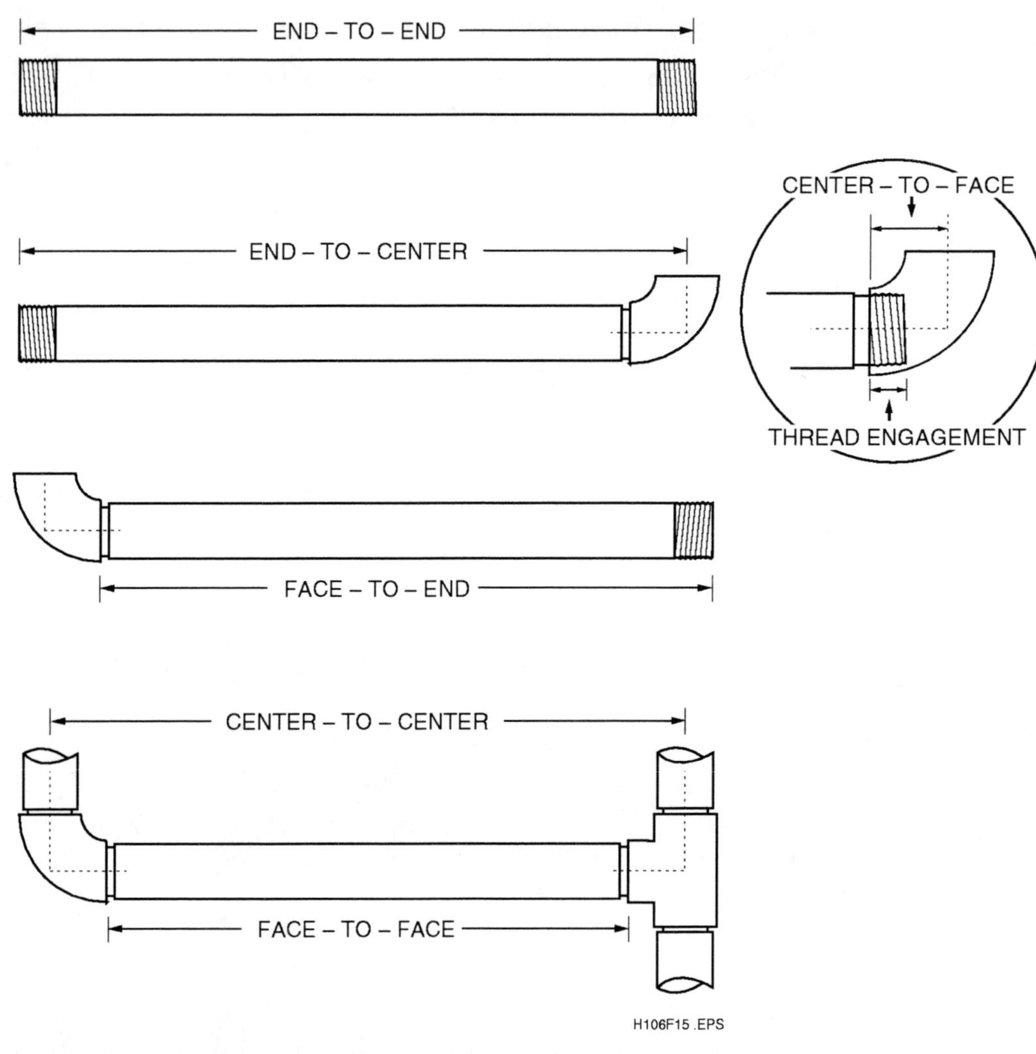

Figure 15.   Pipe Measuring Methods

End-to-end measure is accomplished by measuring the full length of the pipe including the threads at both ends.

End-to-center measure is used for a piece of pipe with a fitting screwed onto one end only. The pipe length is equal to the total end-to-center measurement minus the center-to-face dimension of the fitting plus the length of the thread engagement. Thread engagement accounts for the length of pipe that is threaded into a fitting. See *Table 2*.

Size Of Pipe (In.)	Outside Diameter (In.)	Number Of Threads Per Inch	Total Length Of Threads (In.)	Effective Length (In., Approx.)	Thread Engagement (In., Approx.)
1/4	.54	18	5/8	3/8	3/8
3/8	.675	18	5/8	7/16	3/8
1/2	.84	14	13/16	9/16	1/2
3/4	1.05	14	13/16	9/16	1/2
1	1.315	11-1/2	1	11/16	9/16
1-1/4	1.66	11-1/2	1	11/16	5/8
1-1/2	1.9	11-1/2	1	3/4	5/8
2	2.375	11-1/2	1-1/16	3/4	11/16
2-1/2	2.875	8	1-9/16	1-1/8	15/16
3	3.5	8	1-5/8	1-3/16	1
4	4.5	8	1-3/4	1-5/16	1-1/16
5	5.56	8	1-13/16	1-3/8	1-3/16
6	6.625	8	1-15/16	1-1/2	1-1/4

Table 2.   Pipe Engagement Allowances

Face-to-end measure is used for a piece of pipe with a fitting screwed onto one end only. It differs from end-to-center in that the length of the pipe is equal to the pipe measurement plus the length of the thread engagement.

Center-to-center measure is used to measure pipe with fittings screwed onto both ends. The pipe length is equal to the total center-to-center measurement minus the sum of the two center-to-face dimensions of the fittings, plus two times the length of the thread engagement.

Face-to-face measure can be used under the same conditions as the center-to-center method. It is figured by measuring the length of pipe plus two times the length of the thread engagement.

4.2.0 CUTTING

The pipe cutter (*Figure 9*) is the best tool to use when cutting steel pipe. To operate, revolve the cutter around the pipe and tighten the cutting wheel $\frac{1}{4}$ revolution with each turn. Avoid overtightening the cutting wheel. This will cause a large burr inside the pipe and excessive wear of the cutting wheel.

4.3.0 REAMING

Once the pipe is cut, a reamer (*Figure 10*) removes the burr which forms on the inside of the pipe. Not removing the burr can cause blockage and restrict liquid flow.

4.4.0 THREADING

Threads may be cut by hand with a die and stock or an electric pipe threading machine. See *Figures 11* and *12*.

4.4.1 Using A Hand Die And Stock

To cut threads using a hand die and stock, proceed as follows:

Step 1 Select the correct size die for the pipe being threaded.

Step 2 Inspect the die to see if the cutters are free of nicks and wear.

Step 3 Lock the pipe securely in a vise.

Step 4 Slide the die over the end of the pipe, guide end first.

Step 5 Push the die against the pipe with the heel of one hand. Take three or four short, slow, clockwise turns. Be careful to keep the die pressed firmly against the pipe. When enough thread is cut to keep the die firmly on the pipe, apply some thread-cutting oil. This oil prevents the pipe from overheating due to friction and it lubricates the die. Oil the threading die every two or three downward strokes.

Step 6 Back-off $\frac{1}{4}$ turn after each full turn forward to clear out the metal chips. Continue until the pipe projects one or two threads from the die end of the stock. Too few threads is as bad as too many threads.

Step 7 To remove the die, rotate it counterclockwise.

Step 8 Wipe off excess oil and any chips.

WARNING!　　　　Use a rag, not your bare hands. The chips are sharp and could cause cuts.

4.4.2 Using A Threading Machine

Each threading machine is slightly different. Become familiar with the manufacturer's operating procedures before attempting to operate the machine. Also, become thoroughly familiar with the maintenance and safety instructions of the machine. A poorly maintained machine is a safety hazard.

WARNING! Do not wear loose fitting clothing, long hair, jewelry or loose fitting gloves when using any pipe threading machine.

To thread using a power threading machine:

Step 1 Select the pipe stock and install the correct size die and inspect it for nicks.

Step 2 Mount the pipe stock into the chuck. Long stock must have additional support.

Step 3 Check the pipe and die alignment.

Step 4 Cut threads until two threads appear at the other end of the die. Stop threading action. Apply cutting oil during the threading operation.

Step 5 Back off the die until it is clear of the pipe.

Step 6 Remove the pipe from the machine chuck. Be careful not to mar the threads.

Step 7 Wipe the pipe clean of oil and metal chips.

4.5.0 ASSEMBLING

Apply pipe dope to the male threads before assembling a pipe connection. Do not apply the compound to the threads of the fitting. Either the paste compound or teflon tape may be used.

When using teflon tape, apply the tape in a clockwise direction, the same direction as the fitting turns. Be sure to check the local codes to see if the use of tape is permitted.

Start the fitting onto the threaded pipe by hand. Turn the fitting clockwise. Finish tightening the fitting using pipe wrenches (*Figure 16*).

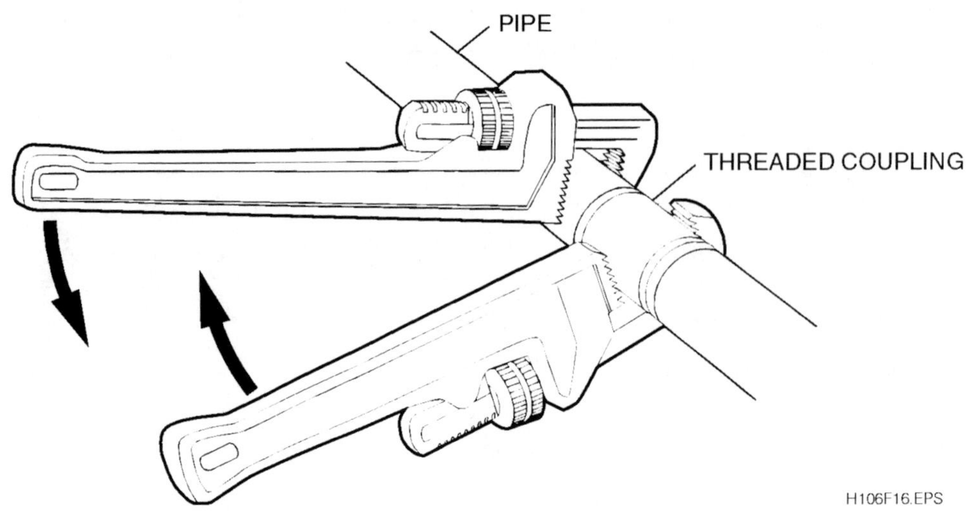

Figure 16.   Tightening Pipe Using Two Pipe Wrenches

4.6.0   HANGING AND SUPPORTING

The various types of hangers and supports used to install steel piping are the same as previously described for non-ferrous metal piping systems (Refer to HVAC Module 03104, Copper and Plastic Piping Practices.)

The main thing to remember is that the method used to install a piping system is generally defined by the builder's specifications. Air conditioning and heating installations planned by architects and engineering consultants include plans and specifications which completely describe the proposed installation. After the job has been awarded to a successful bidder, the engineering drawings are supplemented by the working drawings of the installation contractor. After the working drawings and submittals are approved by the engineer and architect, the work proceeds according to specifications.

Each section of the specification is numbered for ready reference and describes one portion of the work. Specifications are based upon code or ordinances and must be adhered to. A specification for pipe hangers, for example, may read as follows: "All piping shall be supported with hangers spaced not more than 10 feet apart (on center). Hangers shall be the malleable iron split-ring type and shall be as manufactured by XYZ Hangers, Inc. or other approved vendor."

Although piping drawings should be available for every piping system that is installed, this is not always the case. Therefore, the HVAC installer often has to select the best route and install the piping. This is known as field-fabricating or field-routing a pipe run. In this case, the HVAC installer has to determine the placement of the pipe hangers, based on the size, weight, and type of pipe being run. In order for a hanger system to do its job, it must support

HVAC TRAINEE TASK MODULE 03106

the pipe at regular intervals. Evenly spacing the hangers prevents any individual hanger from being overloaded. *Table 3* lists the recommended maximum hanger spacing intervals for carbon steel pipe.

Pipe Size	Rod Diameter	Maximum Spacing
Up to 1-1/4"	3/8"	8'
1-1/2" and 2"	3/8"	10'
2-1/2" to 3-1/2"	1/2"	12'
4 and 5"	5/8"	15'
6"	3/4"	17'
8" to 12"	7/8"	22'

Table 3.   Recommended Maximum Hanger Spacing Intervals For Carbon Steel Pipe

Table 3 only serves as a guide. Always check the job site specifications when determining pipe hanger spacing, and if the pipe sags, add more hangers. The most important thing to remember is that the pipe must run straight without sagging.

5.0.0   GROOVED PIPE

Grooved pipe is so called because grooves instead of welded, flanged, or threaded joints are used for coupling. Each joint in a grooved piping system serves as a union, allowing easy access to any part of the piping system for cleaning or servicing. The leading manufacturer of grooved piping couplings, fittings, and fabrication tools is the Victaulic Company of America. Grooved piping systems have a wide range of applications and can be used with a wide variety of piping materials. The most common uses of grooved piping systems are in oil fields, industrial facilities, mining, municipal systems, and fire protection systems. The following types of piping can be joined by grooved couplings:

- Carbon steel
- Stainless steel
- Aluminum
- PVC plastic
- High-density polyethylene
- Ductile iron

The standard grooved piping coupling consists of a rubber gasket and two housing halves that are bolted together. The housing halves are tightened together until they touch, so no special torquing of the housing bolts is required.

The grooved piping system offers varied mechanical benefits, including the option of rigid or flexible couplings. Rigid and flexible couplings can be incorporated as needed into any system to take full advantage of the characteristics of each. Rigid couplings create a rigid joint useful for risers, mechanical rooms, and other areas where positive clamping with no flexibility within the joints is desired. Flexible couplings provide allowance for controlled pipe movement that occurs with expansion, contraction, and deflection. Flexible couplings may eliminate the need for expansion joints, cold springing, or expansion loops and will provide a virtually stress-free piping system. *Figure 17* shows examples of rigid and flexible grooved pipe couplings.

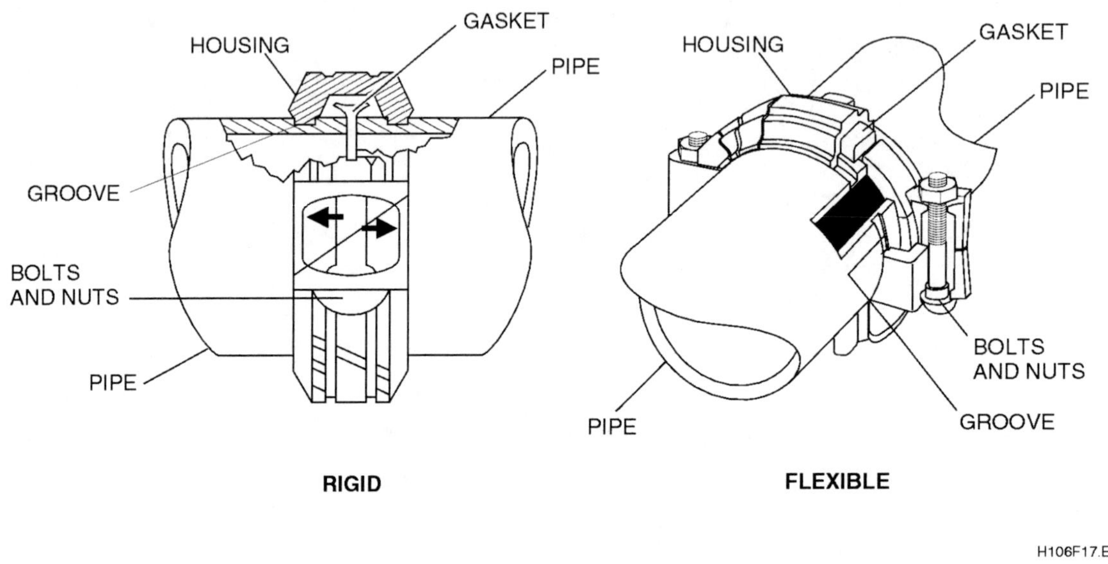

Figure 17. Rigid And Flexible Pipe Couplings

5.1.0 PREPARING PIPE ENDS

Grooved pipe can be delivered to the job site precut to length and grooved, or it can be cut and grooved on the job. Use of the grooved piping method is based on the proper preparation of a groove in the pipe end to receive the coupling housing key. The groove serves as a recess in the pipe with enough depth to secure the coupling housing, yet at the same time, leaves enough wall thickness for a full pressure rating. Groove preparation varies with different pipe materials and wall thicknesses. A variety of tools is available from Victaulic Company of America to properly groove pipe in the shop or in the field. The two methods of forming a groove in pipe include roll grooving and cut grooving. *Figure 18* shows details of pipe grooves.

The dimensions pointed out by the letters in *Figure 18* must comply with engineering specifications at your job site. The A dimension is the distance from the pipe end to the groove and provides the gasket seating area. This area must be free from indentations, projections, or roll marks to provide a leakproof sealing seat for the gasket. The B dimension is the groove width and controls expansion and angular deflection based on its distance from the pipe end

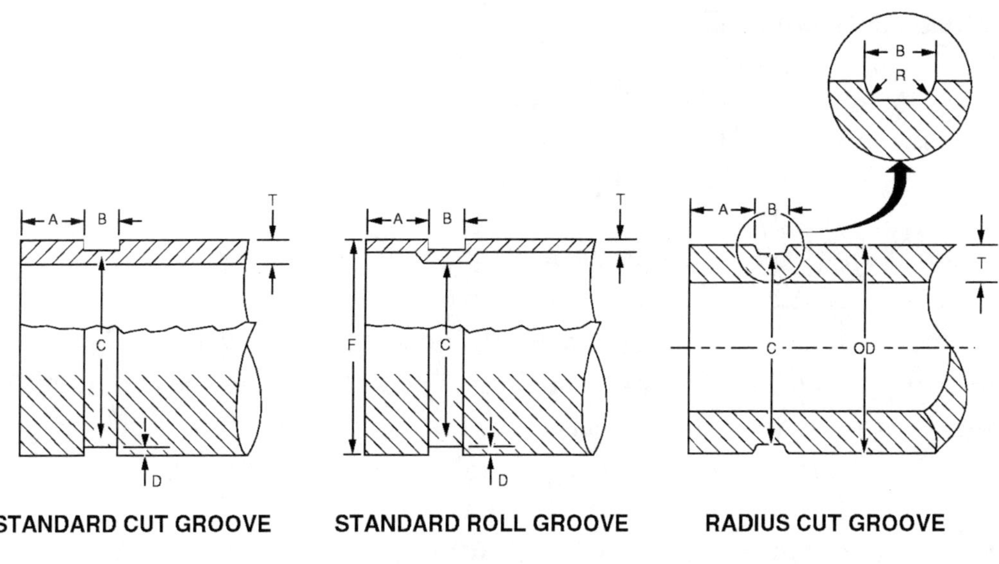

Figure 18. Details Of Pipe Grooves

and its width in relation to the width of the coupling housing key. The C dimension is the proper diameter tolerance and is concentric with the outside diameter of the pipe. The D dimension must be changed if necessary to keep the C dimension within the stated tolerances. The F dimension is used with the standard roll only and gives the maximum allowable pipe end flare. The T dimension is the lightest grade or minimum thickness of pipe suitable for roll or cut grooving. The R dimension is the radius necessary at the bottom of the groove to eliminate a point of stress concentration for cast iron and PVC plastic pipe.

5.1.1 Roll Grooving

Power roll-grooving machines are used to roll grooves at the ends of pipe to prepare the piping for groove-type fittings and couplings. Power grooving machines are available to groove 2" to 16" standard and lightweight steel pipe, aluminum pipe, stainless steel pipe, and PVC plastic pipe.

Roll grooving removes no metal from the pipe but forms a groove by displacing the metal instead. Since the groove is cold-formed, it has rounded edges that reduce the pipe movement after the joint is made up.

5.1.2 Cut Grooving

Cut grooving differs from roll grooving in that a groove is cut into the pipe. Cut grooving is basically intended for standard weight or heavier pipe. The cut removes less than one-half of the pipe wall, which is less depth than thread cuts. Cut-grooving machines are designed to be driven around a stationary pipe. This creates a groove that is of uniform depth and is concentric with the outside diameter of the pipe.

5.2.0   SELECTING GASKETS

There are many types of synthetic rubber gaskets available to provide the option of selecting grooved piping products for the widest range of applications. In order to provide maximum life for the service intended, the proper gasket selection and specification is required.

Several factors must be considered in determining the best gasket to use for a specific service. The foremost consideration is the temperature of the product flowing through the pipe. Temperatures beyond the recommended limits decrease gasket life. Also, the concentration of the product, duration of service, and continuity of service must be considered because there is a direct relationship between temperature, continuity of service, and gasket life. It should also be noted that there are services for which gaskets are not recommended.

5.3.0   INSTALLING GROOVED PIPE COUPLINGS

The procedure used to join grooved pipe is as follows:

Step 1   Make sure that the gasket is suitable for the intended service. Some manufacturers color code their gaskets. Apply a thin coat of lubricant to the gasket lips and the outside of the gasket.

Step 2   Check the pipe ends. To assure a leak-proof seal, they must be free from indentations, projections, or roll marks.

Step 3   Install the gasket over the pipe end. Be sure the gasket lip does not hang over the pipe end.

Step 4   Align and bring the two pipe ends together. Slide the gasket into position and center it between the grooves on each pipe. Be sure that no part of the gasket extends into the groove on either pipe.

Step 5   Assemble the housing segments loosely, leaving one nut and bolt off to allow the housing to swing over the joint.

Step 6   To install the housing, swing it over the gasket and into position in the grooves on both pipes.

Step 7   Insert the remaining bolt and nut. Be sure that the bolt track head engages into the recess in the housing.

Step 8   Tighten the nuts alternately and equally to maintain metal-to-metal contact at the angle bolt pads.

SUMMARY

Steel pipe (both galvanized and black iron) is found on almost every construction site. Its durability, structural strength, low material cost, and low expansion properties make it a dependable plumbing material. Uses for steel pipe include:

- Hot and cold water distribution
- Steam and hot water heating systems
- Gas and air piping systems
- Drainage and vent systems
- Fire protection systems

References

For more advanced study of topics covered in this task module, the following books are suggested:

Modern Plumbing, The Goodheart-Willcox Company, Inc., South Holland, Illinois.
Pipefitter's Handbook, Third Edition, Industrial Press, Inc., 200 Madison Avenue, New York, New York.
Refrigeration & Air Conditioning Technology, Second Edition, Delmar Publishers, Inc., Albany, New York.

SELF CHECK REVIEW / PRACTICE QUESTIONS

1. For which of the following is steel pipe commonly used?
   a. hot and cold water distribution
   b. gas and air piping systems
   c. drainage and vent systems
   d. all of the above

2. The nominal size of pipes 14 inch and larger is based on:
   a. the weight of the pipe.
   b. the outside diameter of the pipe.
   c. the inside diameter of the pipe.
   d. the length of the pipe.

3. The nominal wall thickness of a schedule 80 pipe is the same as that of a _____ weight pipe.
   a. standard (STD)
   b. extra strong (XS)
   c. double extra strong (XXS)
   d. schedule 40

4. The standard number of threads per inch for a pipe whose nominal size is $1/2$ inch is:
   a. 17 threads per inch.
   b. 18 threads per inch.
   c. 14 threads per inch.
   d. $11^{1}/_{2}$ threads per inch.

5. A tee fitting described as 2" x $1^{1}/_{2}$" x 1" has a:
   a. large run size of 2", small run size of $1^{1}/_{2}$" and a branch size of 1".
   b. large run size of 1", small run size of $1^{1}/_{2}$" and a branch size of 2".
   c. large run size of $1^{1}/_{2}$", small run size of 2" and a branch size of 1".
   d. large run size of 2", small run size of 1" and a branch size of $1^{1}/_{2}$".

6. Pipe nipples are used to:
   a. close openings in other fittings.
   b. connect two lengths of pipe when making straight runs.
   c. change direction of the pipe run.
   d. make extensions from a fitting or to join two fittings.

7. Measurement of the full length of pipe including the threads describes a _____ measurement.
   a. end-to-center
   b. end-to-end
   c. face-to-end
   d. face-to-face

8. The threading die should be oiled every _____ when threading pipe.
   a. after the threading of each piece of pipe
   b. 10 downward strokes
   c. two or three downward strokes
   d. once a day

9. Before using a threading machine, you always should:
   a. become familiar with the manufacturer's operating instructions.
   b. become familiar with the safety instructions for the machine.
   c. remove loose fitting clothing and all jewelry.
   d. all of the above.

10. Which of the following joining method(s) is eliminated when grooved piping is used?
    a. threading
    b. flanging
    c. welding
    d. all of the above

ANSWERS TO SELF CHECK REVIEW / PRACTICE QUESTIONS

Answer	**Section Reference**
1. d	1.0.0
2. b	2.3.0
3. b	2.3.0
4. c	2.4.0
5. a	2.5.1
6. d	2.5.5
7. b	4.1.0
8. c	4.4.1
9. d	4.4.2
10. d	5.0.0

ACKNOWLEDGEMENTS

Figures 17 and 18 courtesy of the Victaulic Company of America.

CARBON STEEL PIPE SCHEDULES

PIPE SIZE	OD IN INCHES	OD IN MM	WEIGHTS AND DIMENSIONS OF SEAMLESS AND WELDED STEEL PIPE (P.E.)												
			10	20	30	40	STD.	60	80	XS	100	120	140	160	XXS
1/8	0.405	10.3				0.068 0.24	0.068 0.24		0.095 0.31	0.095 0.31					
1/4	0.540	13.7				0.088 0.42	0.088 0.42		0.119 0.53	0.119 0.53					
3/8	0.675	17.1				0.091 0.57	0.091 0.57		0.126 0.73	0.126 0.73					
1/2	0.840	21.3				0.109 0.85	0.109 0.85		0.147 1.09	0.147 1.09				0.188 1.31	0.294 1.71
3/4	1.050	26.7				0.113 1.13	0.113 1.13		0.154 1.47	0.154 1.47				0.219 1.94	0.308 2.44
1	1.315	33.4				0.133 1.68	0.133 1.68		0.179 2.17	0.179 2.17				0.250 2.84	0.358 3.66
1-1/4	1.660	42.2				0.140 2.27	0.140 2.27		0.191 3.00	0.191 3.00				0.250 3.76	0.382 5.21
1-1/2	1.900	48.3				0.145 2.72	0.145 2.72		0.200 3.63	0.200 3.63				0.281 4.86	0.400 6.41
2	2.375	60.3				0.154 3.65	0.154 3.65		0.218 5.02	0.218 5.02				0.344 7.46	0.436 9.03
2-1/2	2.875	73.0				0.203 5.79	0.203 5.79		0.276 7.66	0.276 7.66				0.375 10.01	0.552 13.69
3	3.500	88.9				0.216 7.58	0.216 7.58		0.300 10.25	0.300 10.25				0.438 14.32	0.600 18.58
3-1/2	4.000	101.6				0.226 9.11	0.226 9.11		0.318 12.50	0.318 12.50					0.636 22.85
4	4.500	114.3				0.237 10.79	0.237 10.79	0.281 12.66	0.337 14.98	0.337 14.98		0.438 19.00		0.531 22.51	0.674 27.54
5	5.563	141.3				0.258 14.62	0.258 14.62		0.375 20.78	0.375 20.78		0.500 27.04		0.625 32.96	0.750 38.55
6	6.625	168.3				0.280 18.97	0.280 18.97		0.432 28.57	0.432 28.57		0.562 36.39		0.719 45.35	0.864 53.16
8	8.625	219.1		0.250 22.36	0.277 24.70	0.322 28.55	0.322 28.55	0.406 35.64	0.500 43.39	0.500 43.39	0.594 50.95	0.719 60.71	0.812 67.76	0.906 74.69	0.875 72.42
10	10.750	273.1		0.250 28.04	0.307 34.24	0.365 40.48	0.365 40.48	0.500 54.74	0.594 64.43	0.500 54.74	0.719 77.03	0.844 89.29	1.000 104.1	1.125 115.6	1.000 104.1
12	12.750	323.9		0.250 33.38	0.330 43.77	0.406 53.52	0.375 49.56	0.562 73.15	0.688 88.63	0.500 65.42	0.844 107.3	1.000 125.5	1.125 139.7	1.312 160.3	1.000 125.5
14	14.000	355.6	0.250 36.71	0.312 45.61	0.375 54.57	0.438 63.44	0.375 54.57	0.594 85.05	0.750 106.1	0.500 72.09	0.938 130.9	1.094 150.8	1.250 170.2	1.406 189.1	
16	16.000	406.4	0.250 42.05	0.312 52.27	0.375 62.58	0.500 82.77	0.375 62.58	0.656 107.5	0.844 136.6	0.500 82.77	1.031 164.8	1.219 192.4	1.438 223.6	1.594 245.3	
18	18.000	457.2	0.250 47.39	0.312 58.94	0.438 82.15	0.562 104.7	0.375 70.59	0.750 138.2	0.938 170.9	0.500 93.45	1.156 208.0	1.375 244.1	1.562 274.2	1.781 308.5	
20	20.000	508.0	0.250 52.73	0.375 78.60	0.500 104.1	0.594 123.1	0.375 78.60	0.812 166.4	1.031 208.9	0.500 104.1	1.281 256.1	1.500 296.4	1.750 341.1	1.969 379.2	
24	24.000	609.6	0.250 63.41	0.375 94.62	0.562 140.7	0.688 171.3	0.375 94.62	0.969 238.4	1.219 296.6	0.500 125.5	1.531 367.4	1.812 429.4	2.062 483.1	2.344 542.1	
26	26.000	660.4	0.312 85.60	0.500 136.2			0.375 102.8			0.500 136.2					
30	30.000	762.0	0.312 98.93	0.500 157.5	0.625 196.1		0.375 118.6			0.500 157.5					
36	36.000	914.4	0.312 118.9	0.500 189.6	0.625 236.1	0.750 282.4	0.375 142.7			0.500 189.6					
42	42.000	1067.0					0.375 166.7			0.500 221.6					
48	48.000	1219.0					0.375 190.7			0.500 253.7					

THE SECOND FIGURE LISTED REPRESENTS THE WEIGHT PER FOOT IN POUNDS

To convert the inch dimensions of outside diameters and wall thickness to millimeters, multiply the inch dimensions by 25.4

CARBON STEEL PIPE WALL THICKNESSES OTHER THAN SCHEDULES

OD — WALL x WALL x 10.68 = WEIGHT PER FOOT OF STEEL PIPE (P.E.)

THE SECOND FIGURE LISTED REPRESENTS THE WEIGHT PER FOOT IN POUNDS

NOMINAL PIPE SIZE	OD IN INCHES															
1	1.315	0.065 0.87	0.109 1.40													
1-1/4	1.660	0.065 1.11	0.109 1.81													
1-1/2	1.900	0.065 1.27	0.109 2.08													
2	2.375	0.065 1.60	0.083 2.03	0.109 2.64	0.120 2.89	0.190 4.43	0.254 5.75	0.281 6.28								
2-1/2	2.875	0.083 2.47	0.109 3.22	0.120 3.53	0.188 5.40	0.217 6.16	0.308 8.44									
3	3.500	0.083 3.03	0.109 3.95	0.120 4.33	0.125 4.51	0.156 5.57	0.188 6.65	0.250 8.68	0.254 8.81	0.281 9.66	0.375 12.52					
3-1/2	4.000	0.083 3.47	0.109 4.53	0.120 4.97	0.125 5.17	0.156 6.40	0.188 7.65	0.250 10.01	0.262 10.46	0.281 11.16						
4	4.500	0.083 3.92	0.109 5.11	0.120 5.61	0.125 5.84	0.141 6.56	0.156 7.24	0.172 7.95	0.188 8.66	0.203 9.32	0.219 10.01	0.224 10.23	0.250 11.35	0.290 13.04	0.312 13.96	0.375 16.52
5	5.563	0.083 4.86	0.109 6.35	0.125 7.26	0.134 7.77	0.156 9.01	0.188 10.79	0.219 12.50	0.281 15.85	0.312 17.50	0.344 19.17					
6	6.625	0.109 7.59	0.125 8.68	0.134 9.29	0.141 9.76	0.156 10.78	0.172 11.85	0.188 12.92	0.203 13.92	0.219 14.98	0.250 17.02	0.312 21.04	0.344 23.08	0.375 25.03	0.500 32.71	0.625 40.05
8	8.625	0.109 9.91	0.125 11.35	0.156 14.11	0.172 15.53	0.188 16.94	0.203 18.26	0.219 19.66	0.264 23.57	0.312 27.70	0.344 30.42	0.375 33.04	0.438 38.30	0.562 48.40	0.625 53.40	
10	10.750	0.156 17.65	0.172 19.43	0.188 21.21	0.203 22.87	0.219 24.63	0.279 31.20	0.344 38.23	0.350 38.88	0.400 44.22	0.438 48.24	0.562 61.15	0.625 67.58	0.812 86.18		
12	12.750	0.172 23.11	0.188 25.22	0.203 27.20	0.219 29.31	0.281 37.42	0.312 41.45	0.344 45.58	0.438 57.59	0.625 80.93	0.750 96.12	0.812 103.5	0.875 111.0	1.500 180.2	1.750 205.6	2.000 229.6
14	14.000	0.188 27.73	0.203 29.91	0.219 32.23	0.281 41.17	0.344 50.17	0.406 58.94	0.469 67.78	0.562 80.66	0.625 89.28	0.688 97.81	0.812 114.4	0.875 122.7	2.000 256.3	2.125 269.5	2.500 307.1
16	16.000	0.188 31.75	0.203 34.25	0.219 36.91	0.281 47.17	0.344 57.52	0.406 67.62	0.438 72.80	0.469 77.79	0.625 102.6	0.750 122.2	0.812 131.7	0.938 150.9	1.125 178.7	1.618 248.5	2.000 299.0
18	18.000	0.188 35.76	0.219 41.59	0.281 53.18	0.344 64.87	0.406 76.29	0.469 87.81	0.625 116.0	0.688 127.2	0.812 149.1	0.875 160.0	1.000 181.6	1.125 202.8	1.250 223.6	1.500 264.3	
20	20.000	0.219 46.27	0.281 59.18	0.312 65.60	0.344 72.21	0.406 84.96	0.438 91.51	0.469 97.83	0.625 129.3	0.750 154.2	0.875 178.7	1.000 202.9	1.250 250.3	1.375 273.5		
24	24.000	0.281 71.18	0.312 78.93	0.344 86.91	0.406 102.3	0.438 110.2	0.469 117.9	0.625 156.0	0.750 186.2	0.875 216.1	1.000 245.6	1.250 303.7	1.312 317.9	1.500 360.5		
26	26.000	0.250 68.75	0.281 77.18	0.344 94.26	0.406 111.0	0.438 119.6	0.469 127.9	0.562 152.7	0.625 169.4	0.656 177.6	0.688 186.0	0.750 202.3	0.875 234.8	1.000 267.0	1.188 314.8	
30	30.000	0.281 89.19	0.344 109.0	0.406 128.3	0.438 138.3	0.469 147.9	0.562 176.7	0.656 205.6	0.750 234.3	0.875 272.2	1.000 309.7	1.250 383.8				
36	36.000	0.281 107.2	0.312 118.9	0.344 131.0	0.406 154.3	0.438 166.4	0.469 178.0	0.562 212.7	0.656 247.6	0.688 259.5	0.875 328.2	1.000 373.8	1.250 463.9	1.500 552.7		
42	42.000	0.312 138.9	0.344 153.0	0.406 180.4	0.438 194.4	0.469 208.0	0.562 248.7	0.625 276.2	0.656 289.7	0.688 303.6	0.750 330.4	0.875 384.3	1.000 437.9	1.125 491.1	1.250 544.0	
48	48.000	0.406 206.4	0.438 222.5	0.469 238.0	0.562 284.7	0.625 316.2	0.656 331.7	0.688 347.6	0.750 378.5	0.812 409.2	0.875 440.4	0.938 471.5	1.000 502.0	1.125 563.2	1.250 624.1	

The NCCER makes every effort to keep these manuals up-to-date and free of technical errors. We appreciate your help in this process. If you have an idea for improving this manual, or if you find an error, a typographical mistake, or an inaccuracy in the NCCER's Craft Training Manuals, please write us, using this form or a photocopy. Be sure to include the exact module number, page number, a description of the problem, and the correction, if possible. Your input will be brought to the attention of the Technical Review Committee. Thank you for your assistance.

Instructors – If you found that additional materials were necessary in order to teach this module effectively, please let us know so that we may include them in the Equipment/Materials list in the Instructor's Guide.

Write: Curriculum and Revision Department
National Center for Construction Education and Research
P.O. Box 141104
Gainesville, FL 32614-1104

Fax: 352-334-0932

Craft

Module Name

Module Number

Page Number(s)

Description of Problem

(Optional) Correction of Problem

(Optional) Your Name and Address

Basic Electricity

Module 03107

BASIC ELECTRICITY

Objectives

Upon completion of this module, the trainee will be able to:

1. State how electrical power is generated and distributed.
2. Describe how voltage, current, resistance, and power are related.
3. Use Ohm's Law to calculate the current, voltage, and resistance in a circuit.
4. Use the power formula to calculate how much power is consumed by a circuit.
5. Describe the differences between series and parallel circuits.
6. Recognize and describe the purpose and operation of the various electrical components used in HVAC equipment.
7. State and demonstrate the safety precautions that must be followed when working on electrical equipment.
8. Make voltage, current, and resistance measurements using electrical test equipment.

Prerequisites

Successful completion of the following Task Modules is required before beginning study of this Task Module: Common Core Curricula, HVAC Modules 03101 through 03106.

Required Student Material

1. Student Module
2. Appropriate Personal Protective Equipment

Course Map Information

This course map shows all of the *Wheels of Learning* task modules in the first level of the HVAC curricula. The suggested training order begins at the bottom and proceeds up. Skill levels increase as a trainee advances on the course map. The training order may be adjusted by the local Training Program Sponsor.

Course Map: HVAC, Level 1

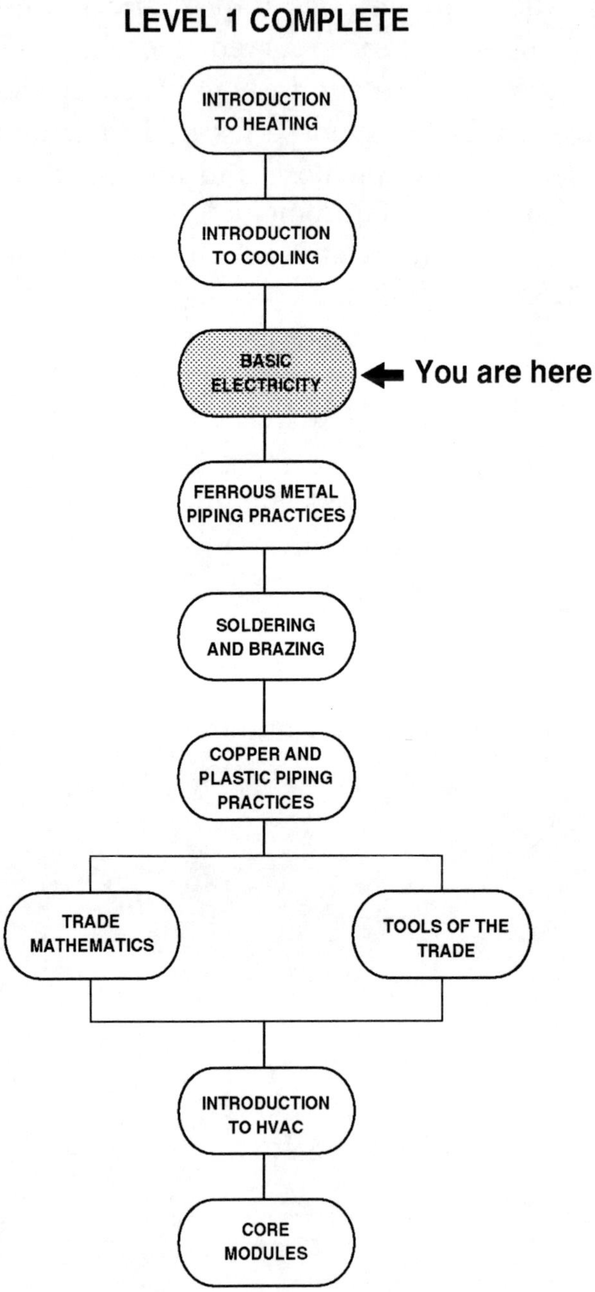

LEVEL 1 COMPLETE

TABLE OF CONTENTS

Trade Terms Introduced In This Module

Alternating current (AC): An electrical current that changes direction on a cyclic basis.

Ammeter: A test instrument used to measure current flow.

Ampere (amp): The unit of measurement for current flow. The magnitude is determined by the number of electrons passing a point at a given time.

Analog meter: A meter that uses a needle to indicate a value on a scale.

Clamp-on ammeter: A current meter in which jaws placed around a conductor sense the magnitude of the current flow through the conductor.

Conductor: A material that readily conducts electricity. Also the wire that connects components in an electrical circuit.

Contactor: A control device consisting of a coil and one or more sets of contacts used as a switching device in high voltage circuits.

Continuity: A continuous current path. Absence of continuity indicates an open circuit.

Current: The rate at which electrons flow in a circuit. Measured in amperes.

Digital meter: A meter that provides a direct numerical reading of the value measured.

Direct current (DC): An electric current that flows in one direction. A battery is a common source of DC voltage.

Electromagnet: A coil of wire wrapped around a soft iron core. When a current flows through the coil, magnetism is created.

HACR circuit breaker: A circuit breaker with a built-in trip delay used specifically in HVAC circuits because of the power surge that occurs with compressor startup.

Induction: To generate a current in a conductor by placing it in a magnetic field.

In-line ammeter: A current-reading meter that is connected in series with the circuit under test.

Insulator: A device that inhibits the flow of current. Opposite of conductor.

Ladder diagram: A simplified schematic diagram in which the load lines are arranged like the rungs of a ladder between vertical lines representing the voltage source.

Line duty: A protective device connected in series with the supply voltage.

Load: A device that converts electrical energy into another form of energy (heat, mechanical motion, light, etc.). Motors are the most common loads in HVAC systems.

Multimeter: A test instrument capable of reading voltage, current, and resistance.

Ohm: The unit of measurement for electrical resistance.

Pilot duty: A protective device that opens the motor control circuit, which then shuts off the motor.

Power: The amount of energy consumed by an electrical load. Measured in watts. Power = Voltage x Current.

Pressurestat: A pressure-sensitive switch used to protect compressors.

Rectifier: A device that converts AC voltage to DC voltage.

Relay: A magnetically-operated device consisting of a coil and one or more sets of contacts. Used in low current circuits.

Resistance: The opposition to the flow of electrons (i.e., load).

Short circuit: A situation where a conductor bypasses the load, causing a very high current flow.

Slow-blow fuse: A fuse with a built-in time delay.

Solenoid: An electromagnetic coil used to control a mechanical device such as a valve or relay contacts.

Solid state: Having to do with semiconductors.

Starter: A magnetic switching device used to control heavy-duty motors.

Transformer: Two or more coils of wire wrapped around a common core. Used to raise and lower voltages.

Volt: The unit of measurement for voltage.

Voltage: A measure of the electrical potential for current flow. Measured in volts.

Watts: The unit of measure for power consumed by a load.

1.0.0 INTRODUCTION

HVAC equipment, like many other things in our lives, runs on electricity. Even a simple gas-fired or oil-fired furnace needs electricity to run the fans that circulate air through the building.

A lot of the problems an HVAC technician faces are in the electrical circuits. In order to determine what is wrong, you must be able to read a circuit diagram and use electrical test equipment to make measurements at key points in the circuit. Your training in electricity will focus on reading electrical circuit diagrams and on using electrical test equipment.

2.0.0 ELECTRICITY

2.1.0 ELECTRICAL POWER GENERATION AND DISTRIBUTION

Electricity comes from electrical generating plants (*Figure 1*) operated by public utilities like your local power company. Steam from coal-burning or nuclear power plants is used to power huge generators called turbines, which generate the electricity. There are also hydroelectric power plants where water flowing over dams is used to drive turbines.

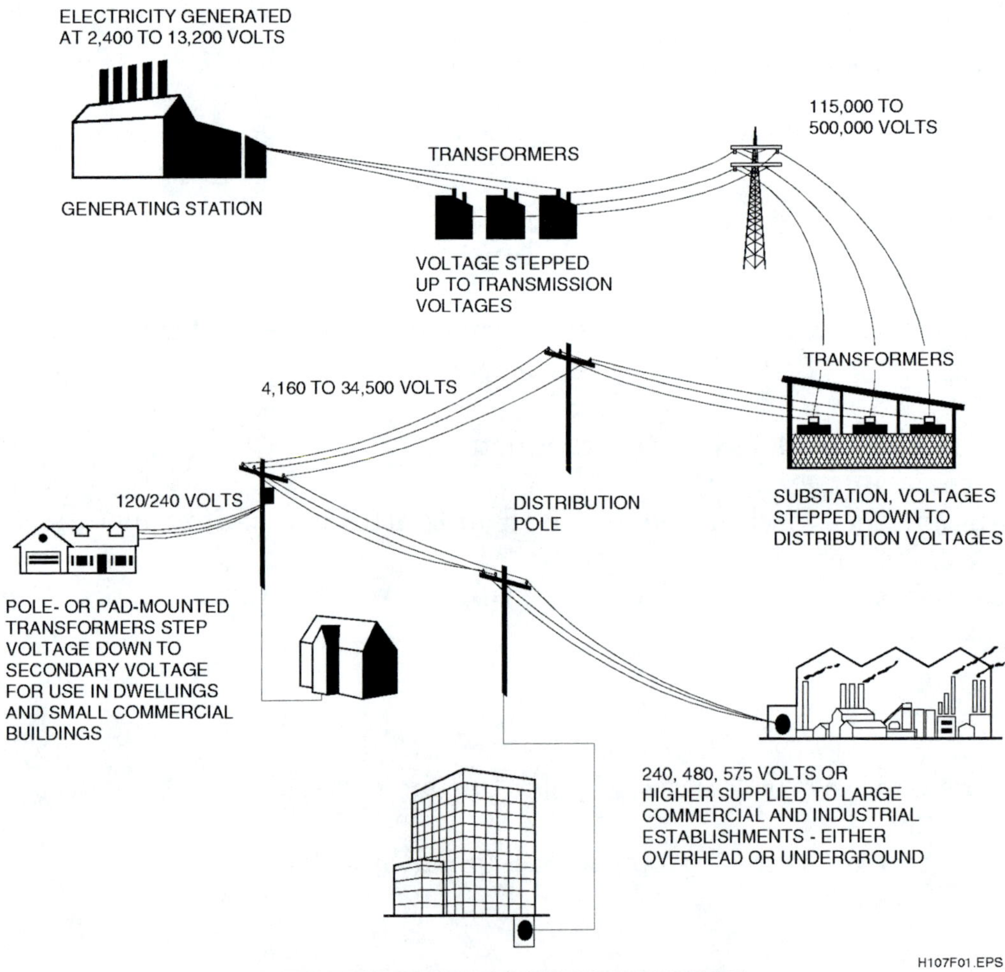

Figure 1. Electrical Power Distribution

The electrical power that travels through long-distance transmission lines may be as high as 500,000 volts. Devices known as **transformers** are used to step the voltage down to lower levels as it reaches electrical substations and eventually our homes, offices, and factories. The voltage we receive at home is usually about 240 volts. At the wall outlet where we plug in small appliances like televisions and toasters, the voltage is about 120 volts (*Figure 2*). Electric stoves, clothes dryers, water heaters, and central air conditioning systems usually require the full 240 volts. Commercial buildings and factories may receive anywhere from 240 to 575 volts. This depends on the amount of power their machines consume.

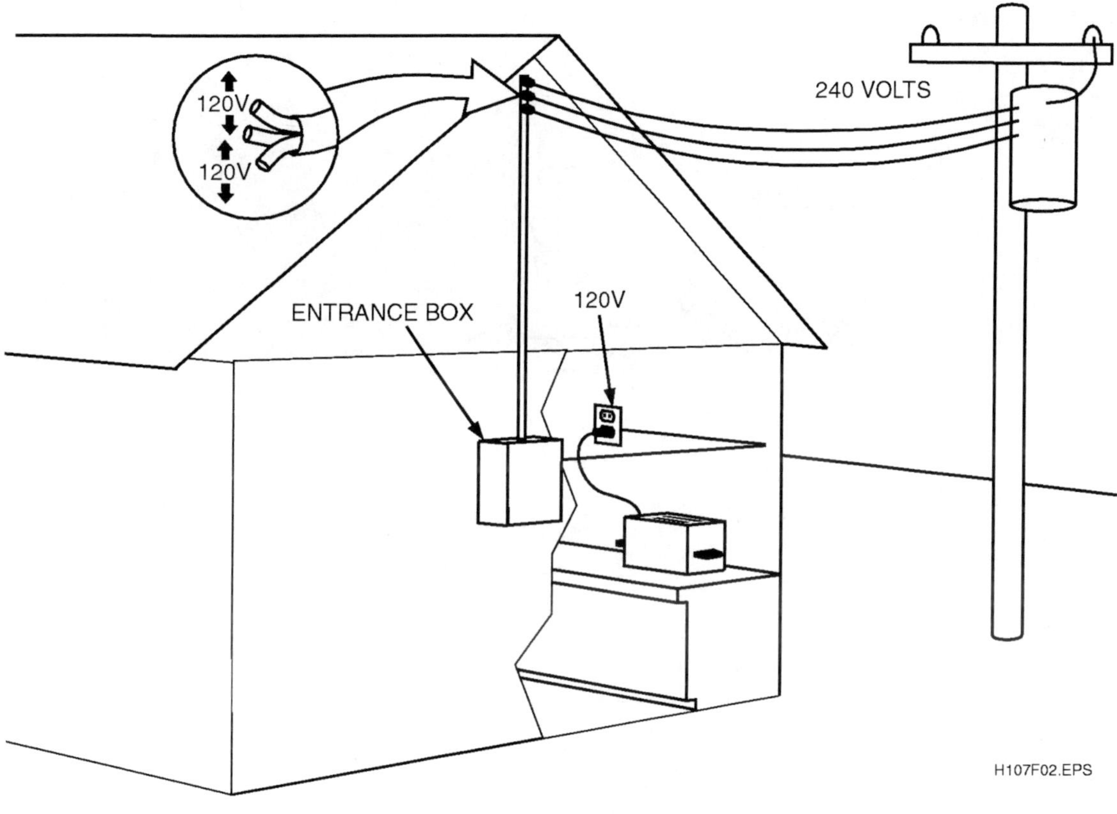

Figure 2.   Internal Power Distribution

2.2.0   CURRENT, VOLTAGE, AND RESISTANCE

An electrical **current** is caused by the movement of electrons from one point to another. Electrons are the negatively-charged particles that exist in all matter. When a difference in the number of electrons exists between two points, electrons will flow from the negative point (the one with more electrons) to the positive point (the one with fewer electrons). The difference in electrical potential between the two points is called **voltage**.

Lightning is a good example of natural electron flow. A storm cloud has a negative charge with respect to the earth. Lightning occurs when the difference in potential between the cloud and the earth becomes so great that the air between them conducts a current.

In the common 12-volt car battery, a chemical reaction causes one of the poles to be negative with respect to the other. If you connect a wire between the negative (-) and positive (+) poles of the battery, electrons will flow from the negative pole to the positive pole.

In order to harness the potential energy of the battery, a **resistance**, such as a light bulb, is connected between the two battery poles (*Figure 3*). Resistance is the property of a material to resist or impede current flow. The resistance in a circuit is also known as a **load**. When the switch closes, the moving electrons flow through the bulb, which converts the electrical energy into light. The bulb consumes the electrical energy in the process. The amount of energy consumed is known as **power**, and is expressed in **watts**. A 60-watt light bulb consumes 60 watts of power.

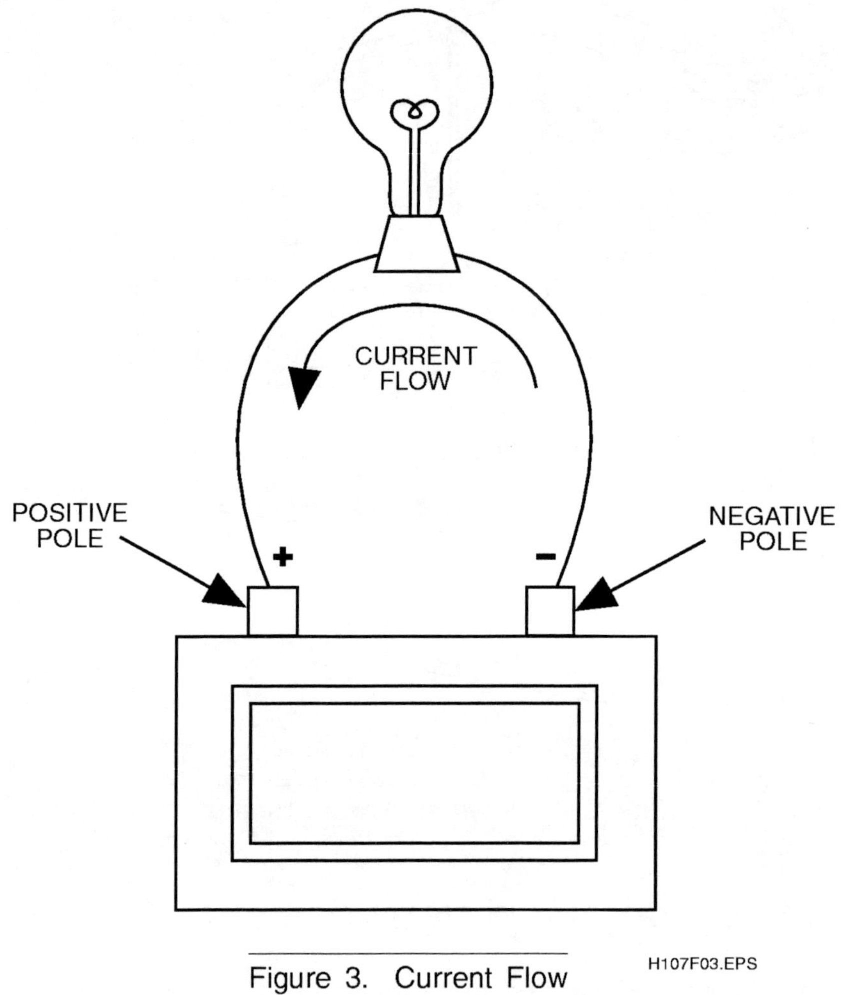

Figure 3. Current Flow

Current is expressed in **amperes** or amps (A), voltage is expressed in **volts** (V), and resistance is expressed in **ohms** (Ω). The AC voltages you will work with range from 24 to 600 volts. The occasional DC voltages you encounter will generally be low voltages in the 5 to 15 volt range. Occasionally, you may encounter a very high voltage circuit, such as the ignitor circuit for a gas furnace, which may produce a 10,000-volt spark.

Currents in HVAC circuits could be as low as a few microamps (millionths of an amp). Currents in the milliamp (thousandths of an amp) range are more common. Some circuits, especially motor circuits, may carry very high currents in the range of 100 amps or more. The important thing to remember is that even currents in the milliamp range can be dangerous under the right conditions.

Resistances are usually large values expressed in thousands of ohms (kilohms or K) or millions of ohms (megohms or M). In some cases, a resistance value may be only a few ohms.

3.0.0   AC AND DC VOLTAGE

The kind of electricity produced by a battery is known as **direct current** or DC. Automobiles, portable stereos, calculators and flashlights are good examples of devices that use DC voltage. Note that all of them are battery-powered. The electricity supplied by your local utility is **alternating current**, or AC. Almost all HVAC devices use AC. Once in a while, you will find a unit that has a DC motor or an electronic circuit board that requires DC voltage. Rather than using batteries, which need to be replaced or recharged, such units contain special circuits called **rectifiers** that convert AC to DC. The device that allows you to plug your calculator or other portable device into a wall socket is also a rectifier.

4.0.0   ELECTRICAL CIRCUIT CHARACTERISTICS

Voltage, current, resistance, and power are closely related. If you know any two of them, you can determine the other two. For example, if you know how much voltage is available and how much power the load consumes, you can figure out how much current the circuit will draw. Why would you want to know that? Let's say you want to add a window air conditioner to a household circuit that is protected by a 20-amp fuse. How do you determine if you can safely add the appliance without overloading the circuit and causing the fuse to blow? One way would be to plug in the air conditioner and see if the fuse blows. A better (and less dangerous way) is to calculate how much current is used or drawn by each appliance connected to the circuit. The material that follows will help you learn how to do that.

4.1.0 OHM'S LAW

Ohm's Law is a formula for calculating voltage, current, or resistance. These values are expressed as E, I, and R, respectively. *Figure 4* shows the three variations of Ohm's Law. The pyramid provides an easy way to remember the formula. If you know two of the values, cover the unknown value with your finger to see how to solve the problem. For example, if you know the voltage and resistance, covering the I shows that you divide E by R.

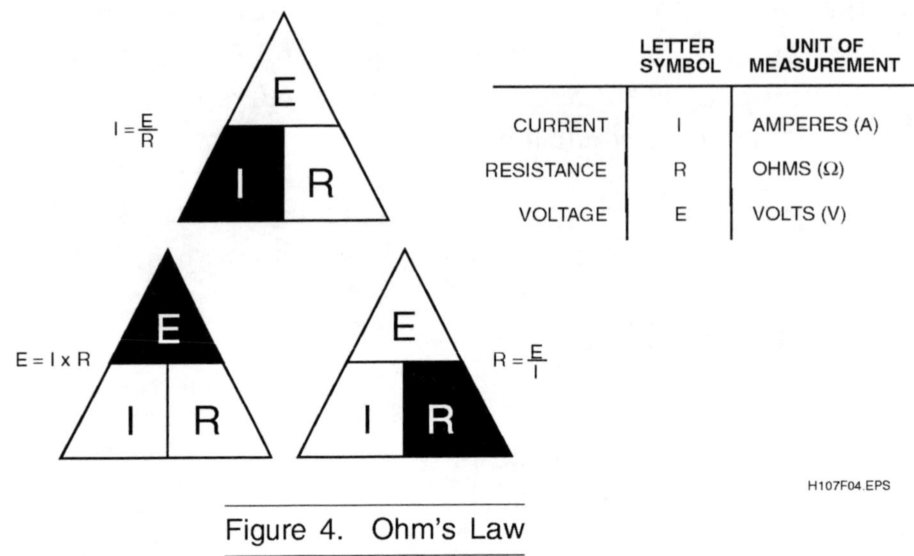

Figure 4.   Ohm's Law

Example: in a 120-volt circuit containing a 60-ohm load, the current flow is 2 amps.

$$I = \frac{E}{R} = \frac{120}{60} = 2$$

4.2.0 ELECTRICAL POWER

Most load devices are rated by the power they consume, rather than the resistance they offer. An electric heater, for example, might have a rating of 2000 watts. In such situations, Ohm's Law will not help you. If you want to know how much current the heater draws, you would need to use the power formula: P = E x I (*Figure 5*). In this case, you would use I = P/E because current is the unknown. In a 120-volt circuit, a 2000-watt heater would draw 16.67 amps (2000/120). One thing you would learn from this calculation is that you would not put this device in a circuit protected by a 15-amp fuse.

Electric motors may be rated in equivalent horsepower, rather than watts. A large swimming pool filter, for example, might contain a 2-horsepower electric motor. There is a simple conversion from horsepower to watts: 1 horsepower = 746 watts. Therefore, a 2-horse motor would consume 1592 watts. In a 120-volt circuit, it would draw more than 13 amps:

$$746 \times 2 = 1592$$

$$I = \frac{P}{E} = \frac{1592\,W}{120\,V} = 13.26 \text{ amps}$$

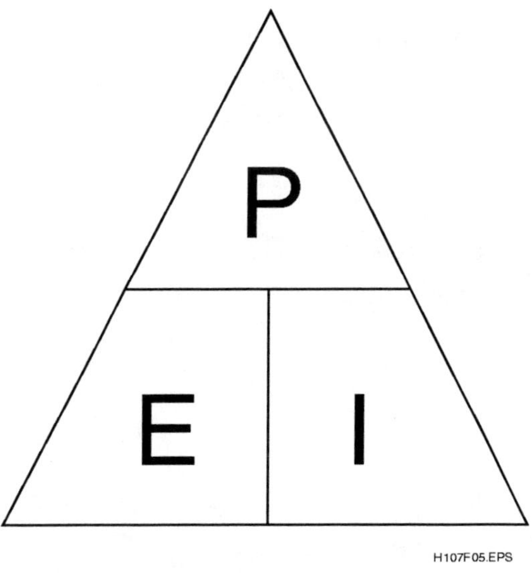

Figure 5. Power Formula

5.0.0 ELECTRICAL CIRCUITS

A basic electrical circuit is shown in *Figure 6*. An electrical circuit is a closed loop that contains a voltage source, a load, and conductors to carry current. It usually contains a switching device to turn the current on and off.

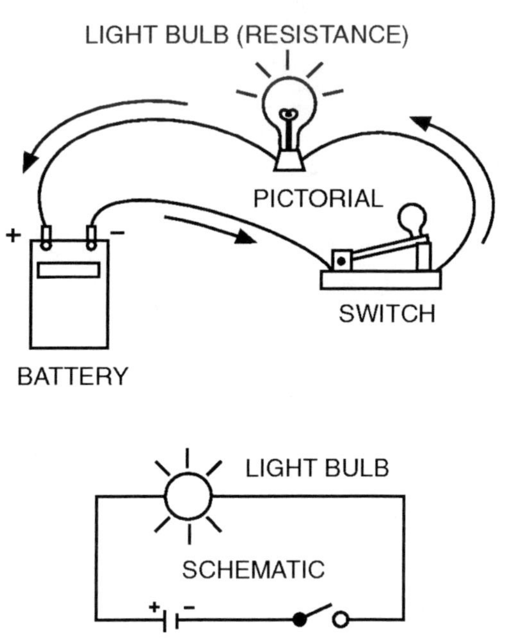

Figure 6. Electrical Circuit

A **conductor** is a material that readily carries an electric current. Most metals are good conductors. Copper is the most common; gold and silver are better conductors, but are too expensive, except in tiny electronic circuits. Water is an excellent conductor. That is why it is dangerous to work with electricity or use electrical appliances in a wet or damp environment.

An **insulator** is the opposite of a conductor. It inhibits the flow of electricity. Rubber is a good insulator. Tools used in electrical trades are often insulated with rubber to prevent electrical shock.

5.1.0   SERIES CIRCUITS

A series circuit provides only one path for current flow. The total resistance of the circuit is equal to the sum of the individual resistances. The 12-volt series circuit in *Figure 7* has two 30-ohm loads. The total resistance is therefore 60 ohms. The amount of current flowing in the circuit is 0.2 amps.

$$I = \frac{E}{R} = \frac{12}{60} = .2$$

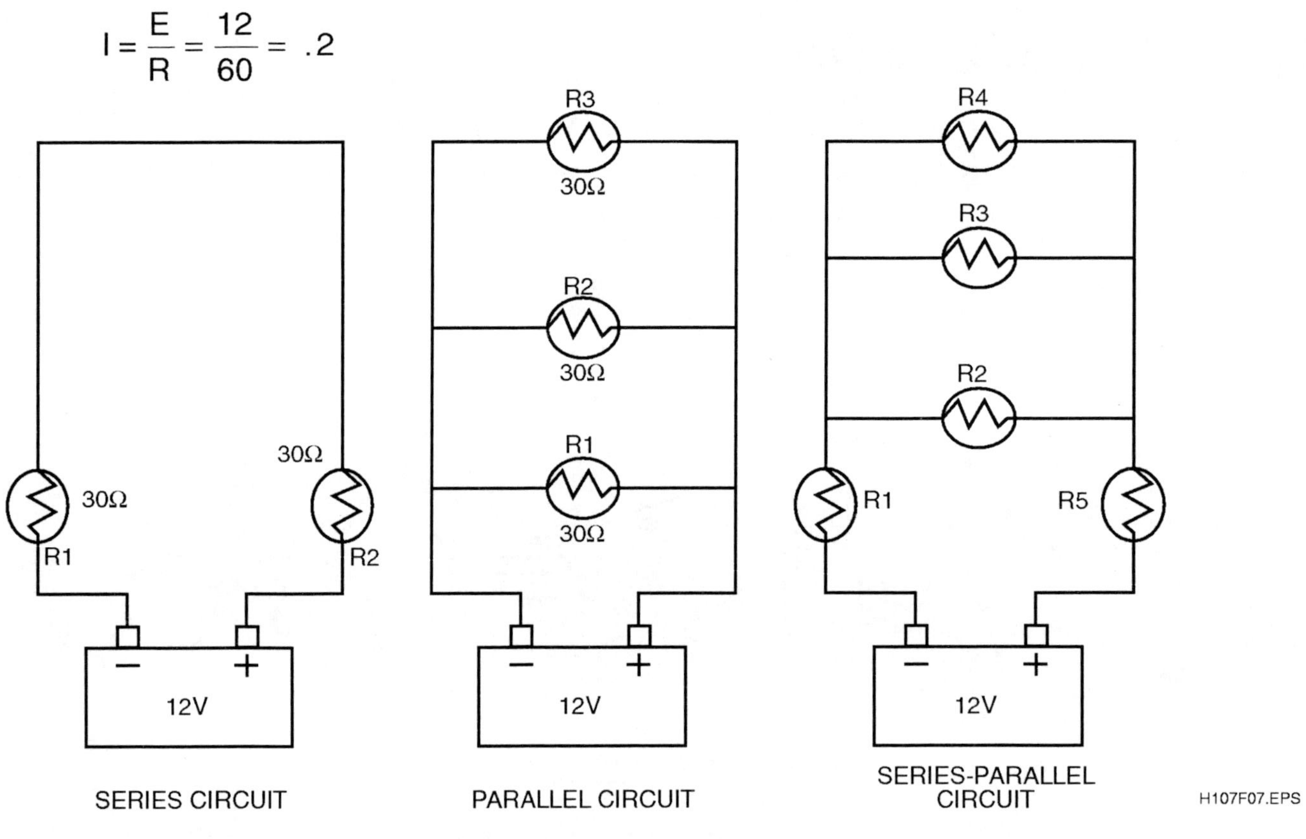

Figure 7.   Types Of Circuits

If there were five 30-ohm loads, the total resistance would be 150 ohms. Circuits containing loads in series are uncommon in HVAC work. An important trait of a series circuit is that if the circuit is open at any point, no current will flow. If you had five light bulbs connected in series and one of them blew out, all five lights would go off.

5.2.0 PARALLEL CIRCUITS

In a parallel circuit, each load is connected directly to the voltage source. The source "sees" the circuit as two or more individual circuits containing one load each. In the parallel circuit in *Figure 7*, the source sees three circuits, each containing a 30-ohm load. The current flow through any load is determined by the resistance of that load. Thus the total current drawn by the circuit is the sum of the individual currents. The total resistance of a parallel circuit is calculated differently than that of a series circuit. In a parallel circuit, the total resistance is less than the smallest of the individual resistances. Follow this example:

For example, each of the 30-ohm loads draws 0.4 amp; therefore the total current is 1.2 amps:

$$I = \frac{E}{R} = \frac{12}{30} = 0.4A \text{ per circuit} \times 3 \text{ circuits} = 1.2A$$

Now, Ohm's Law can be used again to calculate the total resistance:

$$R = \frac{E}{I} = \frac{12}{1.2} = 10 \text{ ohms}$$

This one was simple because all the resistances were the same value. The process is the same when the resistances are different, but the current calculation has to be done for each load. The individual currents are added to get the total current.

Unlike series circuits, if one circuit opens, the remaining circuits continue working. Household circuits are wired in parallel. Almost all the HVAC load circuits you encounter will be parallel circuits.

5.3.0 SERIES-PARALLEL CIRCUITS

Electronic circuits often contain a hybrid arrangement known as a series-parallel circuit (*Figure 7*). It is unlikely, however, that you will ever have to determine the electrical characteristics of one of these circuits. If it becomes necessary, the parallel loads must be converted to their equivalent series resistance. Then the load resistances are added to determine total circuit resistance.

Either of the following formulas can be used to convert parallel resistances to a single resistance value. The first one works only when there are two resistances in parallel. The second is used when there are three or more.

$$\text{Total Resistance} = \frac{R1 \times R2}{R1 + R2} \qquad \text{Total Resistance} = \frac{1}{\dfrac{1}{R1} + \dfrac{1}{R2} + \dfrac{1}{R3}}$$

The operation of many electrical components relies on magnetism. Motors, relays, transformers, and solenoids are examples. Magnetized iron generates a magnetic field consisting of magnetic lines of force, also known as magnetic flux lines (*Figure 8*). Magnetized objects within the field will be attracted or repelled by the magnetic field. The more powerful the magnet, the more powerful the magnetic field around it. Each magnet has a north pole and a south pole. Opposing poles attract each other; like poles repel each other.

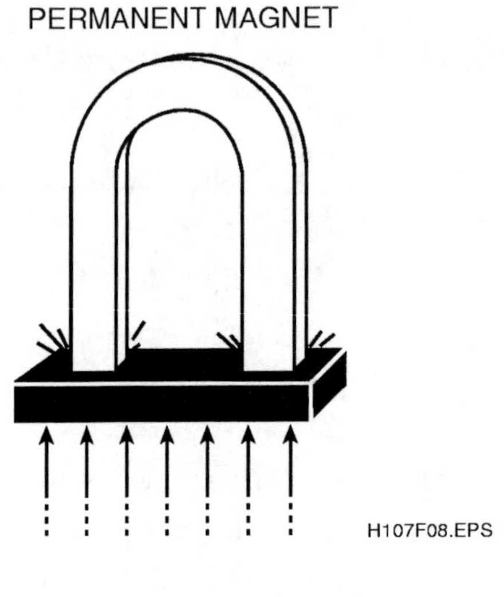

Figure 8.   Magnetism

Electricity also produces magnetism. Current flowing through a conductor produces a small magnetic field around the conductor. If the conductor is coiled around an iron bar the result is a powerful **electromagnet** (*Figure 9*) that attracts and repels other magnetized objects just like an iron magnet does. This is the basis on which electric motors and other components operate.

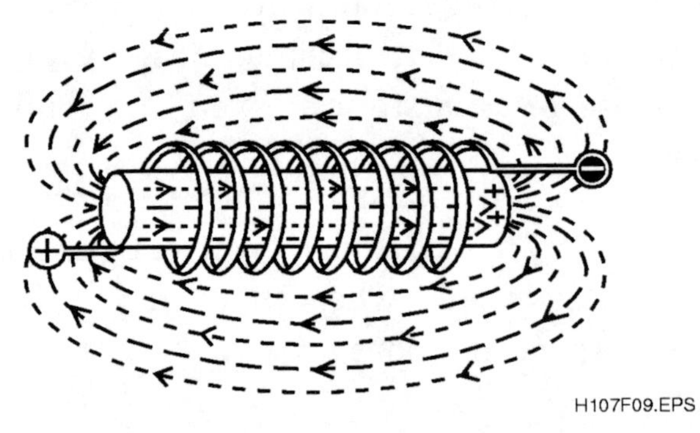

Figure 9.   Electromagnet

Circuit diagrams, often known as wiring diagrams or wiring schematics, use symbols to represent the electrical components and to show how they are connected into the circuit. In this section, you will begin to learn about electrical components and how they are shown on wiring schematics. Electrical components generally fall into two categories: load devices and control devices. Some devices (relays for example) contain both an energy-consuming element (load) and a control element (switch contacts). These are treated as control devices.

7.1.0   LOAD DEVICES

Any device that consumes electrical energy and does work in the process is a load device. Loads convert electrical energy into some other form of energy such as light, heat, or mechanical energy. They have resistance and consume power. An electric motor is a load; so is the burner on an electric stove. As electrical current flows through the burner element, it converts electricity into heat.

7.1.1   Motors

Electric motors convert electrical energy into mechanical energy. They are used primarily to drive compressors and fans. Electric motors are the most common loads found in HVAC systems. *Figure 10* shows some of the ways that motors are depicted on schematic wiring diagrams. Because it is likely that three or more motors will appear on an HVAC circuit diagram, a letter code such as IFM (indoor fan motor) is used to identify the function of a particular motor. The same is true for other components.

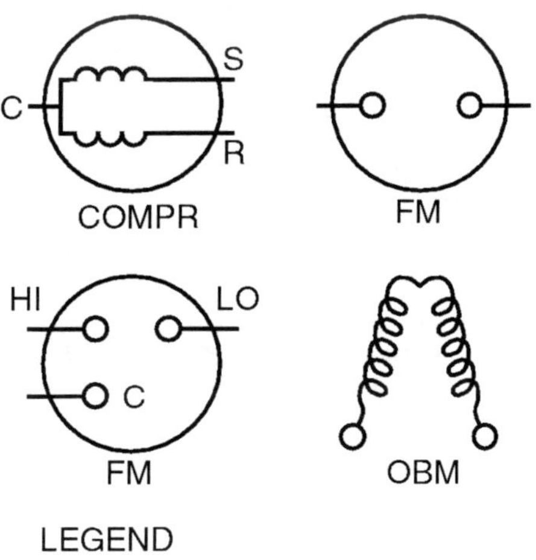

Figure 10.   Electric Motors

7.1.2 Electric Heaters

Electric heaters are also called "resistance" heaters because they are made of high-resistance material that consumes a large amount of energy, converting it into heat. They are somewhat like the burners on an electric stove. The diagram symbol is usually the same as that for a resistor (*Figure 11*).

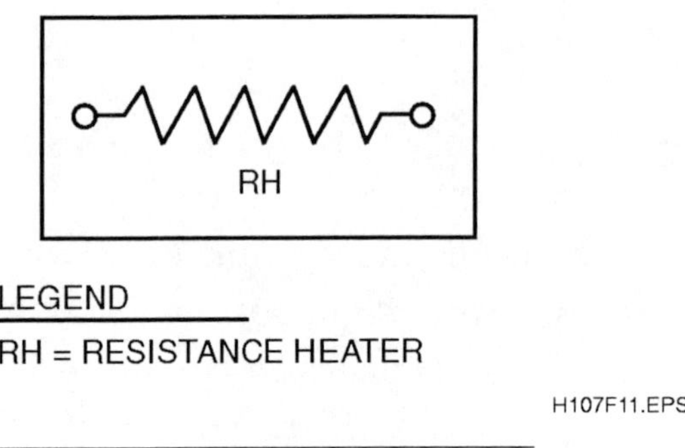

Figure 11. Resistance Heater

7.1.3 Lights

In an HVAC system, a light bulb (*Figure 12*) is classified as a secondary load. It is used to signal an operator about the status of the equipment.

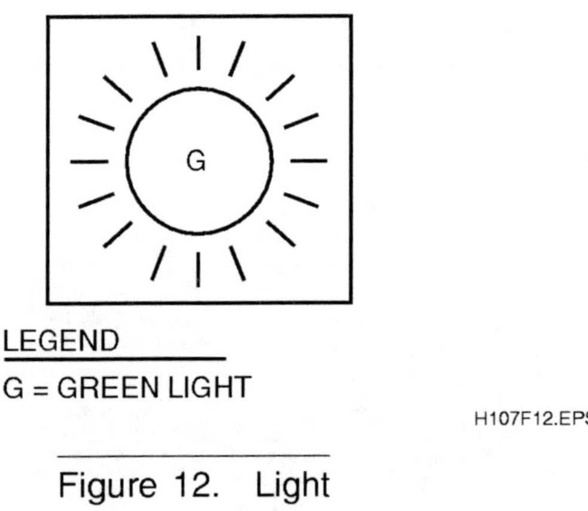

Figure 12. Light

7.2.0 CONTROL DEVICES

There are many types of control devices. Their primary function is to turn loads on and off.

7.2.1 Switches

Switches stop and start the flow of current to a load or to other control devices. Electrical switches are classified according to the force used to operate them; i.e., manual, magnetic, temperature, light, moisture, and pressure.

HVAC TRAINEE TASK MODULE 03107

The simplest type of switch is one that makes (closes) or breaks (opens) a single electrical circuit. More complicated switches control several circuits. The switching action is described by the number of poles (number of electrical circuits to the switch) and the number of throws (number of circuits fed by the switch). *Figure 13* shows some of the common switch arrangements.

1. Single Pole, Single Throw (SPST)
2. Double Pole, Single Throw (DPST)
3. Double Pole, Double Throw (DPDT)

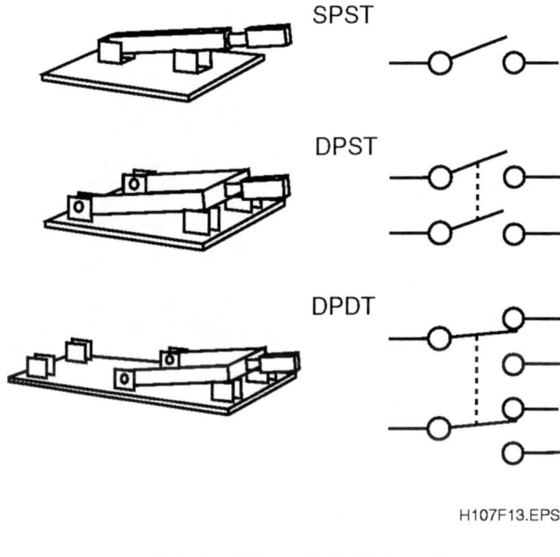

Figure 13.   Switches

An example of a manual switch is the main power disconnect switch (*Figure 14*). It is usually mounted on the unit. It allows the technician to remove power from the system for safety purposes. Some disconnects are fused, so that power is removed at the unit in case of an electrical overload.

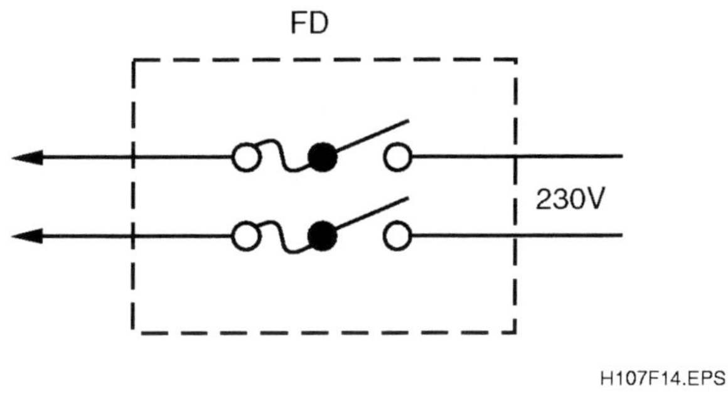

Figure 14.   Fused Disconnect

A thermostatic switch (*Figure 15*) opens and closes in response to changes in temperature. The thermostat shown on the left closes on a temperature drop. It would thus be used to control a heating system. A cooling thermostat would close on a temperature rise. Thermostats are the primary control device in heating and cooling systems. Other temperature-controlled switches provide protection from current or heat overloads. They are commonly used in compressor circuits. Heat-sensing switches are also used as limit switches to shut off furnaces in case of a malfunction.

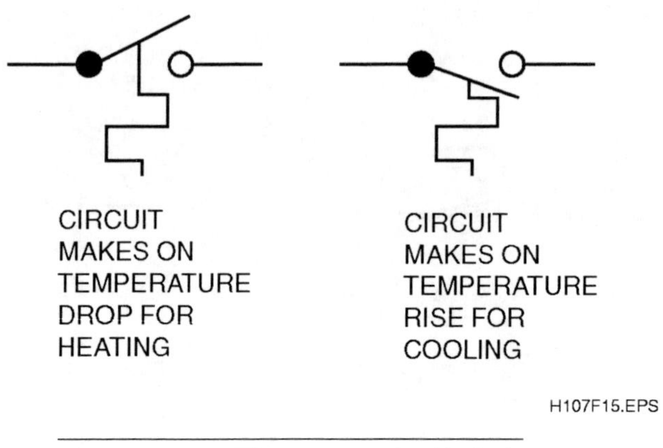

Figure 15.   Thermostatic Switch

Pressure-sensing switches, also known as **pressurestats**, are used in a variety of ways (*Figure 16*). One common use is to shut off an air conditioning compressor if the system pressure becomes too high or too low because of a problem in the system.

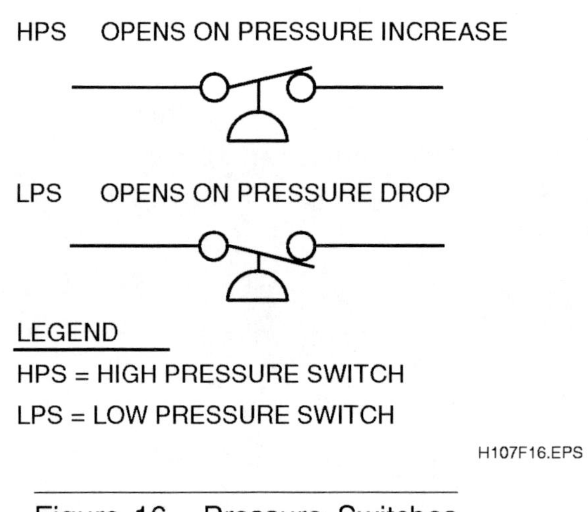

Figure 16.   Pressure Switches

Light-operated switches are sometimes used to sense flame. If the flame in an oil furnace goes out, for example, the light-sensing switch will shut off the flow of oil to the combustion chamber.

HVAC TRAINEE TASK MODULE 03107

Moisture-operated switches are used in humidifiers and dehumidifiers. A strand of hair or nylon, which expands and contracts with changes in moisture, can be used to activate the unit.

7.3.2 Fuses And Circuit Breakers

Fuses and circuit breakers protect components and wiring against damage from current surges or short circuits. By definition, a **short circuit** occurs when current flow bypasses the load. An example would be when a live conductor touches another conducting substance, such as the metal case of a unit. Because there is no load, current flow is uninhibited, and could burn up the wiring, causing a fire. Many home fires are caused by short circuits.

See *Figure 17*. A fuse contains a metal strip which melts when the current exceeds the rated capacity of the fuse; e.g., 20 amps. Because the strip is in series with the circuit, the current flow is cut off when the strip opens. The fuse will blow more quickly in the presence of a short circuit than it will in the presence of a current overload. Once a fuse has blown, it must be replaced.

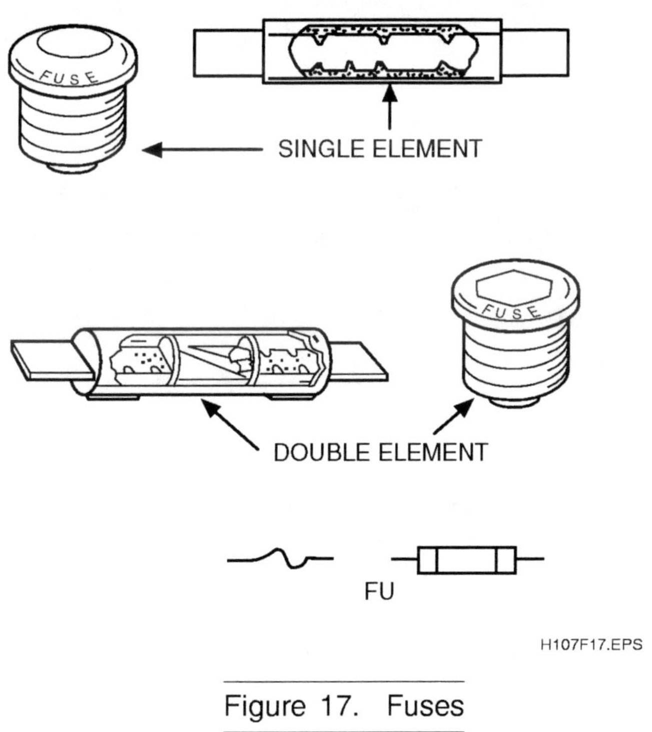

Figure 17. Fuses

Compressor motors require a large startup current. The initial current surge, known as "locked rotor amps" (LRA) may be six times that of its normal running current, known as "full load amps" (FLA). The LRA on many compressors is more than 100 amps. Special delayed-opening fuses (**slow-blow fuses**) are used in air conditioning equipment. These fuses will not blow unless the current surge keeps up.

Circuit breakers (*Figure 18*) serve the same purpose as fuses. Most modern construction uses circuit breakers. The big advantage of circuit breakers is that they operate like switches, and can be reset when they trip. If a circuit breaker trips more than once, however, it is a sign of a circuit problem. Special **HACR circuit breakers** are used for air conditioning equipment for the same reason that slow-blow fuses are used.

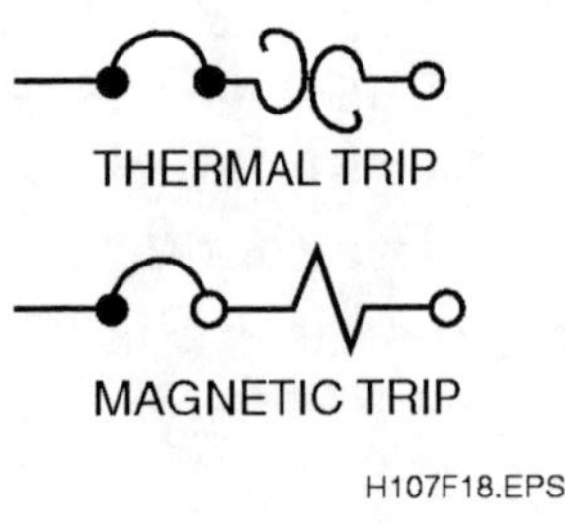

Figure 18.   Circuit Breakers

Thermal trip circuit breakers contain a spring-loaded metal element that opens when high current causes it to overheat. Magnetic trip breakers contain a small coil of wire that magnetically opens the contacts when the current is excessive.

7.3.3   Solenoids

A **solenoid** (*Figure 19*) is an electromagnet. When a current flows through the solenoid, the magnetism produced by the current is used to attract or repel a nearby object. Solenoids are used for a variety of purposes, such as opening and closing valves and switching devices. The "GV" next to the solenoid symbol in *Figure 19* indicates that this solenoid is used to control a gas valve, which feeds natural or LP gas to a gas-fired furnace.

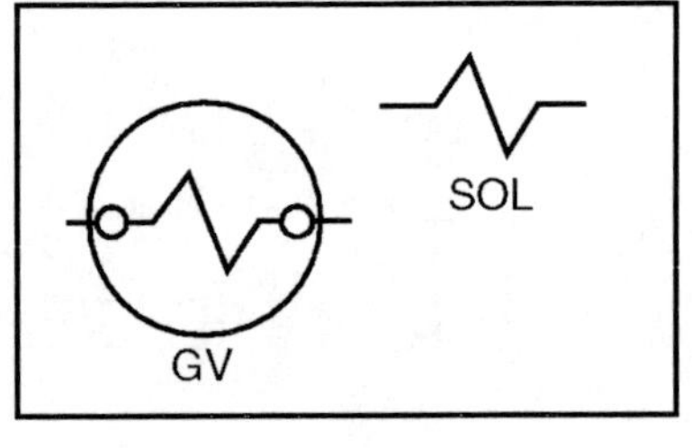

Figure 19.   Solenoid

7.3.4   Relays, Contactors, And Starters

These are magnetically-controlled switching devices consisting of a coil (solenoid) and one or more sets of movable contacts that open and close in response to current flow in the coil. In other words, they are electrically-operated switching devices.

Figure 20 shows different types of **relays**. When a current flows through the relay coil, a magnetic field is created. This field causes the contacts to change position. Some of the contacts close, providing a path for current; these are called normally open, or N.O. contacts because they are open when the relay is deenergized. Other contacts open; these are called normally closed or N.C. contacts because they are closed when the relay is deenergized. Normally closed contacts have a slash through them. The coil and contacts of a relay will be shown at different locations on a diagram because each set of contacts controls a different circuit. You can find them because the coil and contacts will have the same name; in this case, CR for control relay. The same pole-throw descriptions used for switches also apply to relays and contactors.

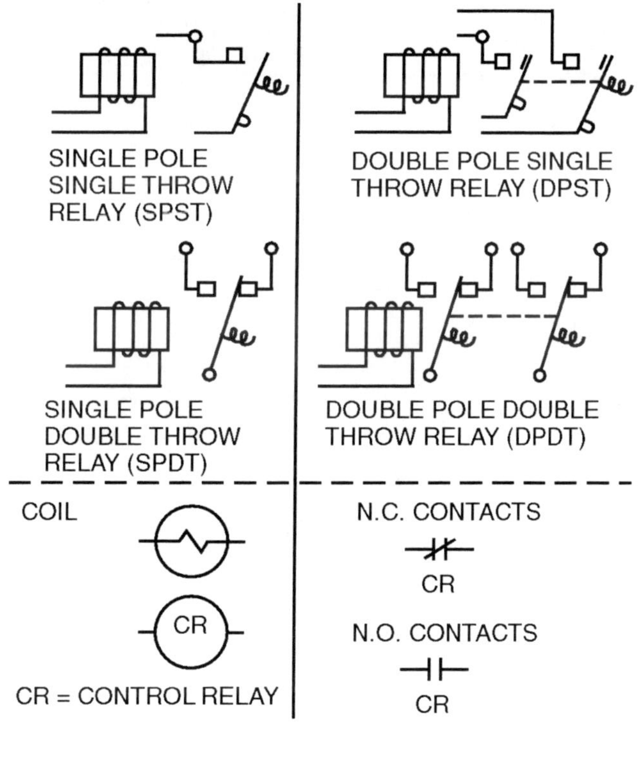

H107F20.EPS

Figure 20.   Relays

Figure 21 shows an example of how a relay might be used in an electrical control circuit. The thermostatic switch on the left closes when the temperature in the space exceeds a selected value (setpoint). The closed thermostat completes a current path to the relay coil. The coil produces a magnetic field that changes the position of the relay contacts. The normally open set of contacts close, completing a path to the fan motor. The normally closed contacts open, causing the FAN OFF light to go out. When the thermostat opens, the current stops and the contacts return to their original positions.

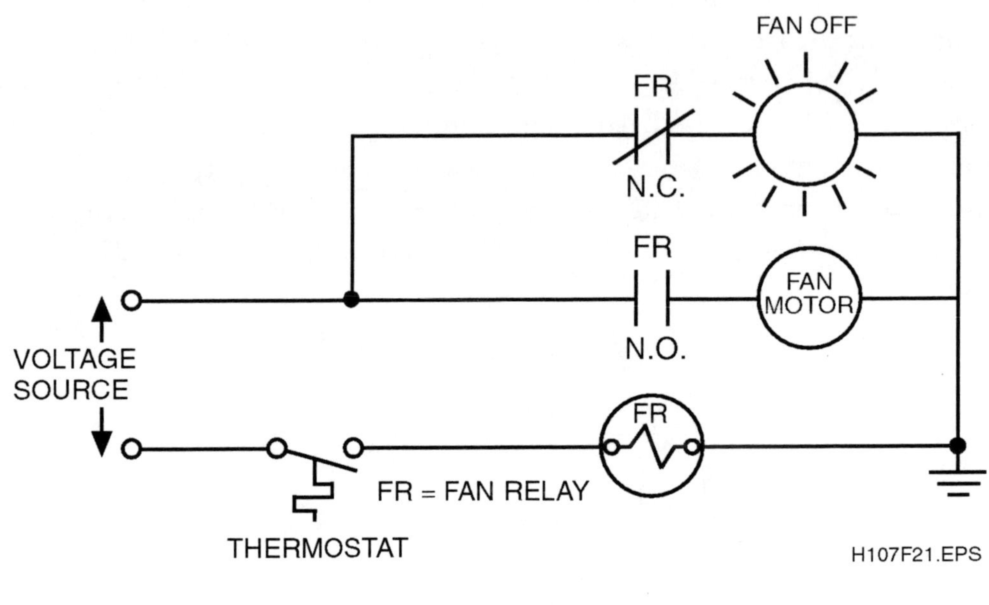

Figure 21. Relay Circuit

A **contactor** is basically a heavy-duty relay designed to carry large currents. They are often used to start and stop compressors.

Starters are used to start large motors. They usually contain overload protection devices, and may include an on-off switch. The National Electric Code (NEC) will specify whether a contactor or starter is required for a motor.

Keep in mind that the contacts of a relay, contactor, or switch will appear in a circuit diagram in their normal state; i.e., the condition they are in with no voltage applied to the circuit. The reader must mentally reposition the contacts to see what happens when the circuit is energized. One way to do that is to imagine a slash in the N.O. contact and no slash in the N.C. contact.

Time delay relays use a thermal element or digital control to delay the energizing or deenergizing of a relay. Delays may range from a few seconds to a few hours. A typical use in HVAC is to keep a furnace fan running for 30 to 60 seconds after the burners have shut off. The fan extracts residual heat from the heat exchangers. This improves heating efficiency.

7.3.5　Transformers

A transformer (*Figure 22*) is used to raise or lower a voltage. The transformer usually consists of two or more coils of wire wound around a common iron core. When a current passes through the primary coil (winding) of the transformer, the resulting magnetic field cuts through the other coil (secondary winding), creating a current in that coil. This process is known as **induction**. Depending on the number of turns of wire in each secondary winding, the voltage induced in the secondary will be stepped down (less than) or stepped up (greater than) that in the primary. The transformer shown in *Figure 22* is a step-down transformer with two secondary windings. Each winding acts as the power source for a circuit.

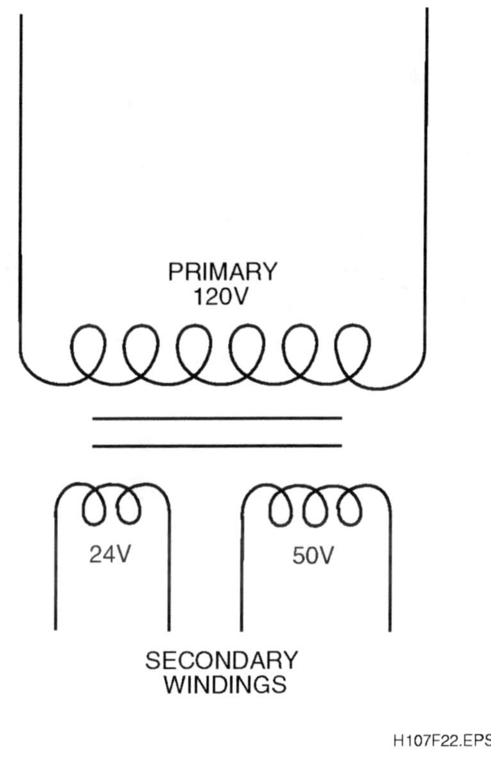

Figure 22.　Transformer

Transformers are used extensively in power distribution systems. In HVAC systems, a transformer is commonly used to step down the 120-volt AC source to 24 volts AC to operate the control circuits. It is common to find 24-volt control circuits in HVAC equipment because low voltage circuits are less dangerous. They are also less expensive to build because low voltage circuits can use lighter-gauge wire and lighter-duty components.

7.3.6 Overload Protection Devices

Overload devices stop the flow of current when safe current or temperature limits are exceeded (see *Figure 23*). Most compressor motor circuits will have one or more of these devices. Thermal overload devices are often embedded in the windings of the motor. Magnetic overloads operate much the same as relays.

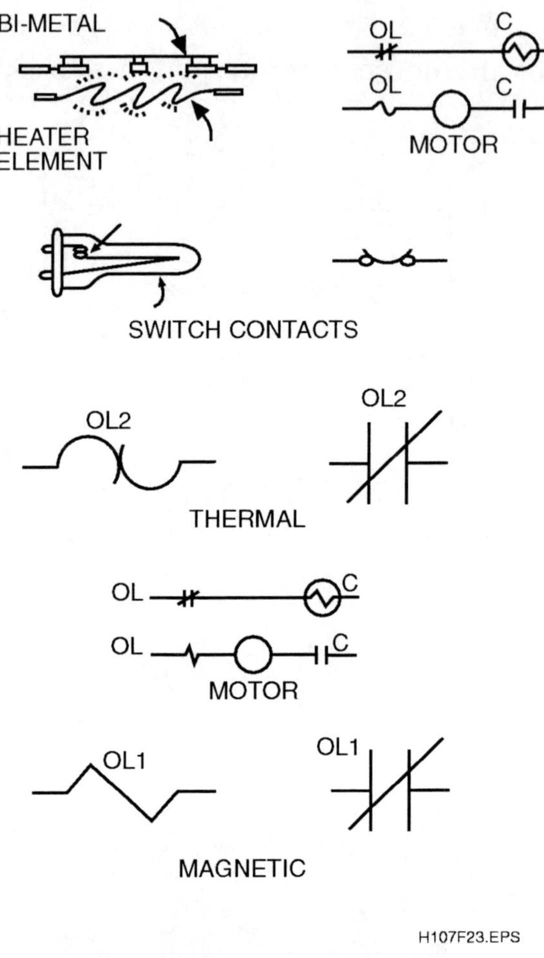

Figure 23.   Overload Devices

Protective devices are placed into two categories. **Line duty** devices are directly in line with the voltage source. In **pilot duty** devices, a small coil of wire senses current and acts as a magnet to open contacts in the motor control circuit when the current is too high.

Some overload devices reset automatically when the cause of the overload has been removed. Others must be manually reset, usually by pressing a button on the device. You would expect the type of overload device that is embedded in the motor windings to be an automatic reset device because it is not accessible.

8.0.0   ELECTRICAL SAFETY

8.1.0   THE EFFECT OF CURRENT

The amount of current that passes through the human body determines the outcome of an electrical shock. The higher the voltage, the greater the chance for a fatal shock. In a one-year study in California, the following results were observed by E. E. Carlton (Supervising Electrical Safety Engineer, Division of Industry Safety, State of California):

- 30% of all electrical accidents were caused by contact with conductors; 34% of these accidents involved conductors carrying 600 volts or more.
- Portable, electrically-operated hand tools caused the second largest number of injuries (15%). Almost 70% of these injuries happened when the frame or case of the tool became energized. These injuries could have been prevented by following proper safety practices and proper grounding.

In the ten years before this study, the same group investigated 9,765 accidental electrical injuries. Over 18% of these injuries involved contact with voltage levels of over 600 volts. A little more than 13% of these high-voltage injuries resulted in death. These high-voltage totals include limited-amperage contacts, which are often found on electronic equipment. When tools or equipment touch high-voltage overhead lines, the chance that a resulting injury will be fatal climbs to 28%. Only 1.4% of the low-voltage injuries were fatal.

WARNING!   High-voltage, defined as 600 volts or more, is almost ten times as likely to kill as low voltage. On the job, you spend most of your time working on or near lower voltages. However, lower voltages can also kill.

Electrical current flows along the path of least resistance to return to its source. If you come in contact with a live conductor, you become a load. *Figure 24* shows how much resistance the human body presents under various circumstances and how this converts to amps or milliamps when the voltage is 110V. Note that the potential for shock increases dramatically if the skin is damp. A cut will also reduce your resistance. Currents of less than 1 amp can severely injure and even kill a person (*Table 1*).

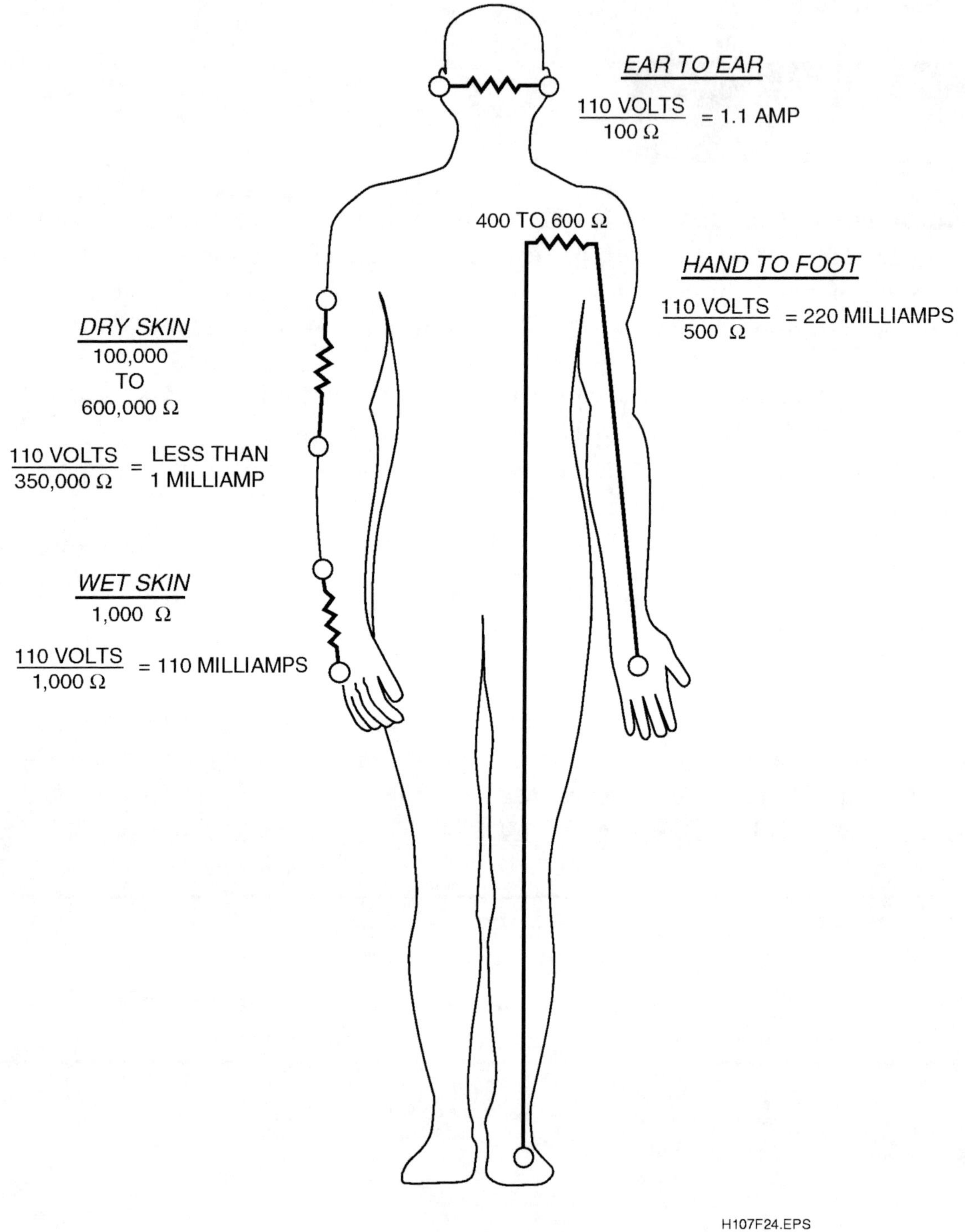

Figure 24. Typical Body Resistances And Currents

CURRENT VALUE	TYPICAL EFFECTS
Less than 1 milliamp	No sensation.
1 to 20 milliamps	Sensation of shock, possibly painful. May lose some muscular control between 10 and 20 milliamps.
20 to 50 milliamps	Painful shock, severe muscular contractions, breathing difficulties.
50 to 200 milliamps	Up to 100 milliamps, same symptoms as above, only more severe. Between 100 and 200 milliamps ventricular fibrillation may occur. This typically results in almost immediate death unless special medical equipment and treatment are available.
Over 200 milliamps	Severe burns and muscular contractions. The chest muscles contract and stop the heart for the duration of the shock.

Table 1. Current Effects On The Human Body

This information has been included in this training module to help you gain respect for the environment where you work and to stress how important safe working habits really are.

8.2.0 SAFETY PRACTICES

Electricians and technicians in many different trades work with potentially deadly levels of electricity every day. They can do so because good safety practices have become second nature to them. Here are some general safety practices to follow whenever you are working with electricity.

* Unless it is essential to work with the power on, always shut off electricity at the source. Lock and tag the power switch in accordance with company or site procedures.
* Use a voltmeter to verify that the power to the unit is actually off. Remember that even though the power may be switched off, there is still potential at the input side of the shutoff switch.

- Use protective equipment such as rubber gloves.
- Use insulated tools.
- Remove metal jewelry such as rings and watches.
- Keep one hand outside the unit when possible.

The National Electrical Code (NEC), when used together with the electrical code for your local area, provides the minimum requirements for the installation of electrical systems. Always use the latest edition of the Code as your on-the-job reference. It specifies the minimum provisions necessary for protecting people and property from electrical hazards. In some areas, different editions of the Code may be in use, so be sure to use the edition specified by your employer.

9.0.0   CIRCUIT DIAGRAMS

A circuit diagram is like a road map. If you know how to read it, you can determine how the electrical circuits are supposed to act when the unit is running properly. You can then check the operation of the circuit to find the portion that is not doing what it is supposed to do.

Diagrams supplied by manufacturers will come in a variety of formats. *Figure 25* is just one example. It contains a wiring diagram and a simplified schematic diagram.

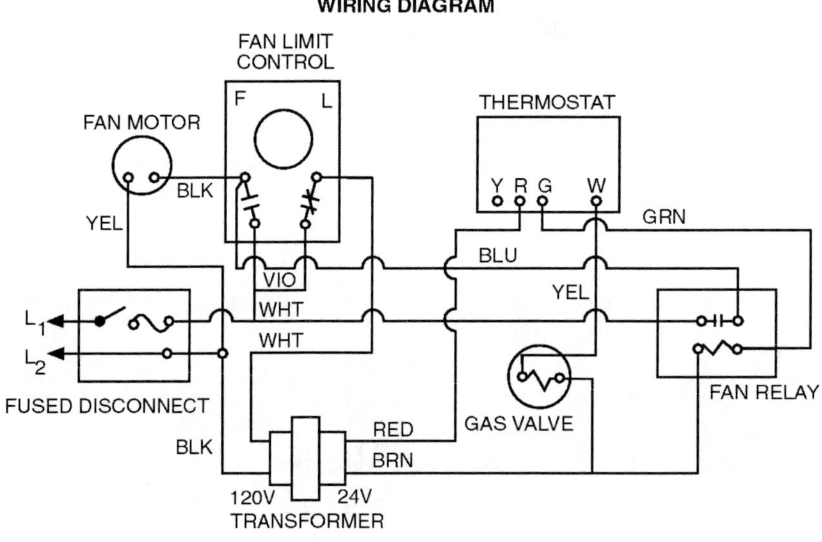

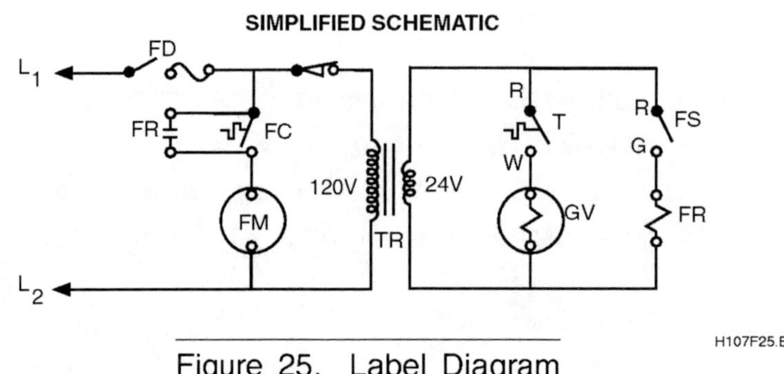

Figure 25. Label Diagram

9.1.0 WIRING DIAGRAM

The wiring diagram shows how the wiring is physically connected. It also shows the color of each wire. The wiring diagram is helpful if it is necessary to rewire a unit. When you are troubleshooting, it will help you find a physical location at which to make a measurement.

9.2.0 SIMPLIFIED SCHEMATIC DIAGRAM

On this diagram, the wire color and physical connection information are removed and the pictorial views of the components are replaced with standard electrical diagram symbols. This simplification makes the diagram easier to read. Also, the simplified schematic is arranged to make it easy to trace circuits. **Ladder diagrams**, such as the one as shown in *Figure 26*, may be provided. The power source is shown as the uprights and each load line and its related control devices is represented as a rung of the ladder.

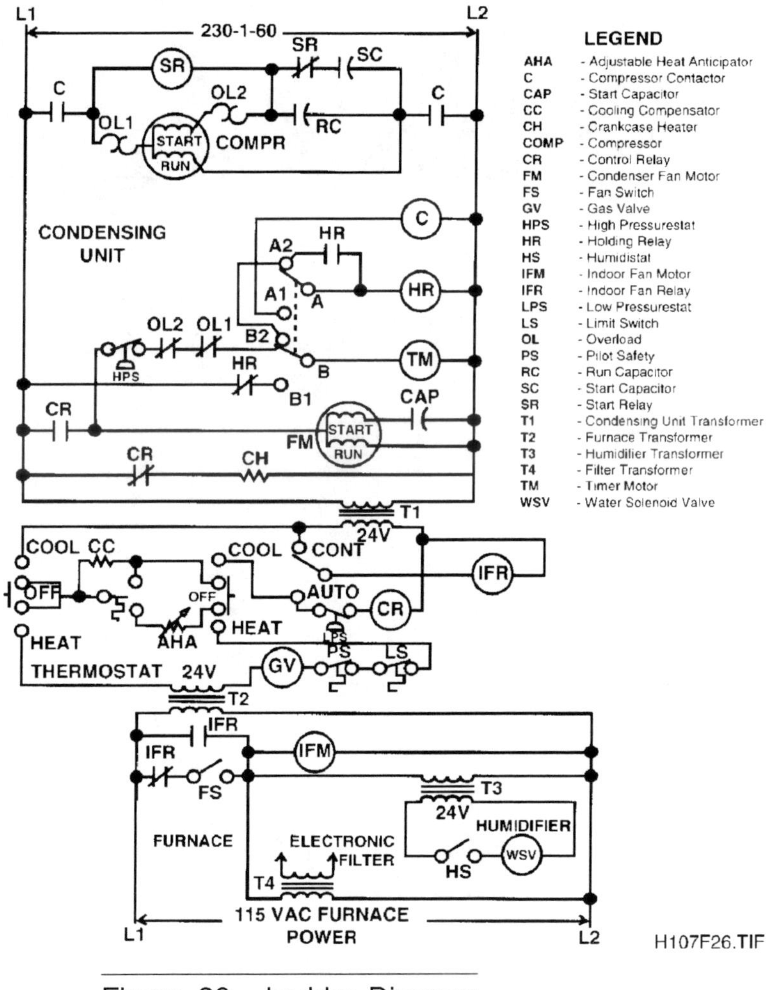

Figure 26. Ladder Diagram

The diagram provided by the manufacturer may also contain a component location diagram. It shows where the electrical components are located in the unit. This will help you make the transition from the schematic diagram, which shows the components in symbol form, but does not provide any information about where they are located.

Electronic circuits use **solid state** timing, switching, and sensing devices to control loads and protective circuits. Electronic circuits operate at much lower voltage and current levels than electromechanical devices such as relays and solenoids. In addition, they are more reliable because they have no moving parts. Most electronic circuits consist of microminiature components mounted on printed circuit boards (*Figure 27*). The components are mounted to the top of the board. Copper runs on the underside of the board serve the same purpose as wiring. Circuits that can be used in a variety of products are often packaged in sealed modules.

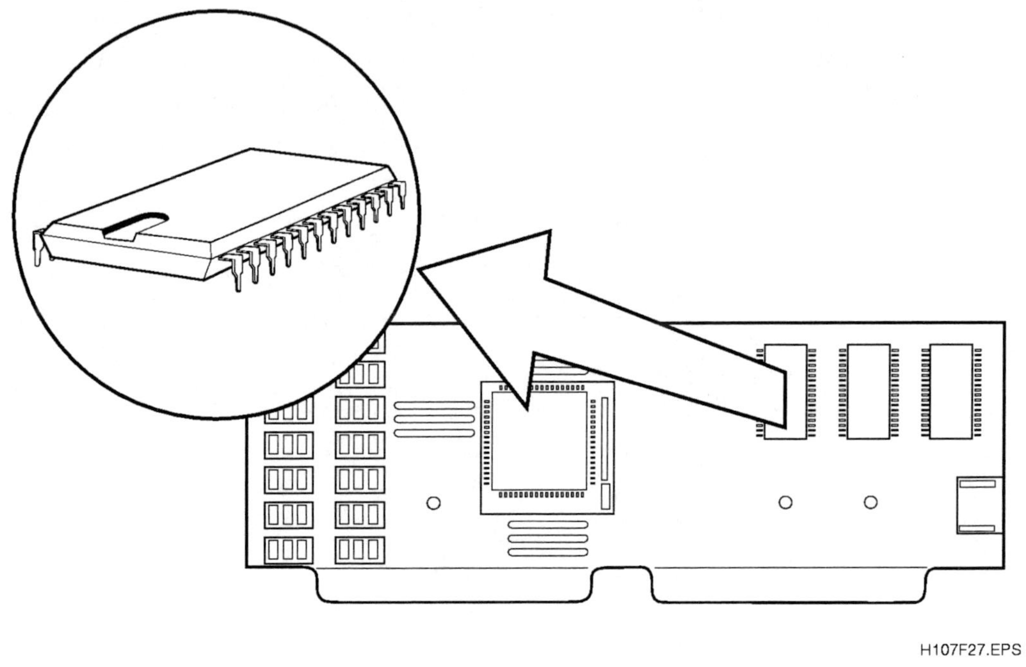

Figure 27. Printed Circuit Board

Microcomputers allow building occupants or managers to custom-tailor the operation of an HVAC system to their needs (*Figure 28*). Programmable thermostats, for example, allow the occupant to "program" a different operating temperature for different times of day. Microcomputers are able to receive a large amount of information and make decisions based on the information and the instructions in their programs. One of the most important features of these circuits is that they offer many more capabilities and more precise control than systems controlled by conventional circuits. Microprocessor-controlled systems are often self-diagnosing. When there is a problem, the microprocessor evaluates information from around the system to determine where the problem is. A digital readout, flashing light code, or other means of communication is then used to tell the technician which component or electronic circuit to check.

Figure 28.   Electronic Control Of HVAC Systems

Electronic circuits are usually treated as "black boxes." In other words, if the technician finds that the control circuit has failed, the entire circuit board or module is replaced. Unlike conventional circuits, it is not necessary to analyze the circuit to figure out which component has failed.

11.0.0   ELECTRICAL MEASURING INSTRUMENTS

When troubleshooting an electrical circuit, it is usually necessary to measure voltage, current, and resistance using electrical meters (*Figure 29*). **Analog meters**, which require the reader to interpret a scale, are still around, but have been largely replaced by direct-reading **digital meters**. **Ammeters** are used to measure alternating current; **multimeters** are commonly used to measure AC and DC voltage, resistance, and direct current. Some can read AC current in the milliamp range.

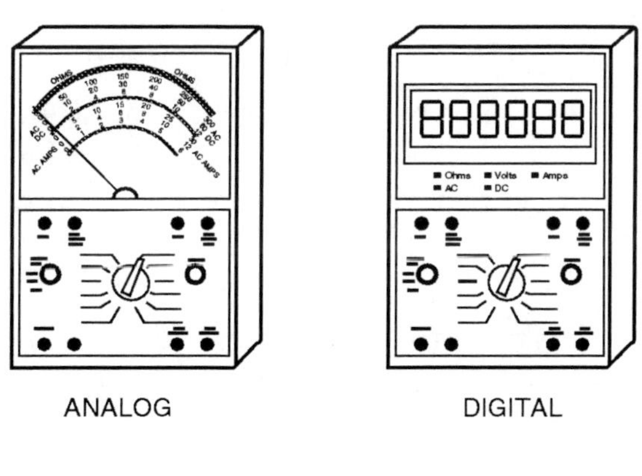

Figure 29.   Analog And Digital Meters

11.1.0 AMMETER

Ammeters are often used to check motor circuits. A **clamp-on ammeter** (*Figure 30*) is normally used to measure AC. The jaws of the ammeter are placed around the wire conductor. Current flowing through the wire creates a magnetic field which induces a proportional current in the ammeter jaws. This current is read by the meter movement and appears as a direct readout or, on an analog meter, as a deflection of the meter needle.

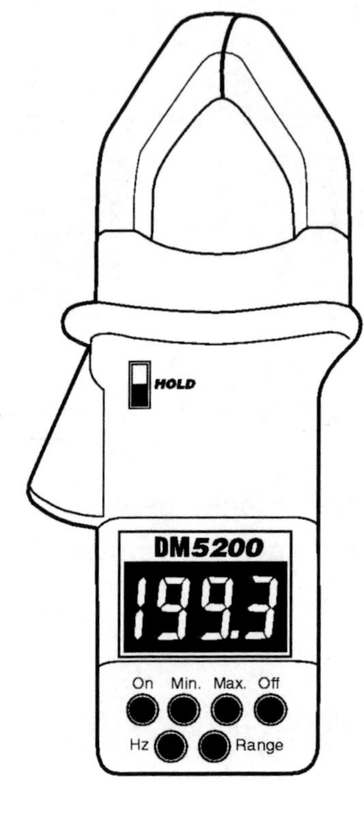

H107F30.EPS

Figure 30. Digital Clamp-On Ammeter

When low currents are being measured, it may be necessary to wrap the conductor around the meter jaws. The reading must then be divided by the number of times the conductor passes through the jaws.

In-line ammeters (*Figure 31*) are less common. This type of meter must be connected in series with the circuit, which means that the circuit must be opened.

Aside from following good safety practices, there are a few things to remember when measuring current:

- If the ammeter jaws are dirty or misaligned, a meter will not read correctly.
- When using an analog meter, always start at a high range and work down to avoid damaging the meter.
- Do not clamp the meter jaws around two different conductors at the same time or an inaccurate reading will result.

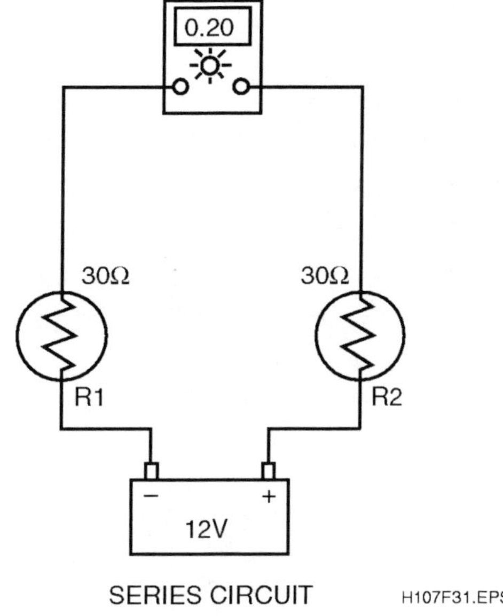

Figure 31. In-Line Ammeter

11.2.0   MULTIMETERS

Multimeters have a selector on the meter that allows the user to select AC or DC voltage (voltmeter), resistance (ohmmeter), or current (ammeter). See *Figure 32*. On an analog meter, the range of values to be read must also be selected.

H107F32.EPS

Figure 32. Digital Multimeter

11.2.1 Voltage Measurements

A voltmeter must be connected in parallel with (across) the component or circuit to be tested (*Figure 33*). If a circuit function is not operating, the voltmeter can be used to determine if the correct voltage is available to the circuit. Voltage must be checked with power applied. When testing a live circuit, it is a good idea to use an insulated alligator clip on the common meter lead. That way, only one hand (the one holding the other probe) is in the unit with power applied. The power must be turned off to connect the alligator clip.

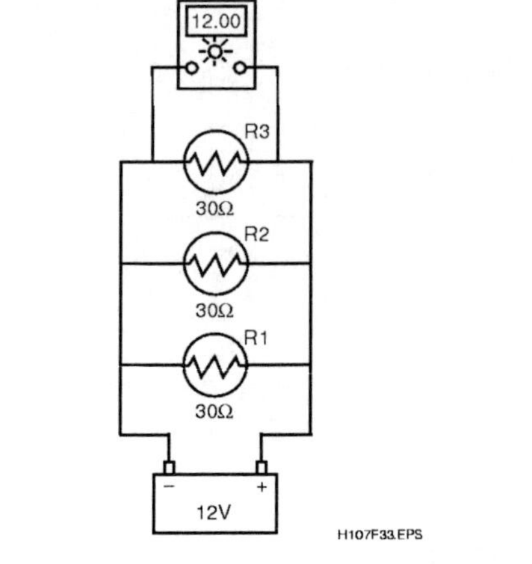

Figure 33.   Voltmeter Connection

11.2.2 Resistance Measurements

An ohmmeter contains an internal battery that acts as a voltage source. Therefore, resistance measurements are always made with the system power shut off. Sometimes, an ohmmeter is used to measure resistance in a load; motor windings are a good example. More often, an ohmmeter is used to check **continuity** in a circuit. A wire or closed switch offers negligible resistance. With the ohmmeter connected as in *Figure 34*, and the three switches closed, the current produced by the ohmmeter battery will flow unopposed and the meter will show zero resistance. The circuit has continuity; i.e., it is continuous. If a switch is open, however, there is no path for current and the meter will see infinite resistance (∞); that is, a lack of continuity.

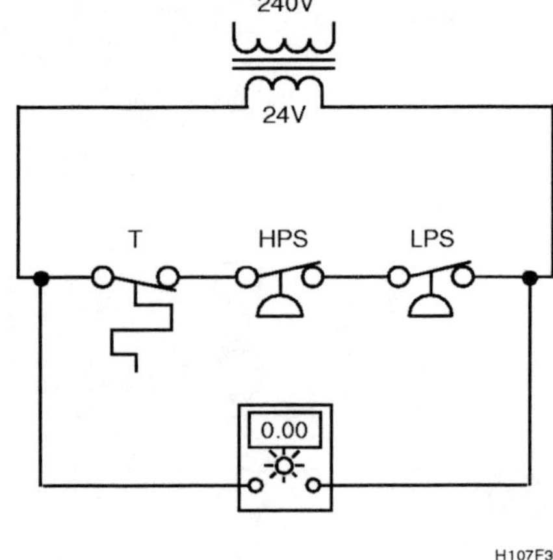

Figure 34.   Ohmmeter Connection For Continuity Testing

HVAC TRAINEE TASK MODULE 03107

SUMMARY

All HVAC systems contain electrical circuits. In order to install and service HVAC systems, the HVAC technician must know how electrical components work, how to read circuit diagrams, and how to use electrical test equipment. In working with electricity, knowing and following proper safety practices is extremely important. Failure to develop good safety habits and follow safety rules can result in injury or death to yourself or your co-workers.

References

For advanced study of topics covered in this task module the following books are suggested:

Modern Refrigeration and Air Conditioning, The Goodheart-Willcox Company, Inc., South Holland, Illinois.

Refrigeration & Air Conditioning Technology, Second Edition, Delmar Publishers, Inc., Albany, New York.

Vest Pocket Guide to the National Electrical Code (available from the Construction Education Foundation).

SELF CHECK REVIEW/PRACTICE QUESTIONS

1. A transformer is used to:
   a. transform electrical energy into another form of energy.
   b. raise or lower a voltage.
   c. consume power.
   d. increase the resistance of a circuit.

2. The difference in potential between the two poles of a car battery is caused by:
   a. the current flowing between the two poles.
   b. the circuit resistance.
   c. the amount of power consumed by the load.
   d. a chemical reaction.

3. Current is measured in:
   a. milliwatts per square meter.
   b. volts.
   c. amps.
   d. miles per hour.

4. A _____ converts AC voltage to DC voltage.
   a. voltmeter
   b. inverter
   c. rectifier
   d. transformer

5. How much current is drawn by a 120-volt circuit containing a 1500-watt load?
   a. 125 milliamps
   b. 0.08 amps
   c. 180,000 amps
   d. 12.5 amps

6. How much power is consumed by a 120-volt parallel circuit that contains a 100-ohm load and an 80-ohm load and draws 2.7 amps?
   a. 324 watts
   b. 44 watts
   c. 80 watts
   d. 120 watts

7. How much current will be drawn by a 120-volt series circuit containing a 100-ohm load and a 50-ohm load?
    a. 0.8 watts
    b. 800 milliamps
    c. 8 amps
    d. 1.25 amps

8. A DPDT switch can control _____ circuits.
    a. 4
    b. 1
    c. 3
    d. 2

9. Fuses are rated on the basis of:
    a. how fast they blow.
    b. how much current they can withstand before they blow.
    c. full load amps (FLA).
    d. locked rotor amps (LRA).

10. It is okay to work in a unit with the power on, as long as you:
    a. wear metal jewelry.
    b. have both hands in the unit.
    c. wear a hard hat.
    d. none of the above.

ANSWERS TO SELF CHECK REVIEW/PRACTICE QUESTIONS

Answers	Section Reference
1. b	2.1.0
2. d	2.2.0
3. c	2.2.0
4. c	3.0.0
5. d	4.2.0
6. a	4.2.0
7. b	5.1.0
8. d	7.2.1
9. b	7.3.2
10. d	8.2.0

ACKNOWLEDGEMENTS

Figure 26 courtesy of Carrier Corporation.

The NCCER makes every effort to keep these manuals up-to-date and free of technical errors. We appreciate your help in this process. If you have an idea for improving this manual, or if you find an error, a typographical mistake, or an inaccuracy in the NCCER's Craft Training Manuals, please write us, using this form or a photocopy. Be sure to include the exact module number, page number, a description of the problem, and the correction, if possible. Your input will be brought to the attention of the Technical Review Committee. Thank you for your assistance.

Instructors – If you found that additional materials were necessary in order to teach this module effectively, please let us know so that we may include them in the Equipment/Materials list in the Instructor's Guide.

Write: Curriculum and Revision Department
National Center for Construction Education and Research
P.O. Box 141104
Gainesville, FL 32614-1104
Fax: 352-334-0932

Craft ___________________________ Module Name ___________________________

Module Number ___________________ Page Number(s) ___________________

Description of Problem ___________________________

____

____

____

(Optional) Correction of Problem ___________________________

____

____

(Optional) Your Name and Address ___________________________

____

Introduction to Cooling

Module 03108

N^{ATIONAL} C^{ENTER FOR} C^{ONSTRUCTION} E^{DUCATION AND} R^{ESEARCH}

INTRODUCTION TO COOLING

Objectives

Upon completion of this module, the trainee will be able to:

1. Explain how heat transfer occurs in a cooling system, demonstrating an understanding of the terms and concepts used in the refrigeration cycle.
2. Calculate the temperature and pressure relationships at key points in the refrigeration cycle.
3. Under supervision, use temperature and pressure measuring instruments to make readings at key points in the refrigeration cycle.
4. Identify commonly used refrigerants and demonstrate the procedures for handling these refrigerants.
5. Recognize the major components of a cooling system and explain how each type works.
6. Recognize the major accessories available with cooling systems and explain how each type works.
7. Recognize the control devices used in cooling systems and explain how each type works.
8. Under supervision, perform basic power-off maintenance procedures applicable to cooling systems.
9. State the correct methods to be used when piping a refrigeration system.

Prerequisites

Successful completion of the following Task Modules is required before beginning study of this Task Module: Common Core Curricula, HVAC Modules 03101 through 03107.

Required Student Material

1. Trainee Task Module
2. Appropriate Personal Protective Equipment

Course Map Information

This course map shows all of the *Wheels of Learning* task modules in the first level of the HVAC curricula. The suggested training order begins at the bottom and proceeds up. Skill levels increase as a trainee advances on the course map. The training order may be adjusted by the local Training Program Sponsor.

Course Map: HVAC, Level 1

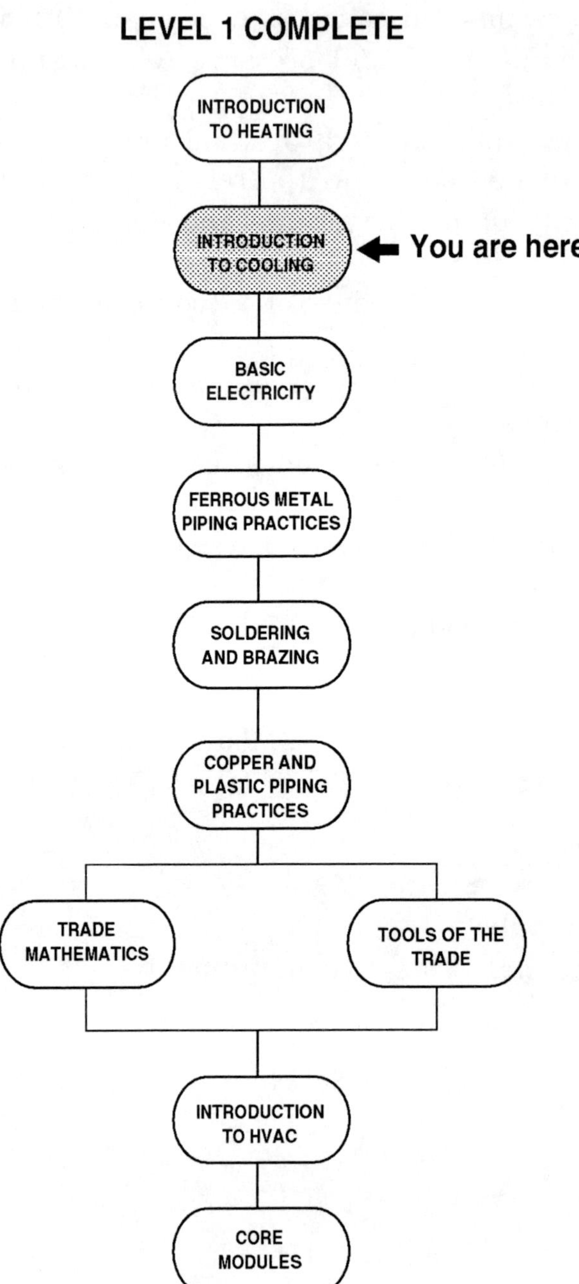

TABLE OF CONTENTS

Table of Contents (Continued)

Table of Contents (Continued)

Trade Terms Introduced In This Module

Absolute pressure: Positive pressure measurements that start at zero (no pressure at all). Also gauge pressure plus the pressure of the atmosphere, 14.7 psi at sea level at 70°F.

Atmospheric pressure: The pressure exerted on all things on the earth's surface as a result of the weight of the atmosphere. It is 14.7 psi at sea level at 70°F.

Azeotropes: Refrigerants made by combining two different refrigerants. When mixed in certain proportions, a new compound is formed that has a boiling point different than either of the base refrigerants.

British thermal unit (Btu): The amount of heat needed to raise the temperature of one pound of water one degree Fahrenheit.

Cold: A relative term for temperature. Cold means having less heat energy than another object against which it is being compared.

Conduction: A means of heat transfer in which heat is moved from one substance to another by means of direct contact.

Conductor: A material in which the transfer of heat by conduction occurs easily.

Convection: The transfer of heat by the flow of liquid or gas.

Energy Efficiency Ratio (EER): The ratio of the rated cooling capacity in Btu's per hour divided by the amount of electrical power used in watts at any given set of conditions.

Fluorocarbons: Halocarbons in which at least one or more of the hydrogen atoms has been replaced with fluorine.

Gauge pressure: The pressure measured on a gauge, expressed as psig or in. Hg. vac. Also pressure measurements that are made in comparison to atmospheric pressure.

Halocarbons: Hydrocarbons, like methane and ethane, that have most or all of their hydrogen atoms replaced with the non-metallic elements fluorine, chlorine, bromine, astatine or iodine.

Halogens: Substance containing chlorine, fluorine, bromine, astatine or iodine.

Heat: A form of energy. It causes molecules to be in motion and raises the temperature of a substance. Other forms of energy like electricity, light, and magnetism deteriorate into heat.

Heat content: The amount of heat energy contained in a substance. Measured in Btu's.

Hydrocarbons: Compounds containing only hydrogen and carbon atoms in various combinations.

Insulators: Materials that resist heat transfer by conduction.

Latent heat: The heat energy absorbed or rejected when a substance is changing state (solid to liquid, liquid to gas, or vice-versa) and there is no change in the measured temperature.

Latent heat of condensation: The heat given up or removed from a gas in changing back to a liquid state (steam to water).

Latent heat of fusion: The heat gained or lost in changing to or from a solid (ice to water or water to ice).

Latent heat of vaporization: The heat gained in changing from a liquid to a gas (water to steam).

Occupational Health and Safety Administration (OSHA): A department of the U.S. government concerned with occupational safety.

Pressure: Force per unit of area.

Radiation: The movement of heat in the form of invisible rays or waves, similar to light.

Refrigerant: A fluid (liquid or gas) that picks up heat by evaporating at a low temperature and pressure and gives up heat by condensing at a higher temperature and pressure.

Refrigeration: The transfer of heat from a space or object where it is not wanted to a space or object where it is not objectionable.

Seasonal Energy Efficiency Ratio (SEER): The total cooling of an air conditioner or heat pump in Btu's during its normal annual usage period for cooling divided by the total electrical energy input in watt-hours during the same period.

Sensible heat: Heat that can be measured by a thermometer or sensed by touch. The energy of molecular motion.

Specific heat: The amount of heat required to raise one pound of a substance one degree Fahrenheit. Expressed as Btu/lb./°F.

Specific heat capacity: The amount of heat, measured in Btu's, added or removed to change the temperature of one pound of a substance one degree Fahrenheit.

Subcooling: The temperature of a liquid when it is cooled below its condensing temperature.

Superheat: The measurable heat added to the vapor or gas produced after a liquid has reached its boiling point and completely changed into a vapor.

Thermistor: A semiconductor device that changes resistance with a change in temperature.

Thermocouple: A device made of two different metals that generates electricity when there is a difference in temperature from one end to the other.

Ton of refrigeration: Large unit for measuring the rate of heat transfer. One ton is defined as 12,000 Btu's per hour or 12,000 Btuh.

Total heat: Sensible heat plus latent heat.

1.0.0 INTRODUCTION

A HVAC technician works with air conditioning equipment. The term air conditioning includes many methods for conditioning air. In this module, the focus is only on the cooling aspect of air conditioning. In this regard, air conditioning is a form of refrigeration.

To prepare yourself for working on cooling equipment, you must understand the basic concepts of refrigeration. You must also understand the operation of the mechanical refrigeration system: its purpose, function, components and conditions. There are many types of mechanical refrigeration systems. If you try to learn refrigeration by learning how each one works, it would be a very long and difficult task. However, if you learn the basics of refrigeration given in this module, you should be able to understand most systems. This is because the principles of mechanical refrigeration and the basic parts used in a system are the same, no matter how big or small the system or how the parts are packaged.

From this study, your ability to install and service all sorts of cooling equipment will be enhanced. This is true whether the cooling equipment is used for personal comfort air conditioning, food preservation, or in industrial processes.

In nature, there is nothing from which heat or temperature is totally absent. **Cold** is a relative term for temperature. It means an object has less heat energy (making it colder) than another object to which it is being compared. For ease of understanding, it is helpful to define cold as being the absence of heat.

Refrigeration is the transfer of heat from a place or object where it is not wanted to a place or object where it is not objectionable. Simply defined, refrigeration is cooling by the removal of heat. Air conditioners and refrigerators do not "pump cold" into a space. They take heat out of the space or object to be cooled and move it outside (*Figure 1*). A chemical fluid known as refrigerant circulates through a refrigerator or air conditioning system. The refrigerant absorbs heat from the refrigerated space, then carries it to a location outside the space. You will learn more about refrigerant later in this module.

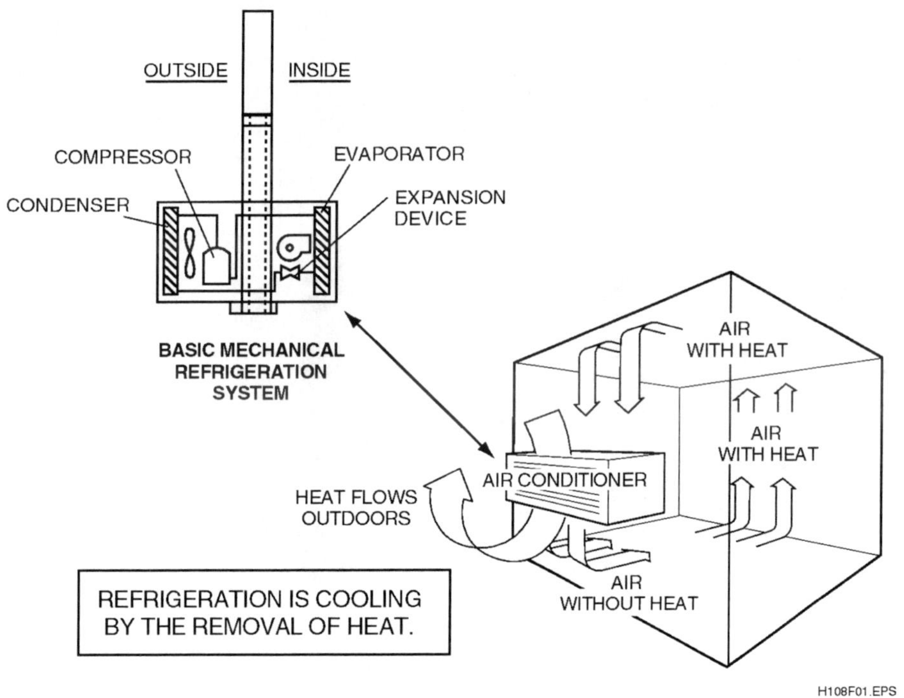

Figure 1.   Refrigeration – Transfer Of Heat

2.1.0 HEAT

To understand refrigeration you must understand heat. Like light, electricity, and magnetism, **heat** is a form of energy. Heat can be measured and controlled. Like other forms of energy, it can do work. Its ability to do work depends on two characteristics: temperature and heat content (quantity).

2.1.1 Temperature

Temperature compares the degree of hotness or coldness of any object or substance. The intensity of heat is measured in degrees (°) with a thermometer. Two temperature scales are normally used for measuring temperature. The one you will use most often is called the Fahrenheit scale (*Figure 2*). The other scale commonly used worldwide, and for scientific work in the U.S., is the Celsius or Centigrade scale.

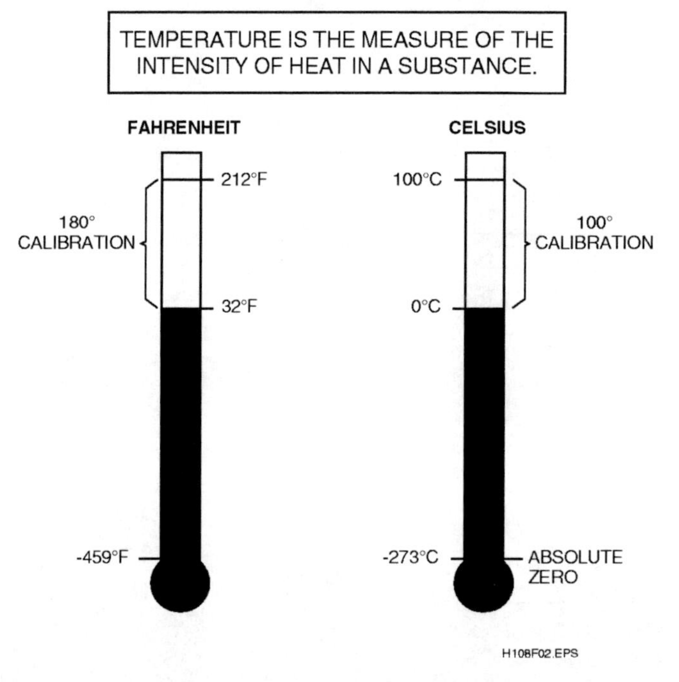

Figure 2. Fahrenheit And Celsius Temperature Scales

On the Fahrenheit scale, abbreviated °F, water boils at 212°F and freezes at 32°F. The distance between the two points is divided into 180 equal parts. On the Celsius scale, abbreviated °C, water boils at 100°C and freezes at 0°C. The distance between the two points is divided into 100 equal parts. The absolute zero point shown on the temperature scale is the theoretical point where all molecular motion stops, resulting in zero heat content.

2.1.2 Heat Content

Heat content or the quantity of heat is the amount of heat energy contained in a substance. Heat content is measured by the **British thermal unit (Btu)**. One Btu is the amount of heat needed to raise the temperature of one pound of water one degree Fahrenheit. (One pound of water is equal to about a pint of water.) For example, by heating 10 pounds of water from 40°F to 50°F (difference of 10°F) 100 Btu's of heat are added to the water. The opposite is also true. If the 10 pounds of water had been cooled 10°F, 100 Btu's would have been removed.

2.1.3 Sensible And Latent Heat

Depending on the heat content and temperature, materials or substances can exist in three states: solid, liquid, and gas.

Using water as an example, the three states are ice (solid), water (liquid), and steam or vapor (gas). In a refrigeration system, the refrigerant exists in either a liquid or gas state, depending on its heat content. For this reason, these two states are the ones you will be most concerned with in your study of refrigeration.

When a substance such as water changes from one state to another, something peculiar happens (*Figure 3*). If heat is added to one pound of ice at 0°F, a thermometer would show a rise in temperature until the reading reaches 32°F. This is the point where the ice starts changing into water. If heat is continually added, the thermometer reading remains fixed at 32°F instead of rising as expected. It continues to read 32°F until the entire pound of ice is melted. The increase in temperature from zero to 32°F registered by the thermometer is called **sensible heat**. The heat that was added to the ice and caused its change in state from a solid to a liquid, but did not register on the thermometer is called **latent heat**.

- Sensible heat is heat that can be sensed by a thermometer or by touch.
- Latent heat is the heat energy absorbed or rejected when a substance is changing state (solid to liquid, liquid to gas, or vice-versa) without a change in the measured temperature.
- Sensible heat plus latent heat equals **total heat**.

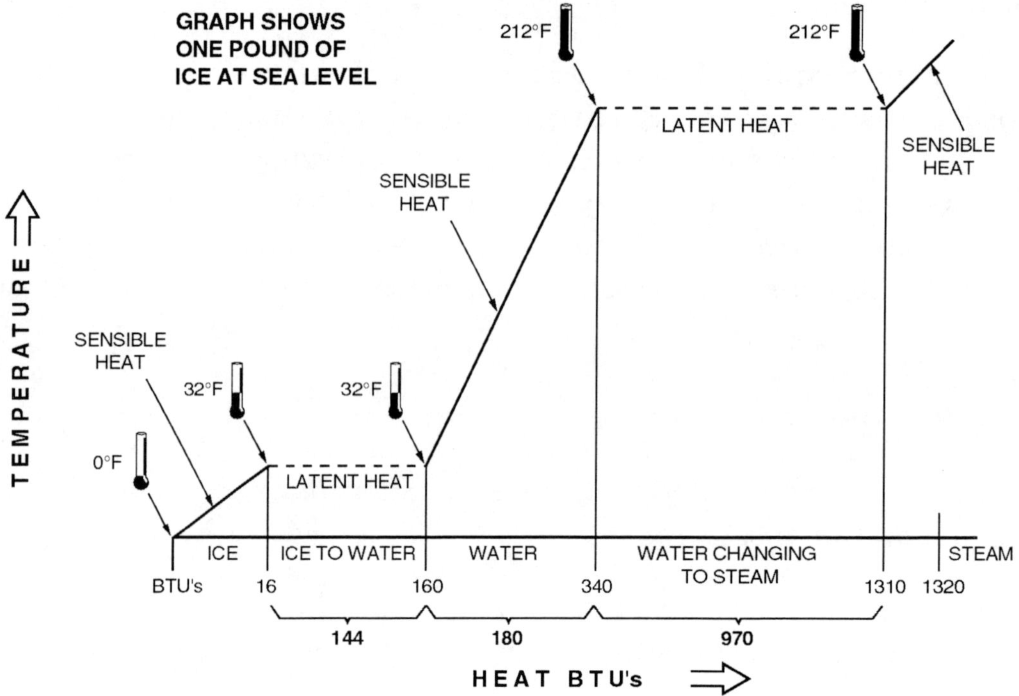

SENSIBLE HEAT:

HEAT THAT CAUSES A TEMPERATURE CHANGE AND CAN BE SENSED BY A THERMOMETER OR BY TOUCH.

LATENT HEAT:

HEAT THAT PRODUCES A CHANGE IN STATE WITHOUT A CHANGE IN TEMPERATURE.

H108F03.EPS

Figure 3. Changing States Of Water

If we continue to add heat to the pound of water after all the ice has melted, the thermometer will once again show an increase in temperature until the temperature reaches 212°F. At this point, the water starts boiling and changes state from water into steam or water vapor. As even more heat is added, the thermometer reading remains at 212°F until all the water has turned into steam. If we continue to add heat, the thermometer will once again register sensible heat called superheat. **Superheat** is the measurable heat added to the vapor or gas once a liquid has reached its boiling point and has been completely changed into a vapor.

As shown in *Figure 3*, it takes a great deal more heat to cause a change in state than is needed for a degree change in temperature. It required 144 Btu's of latent heat to melt the pound of ice before the temperature began to rise. This is 144 times as much heat as is needed to raise the temperature of water one degree. The change from water to steam required an even greater amount of latent heat. It took 970 Btu's to change the water to steam. It is important to remember that none of this latent heat registered on the thermometer.

The latent heat added or removed in changing to and from the solid, liquid and vapor states
has several special names as shown in *Figure 4*.

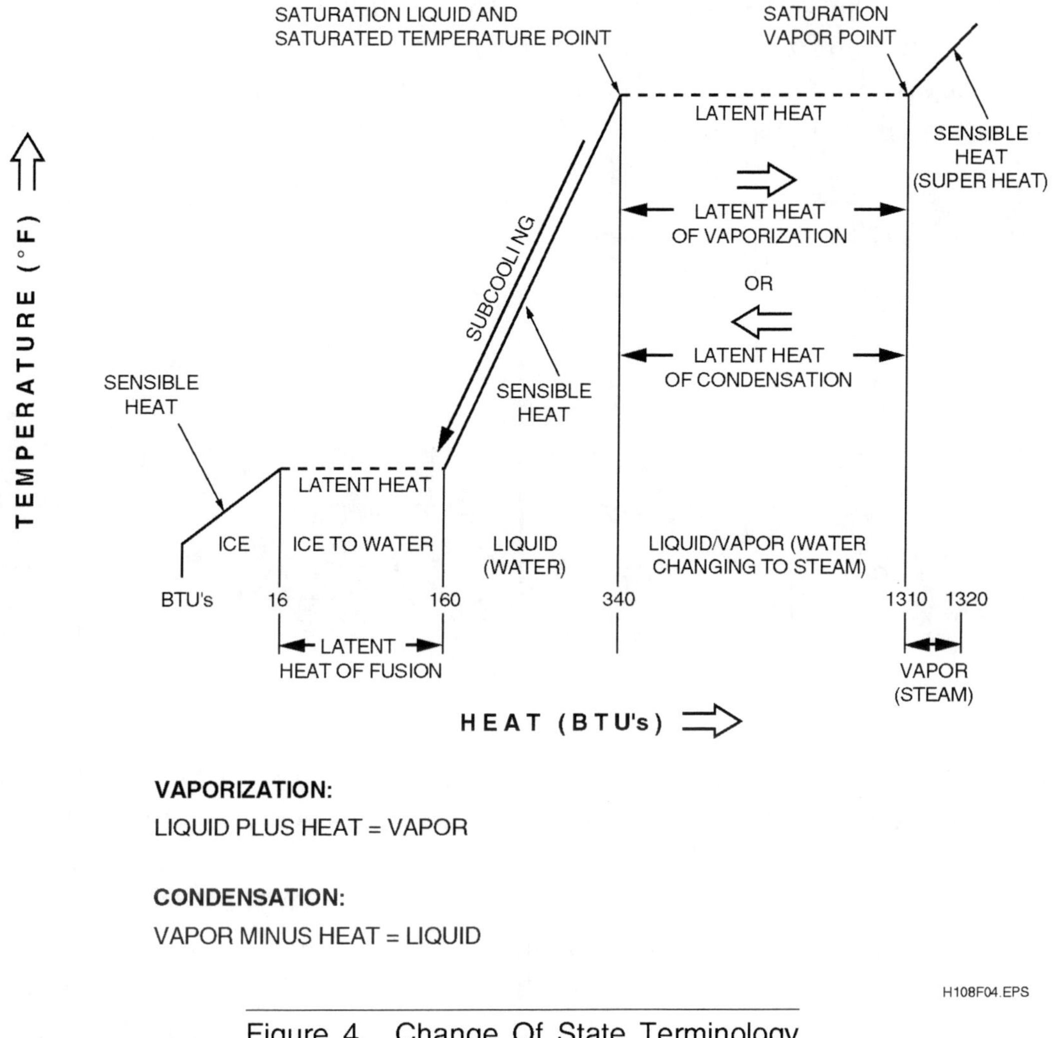

Figure 4.   Change Of State Terminology

- **Latent heat of fusion:** The heat gained or lost in changing to or from a solid (ice to
  water or water to ice).
- **Latent heat of vaporization:** The heat gained in changing from a liquid to a gas (water
  to steam). The temperature at which this occurs is known as the boiling point. In
  refrigeration work, the boiling point is also known as the saturation temperature.
- **Latent heat of condensation:** The heat given up or removed from a gas in changing
  back to a liquid state (steam to water).

As mentioned previously, the two states you are most concerned with in learning about
refrigeration are the liquid and gas states. Another term used in refrigeration work
concerning changes in the state of matter is subcooling. **Subcooling** is the reverse of
superheat. It is the temperature of a liquid when it is cooled below its condensing
temperature. For example, the condensing temperature of water is 212°F. If the water is
subcooled 10 degrees, the temperature is cooled to 202°F.

2.1.4 Specific Heat Capacity

Specific heat is the amount of heat required to raise one pound of a substance one degree Fahrenheit. In the previous examples we saw that water in the liquid state has a specific heat of one Btu per pound of water per each degree Fahrenheit of temperature change (1 Btu/lb./°F). The specific heat of each substance is different. *Table 1* shows the specific heat values for some common substances.

Water	1.00	Iron	0.10
Ice	0.50	Mercury	0.03
Air (dry)	0.24	Copper	0.09
Steam	0.48	Alcohol	0.60
Aluminum	0.22	Kerosene	0.50
Brass	0.09	Olive Oil	0.47
Lead	0.03	Pine	0.67

Table 1.   Specific Heats

Specific heats vary from substance to substance, but also from one state of a substance to another state. As shown in *Table 1*, liquid water has a specific heat of 1.00. Ice, which is solid water, has a specific heat of 0.50. Substances with lower specific heat numbers are the most easily heated. Those with higher numbers require more heat. The result is that it takes twice as many Btu's of heat to raise one pound of water one degree as it does to raise one pound of ice one degree. The effect of specific heat is shown in *Figure 3*. Because the specific heat of ice is 0.50, it took only 16 Btu's of heat to raise the temperature of the ice from 0°F to 32°F. Water with a specific heat of 1.00 took 180 Btu's to raise the temperature from 32°F to 212°F, or 180 degrees.

The **specific heat capacity** of a substance is the amount of heat, measured in Btu's, added or removed to change the temperature of one pound of a substance one degree Fahrenheit. Specific heat capacity can be calculated using the formula:

$$Btu = S \times W \, (\Delta T)$$

Where:
S = Specific heat of the substance (Btu/lb./°F)
W = Weight of the substance (lb.)
ΔT = Change in the temperature of the substance (°F)

As previously discussed, heating one pound of water from 32°F to 212°F required 180 Btu's. Using the formula for specific heat capacity, this can be proved as follows:

$$
\begin{aligned}
\text{Btu} &= \text{S} \times \text{W} \, (\Delta T) \\
&= 1.00 \text{ Btu/lb/°F} \times 1.0 \text{ lb.} \times 180°F \, (212°F - 32°F) \\
&= 1 \times 1 \times 180 \\
&= 1 \times 180 \\
&= 180
\end{aligned}
$$

Likewise, heating one pound of ice from 0°F to 32°F required 16 Btu's. This can also be proved as follows:

$$
\begin{aligned}
\text{Btu} &= \text{S} \times \text{W} \, (\Delta T) \\
&= 0.50 \text{ Btu/lb/°F} \times 1.0 \text{ lb.} \times 32°F \, (32°F - 0°F) \\
&= 0.5 \times 1 \times 32 \\
&= .5 \times 32 \\
&= 16
\end{aligned}
$$

2.2.0 HEAT TRANSFER

Heat transfer is the movement of heat from one place to another, either within a substance or between substances. Heat always flows from a warmer location to a cooler location, like water running down hill. It's important to remember that to have heat flow there must be a difference in temperature. There are three ways to move or transfer heat. These are listed below and shown in *Figure 5*.

*   Conduction
*   Convection
*   Radiation

2.2.1 Conduction

Conduction is a means of heat transfer in which heat is moved from molecule to molecule within a substance. When these molecules are heated, they move about, colliding with one another. These collisions continue in a direction towards the cooler part of the material, causing the movement of heat in the same direction. For example, when copper tubing is being heated by a torch, the molecules in the tubing nearest the torch get heated first and begin to move and collide with the nearby molecules in the tubing. These molecules then collide with other molecules, causing the tubing to become heated. The result is that the heat is carried by conduction from the end being heated by the torch toward the cold end.

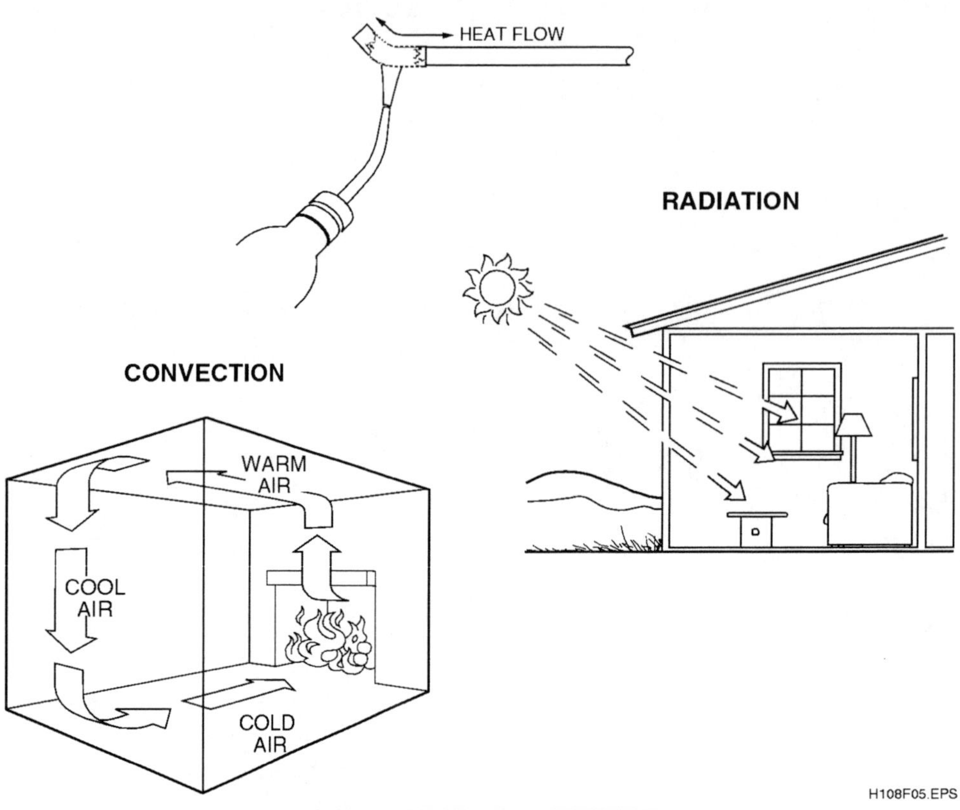

Figure 5. Heat Movement

2.2.2 Convection

Convection is the transfer of heat by the flow of liquid or gas. As shown in *Figure 5*, air near the fireplace is heated by conduction and becomes warmer than the air in the rest of the room. Since warm air rises, the heated air moves upward toward the ceiling. In doing so, it gives up heat as it goes and then settles back down to the floor as it cools. The cooler air at the floor level moves towards the fireplace to replace the rising warm air. It too will warm, rise, give off heat, and settle back down. This circulation of air is accomplished via convection. Convection can be either natural or forced. Natural convection is shown by the example of the fireplace. Forced convection uses fans or pumps to speed up the circulation process, such as those found in home heating and air conditioning systems.

2.2.3 Radiation

Radiation is the movement of heat in the form of invisible rays or waves, similar to light. Like light, it needs no medium on which to travel. Radiation takes place free of convection. It travels in straight lines from the heat source to the point where it is absorbed without heating the space in between. Heat from the sun travelling through space and warming our homes is a good example of heat transfer by radiation. The solar radiation comes through the windows in a building, strikes the walls, floors, furniture and people and is absorbed by them.

HVAC TRAINEE TASK MODULE 03108

2.2.4 Conductors And Insulators

The rate of heat conduction varies for different substances. Some support the transfer of heat, while others restrict it. Materials in which the transfer of heat by conduction occurs easily are called **conductors**. **Insulators** are materials that resist heat transfer by conduction. Cork, fiberglass, and polyurethane foam are examples of insulators. Most metals are good conductors of heat. Copper and aluminum are used in refrigeration systems because of their good heat conduction ability.

2.2.5 Rate Of Heat Transfer

The rate of heat transfer describes how fast heat can be added to or removed from an object or between objects. It usually is expressed in two ways. One way is in Btu's per hour or Btuh. Another way is by the ton. This is a much larger unit of measure. The ton is commonly used in refrigeration work to describe the heat or cooling load for a space, or the capacity of a piece of equipment or system.

One **ton of refrigeration** is defined as 12,000 Btu's per hour or 12,000 Btuh (*Figure 6*). The ton is based on the amount of heat required to melt one ton of ice in a 24-hour period. As learned earlier, one pound of ice at 32°F absorbs 144 Btu's of heat while melting. Assuming that it takes one hour to melt, the rate of heat transfer is 144 Btuh. Since a ton of ice contains 2000 pounds, a total of 288,000 Btu's per day, or 12,000 Btu's per hour (288,000 ÷ 24) is the amount of heat required to melt the ice.

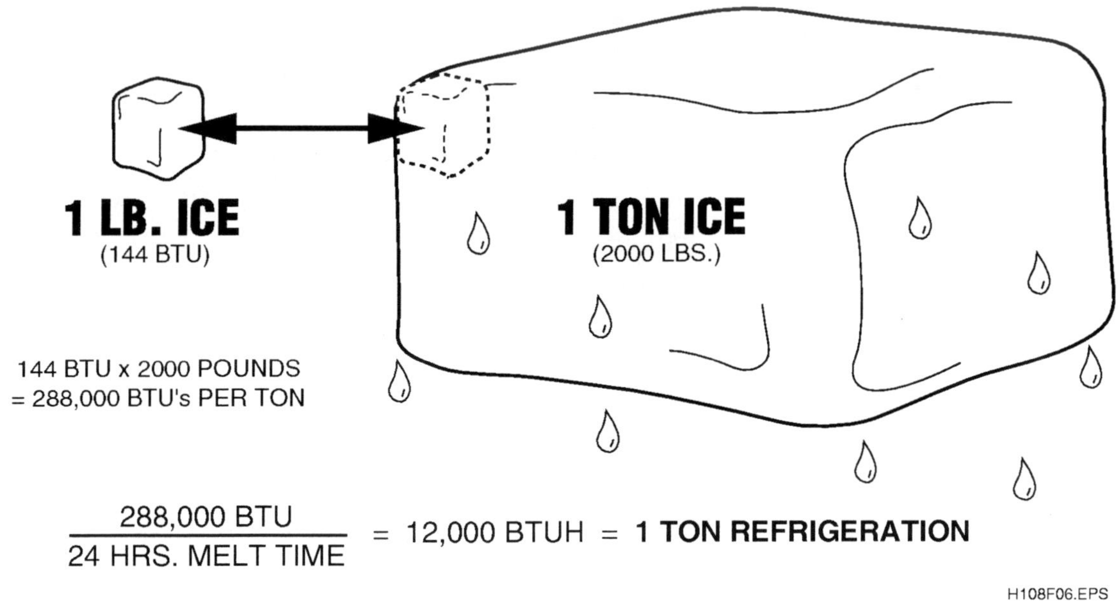

Figure 6.   One Ton Of Refrigeration

2.3.0 PRESSURE

Pressure is defined as force per unit area. This is normally expressed in pounds per square inch, inches of mercury, or for very low pressures, inches of water. Depending on the state of a substance, pressure may be exerted in one direction, several directions, or all directions (*Figure 7*). Using the three states of water as an example, ice (a solid) exerts pressure only in a downward direction. The same is true for all solid materials. As a liquid, water exerts pressure against all sides of the container in contact with it. As a gas (water vapor), it exerts pressure on all the surfaces of the container because it completely fills the container. In refrigeration work, the term fluid is generally used when describing pressure. Fluid means the liquid or gaseous state of a material such as a refrigerant. Fluids tend to exert pressure equally in all directions like the water vapor given in the example.

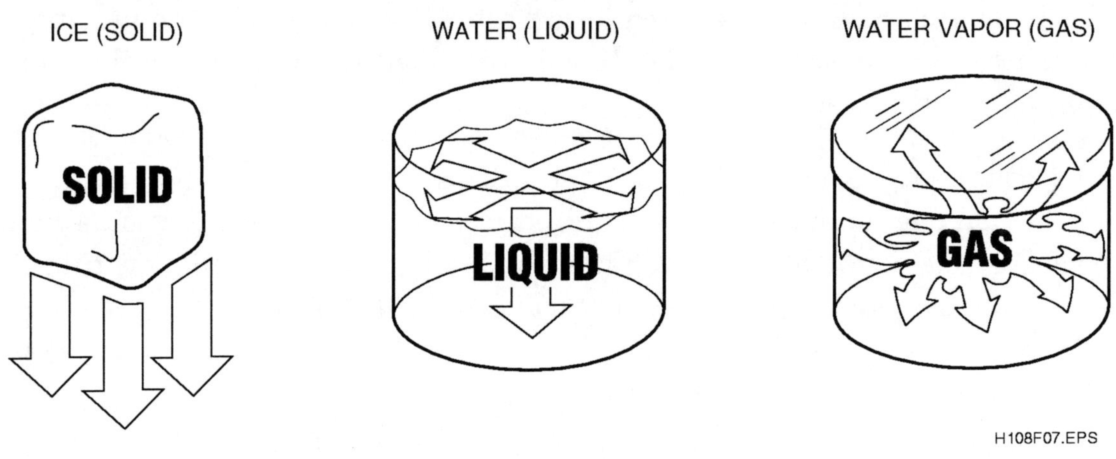

Figure 7.   The Directions Of Pressure

2.3.1   Atmospheric Pressure

The earth is surrounded by a blanket of air called the atmosphere. Air is matter consisting of oxygen, nitrogen and water vapor. It has weight and exerts a force called **atmospheric pressure** on all things on the earth's surface. Atmospheric pressure can be measured with a barometer. For this reason, it is often referred to as barometric pressure. The barometer compares the pressure of the atmosphere to the pressure of no atmosphere.

Figure 8 shows a simple mercury tube barometer. The top of the barometer tube is sealed while the open end at the bottom rests in a container of mercury. Air pressure pushing down on the mercury in the container causes the column of mercury in the tube to rise. The extent of the rise is determined by the amount of pressure applied to the mercury.

   HVAC TRAINEE TASK MODULE 03108

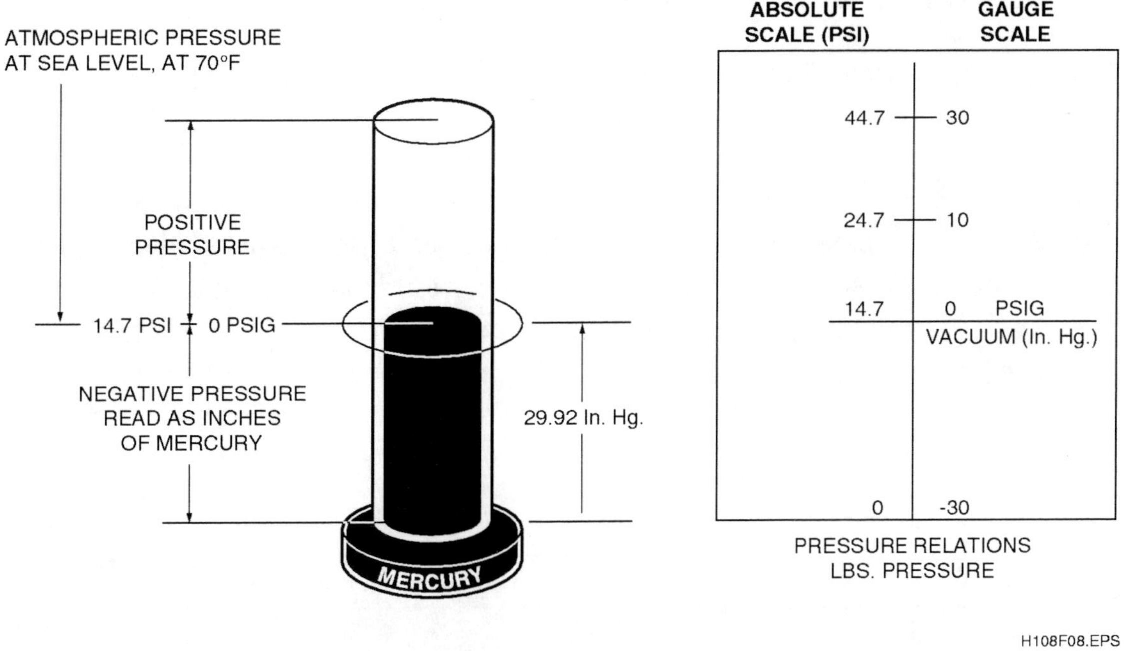

Figure 8.   Absolute And Gauge Pressure Scale Comparison

Visualize a column of air with a cross-sectional area of one square inch and extending from the earth's surface at sea level to the limits of the atmosphere. Also, assume that the temperature at sea level is 70°F. If this column of air is applied to the mercury tube barometer, the height of the mercury in the tube will be slightly less than 30 inches (29.92 inches). Its weight will be 14.7 pounds. This means that every square inch of any surface at sea level has 14.7 pounds of air pressure pushing down on it. These values of 29.92 inches of mercury and 14.7 pounds per square inch at sea level at 70°F are standards that are used frequently in refrigeration work. A pressure scale, called the **absolute pressure** scale, is based on the barometer measurements just described. On this scale, pressures are expressed as pounds per square inch (psi) or pounds per square inch absolute (psia), starting from zero, which represents a complete absence of pressure.

2.3.2   Gauge Pressure

Another scale, called **gauge pressure**, is normally used for refrigeration work. Gauge pressure scales use atmospheric pressure as their zero starting point. Positive gauge pressures, those above zero (14.7 psi), are expressed in pounds per square inch gauge or psig. Negative pressures, those below 0 psig, are expressed in inches of mercury vacuum or in. Hg. vac. Gauge pressures can easily be converted to absolute pressures by adding 14.7 to the gauge pressure value. For example, a gauge pressure of 10 psig equals an absolute pressure of 24.7 (10 + 14.7). A comparison of the gauge and absolute pressure scales is shown in *Figure 8*.

2.3.3   Pressure/Temperature Relationships

Pressure and temperature have a special relationship. Two things are important to remember. First, the temperature at which a liquid or gas changes state is dependent on the pressure. Second, the boiling temperature of a liquid will drop as the pressure on it decreases. It will rise as the pressure increases. Using our water example, water boils at 212°F at sea level (14.7 psia). With a lower atmospheric pressure of about 11.6 psia that exists at 5000 feet above sea level, the same water would boil at the lower temperature of about 203°F. At the higher pressure of about 29.7 psia (15 psig + 14.7 atmospheric pressure) such as can be reached in a pressure cooker, the water boils at the higher temperature of about 250°F. *Figure 9* shows the temperature/pressure relationship of water.

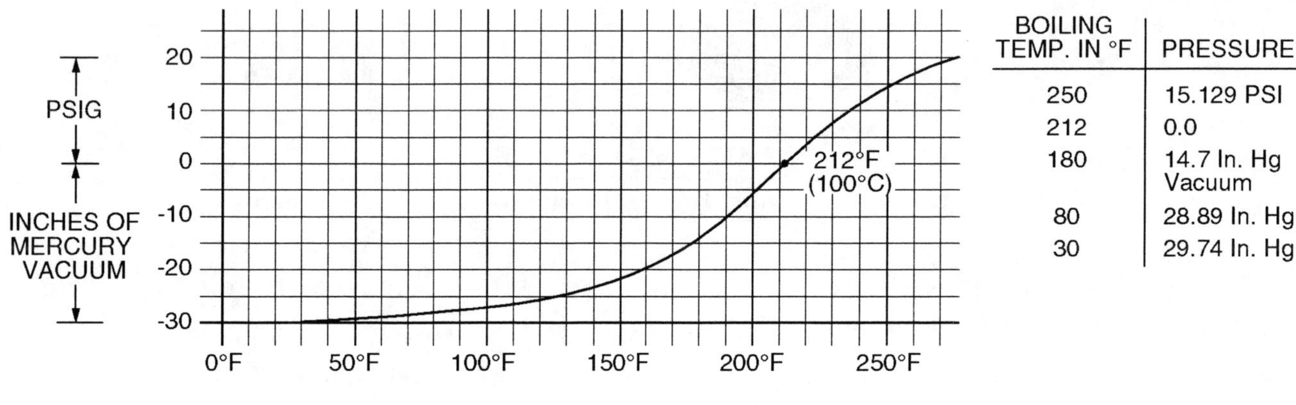

BOILING TEMP. IN °F	PRESSURE
250	15.129 PSI
212	0.0
180	14.7 In. Hg Vacuum
80	28.89 In. Hg
30	29.74 In. Hg

Figure 9.   Temperature/Pressure Relationship Of Water

If the pressure on a liquid can be lowered enough, its boiling point need not be a high temperature. This relationship is basic to our work in refrigeration. Each refrigerant used with cooling equipment has its own temperature/pressure relationship. Like water, the boiling temperature of refrigerants rises as the pressure is increased, and drops as the pressure is decreased.

2.3.4   Movement Of Fluids

Differences in pressure cause the flow of fluids. This flow is always from a higher pressure to a lower pressure. Just as heat moves from a higher temperature to a lower temperature, so too does a liquid or gas move from a higher pressure to a lower pressure. The result is that liquids and gases can be made to move by adjusting or changing the pressure around them. In refrigeration, a compressor is used to create the pressure differential that causes refrigerant to flow in a system.

2.4.0 TEMPERATURE AND PRESSURE MEASUREMENT INSTRUMENTS

2.4.1 Thermometer

The dial and electronic thermometers are two types of thermometers most commonly used for measuring temperatures in HVAC equipment. Many other special-purpose thermometers are also available. Thermometers come in a variety of temperature ranges based on their intended use. *Figure 10* shows dial and electronic thermometers.

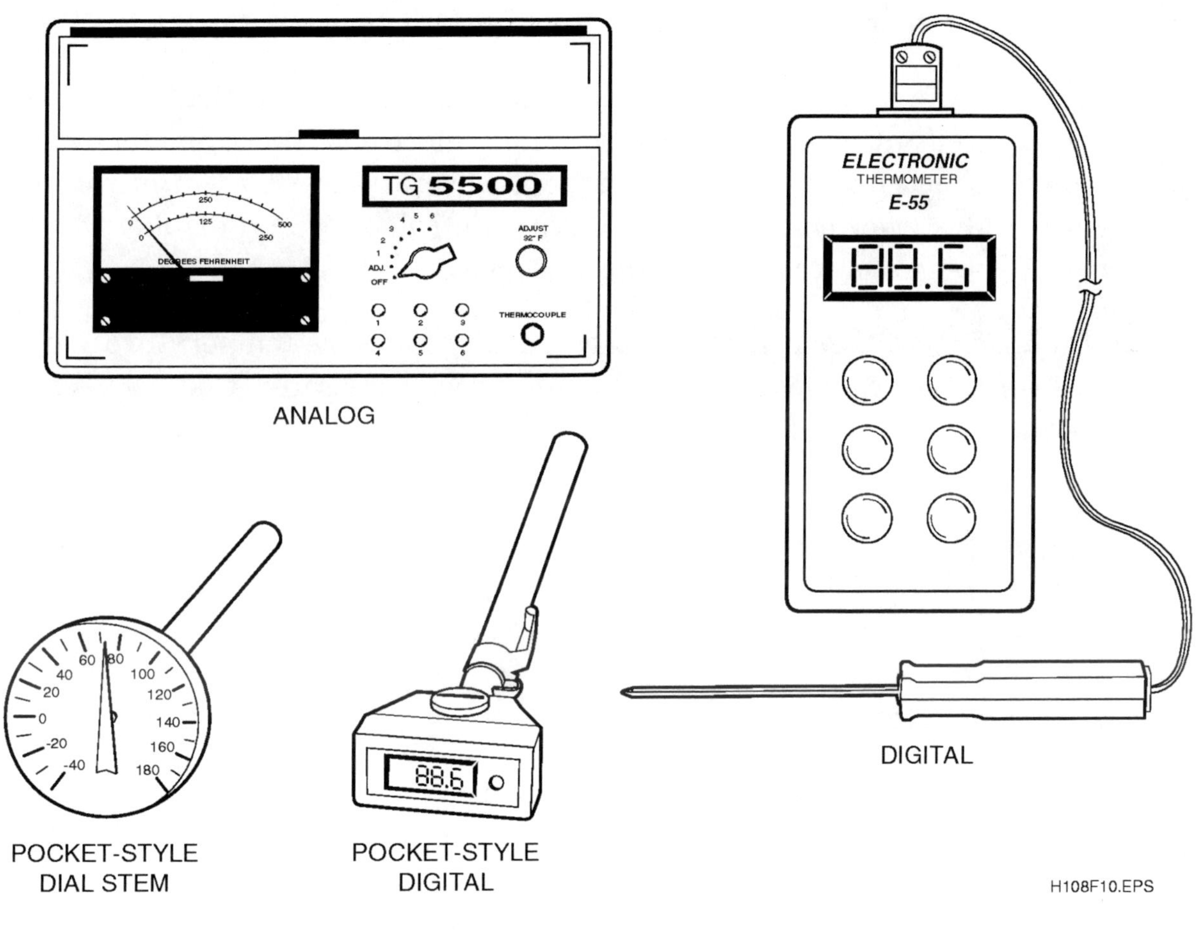

Figure 10.   Thermometers

Dial thermometers are available in various forms. They are popular because they are rugged, small, and inexpensive. They come in many stem lengths and dial sizes. Because pocket-style dial thermometers have smaller-size calibrated scales, their use sometimes makes it difficult to get accurate readings. When accuracy is needed, electronic thermometers should be used.

Electronic thermometers display temperature on either an analog meter or digital liquid crystal display (LCD) readout. They generally use either a thermocouple or thermistor-type temperature probe, or both, to sense the heat and generate the temperature reading. Thermocouple probes use a sensing device (**thermocouple**) made of two dissimilar wires welded together at one end called a junction. When the junction is heated, it generates a low level DC voltage that produces a temperature reading on the electronic thermometer indicator. Thermocouple probes tend to be rugged and inexpensive. Thermistor probes use a semiconductor electronic element (**thermistor**) whose resistance changes with a change in temperature. As the temperature being measured varies, the thermistor resistance varies, causing the electronic thermometer indicator reading to change accordingly.

Often, several different probes are used with the same electronic thermometer to allow temperature measurement over a wide range. Many electronic thermometers have two or more probes so that measurements can be made at several locations within the equipment at the same time. Most thermometers of this type can calculate and display the difference in temperature between the locations being measured.

Many digital multimeters (DMM's) can also be used to measure temperature. This feature requires the use of thermocouple and/or thermistor probe accessories that convert the DMM into an electronic thermometer. Some DMM's can measure temperature using a non-contact infrared probe.

Electronic thermometers are precise measuring instruments. Be sure to read and follow the manufacturer's instructions for operating electronic thermometers. Also, be sure to follow the manufacturer's instructions for calibration of the instrument.

2.4.2   Gauge Manifold Set

The gauge manifold set is the most frequently used item of service equipment. It is used to measure low-side and high-side pressures in an operating system in order to evaluate system performance. The gauge manifold set is regularly used to route and control the flow of refrigerant, refrigerant oil, or other acceptable gases to and from the system in support of many other servicing tasks.

Figure 11 shows a standard two-valve gauge manifold set. It consists of two pressure gauges mounted on a manifold assembly. A compound gauge is mounted on the left side of the manifold and a high-pressure gauge is mounted on the right. A compound gauge allows measurement of pressures both above and below atmospheric pressure. Most compound gauges used with the gauge manifold set can measure pressures above atmospheric pressure in the range from 0 to 150 pounds per square inch gauge (psig). Below atmospheric pressure they can measure from 0 to 30 in. Hg vacuum. Zero, which represents atmospheric pressure, is the starting point for both scales. The compound gauge is used to measure system low-side (suction) pressures, including any vacuum that exists in a system. The high-pressure

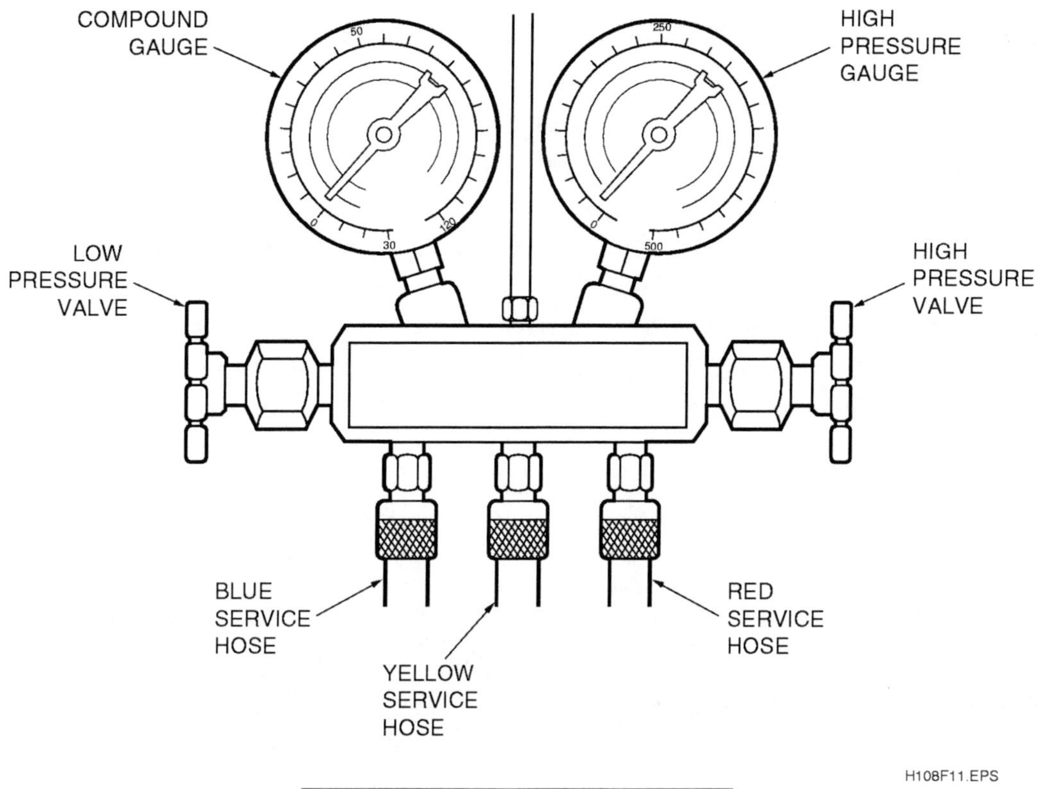

Figure 11. Gauge Manifold Set

gauge on the right side of the manifold is used to measure system high-side (discharge) pressures. High-pressure gauges can usually measure system pressures in the range from 0 to 500 psig.

As shown, a two-valve gauge manifold set has two hand valves and three hose ports. The hand valves are adjusted to monitor system pressures on the compound and high-pressure gauges and/or route the flow of refrigerant to and from the system during servicing activities. The gauge manifold set hose ports are connected to the system being serviced and/or other service instruments through a set of environmentally safe, high-vacuum, high-pressure service hoses. It is recommended that these hoses be equipped with fast, self-sealing type fittings that immediately trap refrigerant in the hoses when disconnected. Their use helps to meet the clean air non-venting regulations and also greatly reduces the amount of air that can enter and contaminate the hose after it has been disconnected.

Most gauge manifold sets and service hoses are color coded. The low-pressure compound gauge, hand valve, and low-pressure hose port are blue. A blue service hose is used to connect the manifold low-pressure hose port to the equipment suction service valve. Red marks the high-pressure gauge, hand valve, and hose port. A red service hose is used to connect the high-pressure hose port to the equipment discharge service valve or liquid line valve. The center hose port is the utility port. This port is normally connected through a yellow service hose to other service instruments or devices. When not in use, the utility port should be capped.

Gauge manifold sets are also available with four hand valves and related hose ports. Use of this type of manifold can reduce service time by eliminating the need to switch a single utility hose between different service devices. Four-valve manifolds and related service hose sets are color-coded as follows: blue (low pressure), red (high pressure), yellow (charging) and black (vacuum). Digital electronic gauge manifold sets that display system pressures on liquid crystal display (LCD) readouts are also available.

The gauge manifold set is a precise measuring instrument. Its accuracy is critical to correct servicing. The technician must ensure that the gauge manifold set is always handled with care both during use and in transport. The calibration of the gauge manifold set should be checked regularly. If necessary, it should be calibrated according to the manufacturer's instructions.

2.4.3   Manometer

Another instrument that is used to measure pressure is the manometer. Both electronic and non-electronic manometers are in common use. Manometers can be used to measure pressures within a refrigeration system, but are seldom used for this purpose. Manometers are most often used to measure gas pressure in heating systems and velocity or static air pressure in air distribution systems. You will learn more about manometers and their use later in your training.

3.0.0   MECHANICAL REFRIGERATION SYSTEM

3.1.0   SYSTEM COMPONENTS

There are many types of systems used to provide cooling for personal comfort, food preservation, and industrial processes. Each of these uses a mechanical refrigeration system. *Figure 12* shows a basic system. Its components are:

- Evaporator — A heat exchanger where the heat from the area or item being cooled is transferred to the refrigerant.
- Compressor — Creates the pressure differences in the system needed to make the refrigerant flow and the refrigeration cycle work.
- Condenser — A heat exchanger where the heat absorbed by the refrigerant is transferred to the cooler outdoor air or another cooler substance.
- Expansion device — Provides a pressure drop that lowers the boiling point of the refrigerant just before it enters the evaporator. This is also known as the metering device.

Also shown in *Figure 12* is the piping, called lines, used to connect the basic components in order to provide the path for refrigerant flow. Together, the components and lines form a closed refrigerant system. The lines are:

- Suction line — The tubing that carries heat-laden refrigerant gas from the evaporator to the compressor.
- Hot gas line (also called the discharge line) — The tubing that carries hot refrigerant gas from the compressor to the condenser.
- Liquid line — The tubing that carries the liquid refrigerant formed in the condenser to the expansion device.

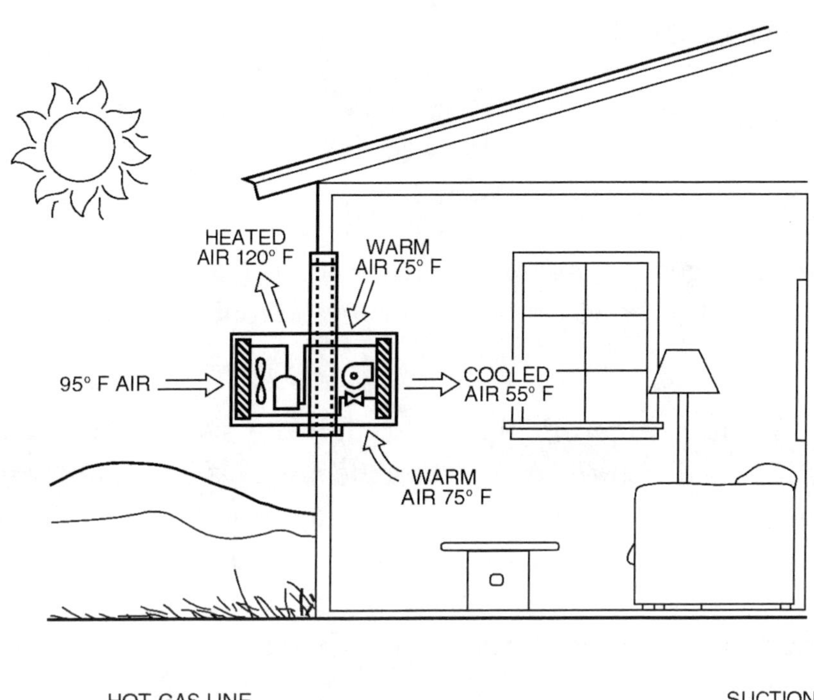

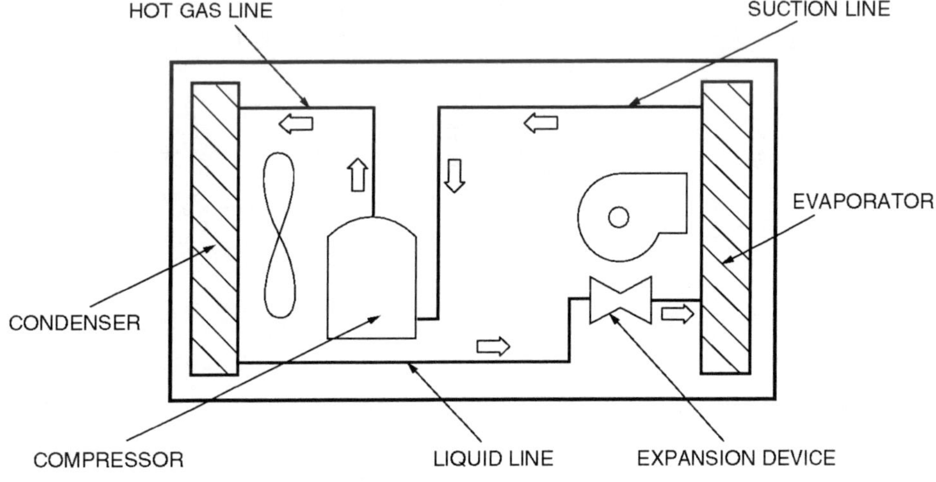

Figure 12. Basic Refrigeration Cycle

The arrows on *Figure 12* show the direction of flow through the system. The purpose of the refrigerant is to move heat. It is the medium by which heat can be moved into or out of a space or substance. A **refrigerant** is any fluid (liquid or gas) that picks up heat by evaporating at a low temperature and pressure and gives up heat by condensing at a higher temperature and pressure. Refrigerants often boil at very low temperatures (as low as -21°F).

Remember that the operation of a mechanical refrigeration system is the same for all systems. Only the type of refrigerant used, the size and style of the components, and the installed locations of the components and the lines will change from system to system. Other devices, called accessories, may be used in some systems to gain the desired cooling effect. The events that take place within the system happen again and again in the same order. This repeating series of events is called the "refrigeration cycle."

3.2.0   REFRIGERATION CYCLE

3.2.1   Basic Operation

The refrigeration cycle is based on two principles:

1. As liquid changes to a gas or vapor, it is capable of absorbing large quantities of heat.
2. The boiling point of a liquid can be changed by changing the pressure exerted on the liquid.

As shown in *Figure 12*, the refrigerant flows through the system in the direction indicated by the arrows. We will begin with the evaporator. It receives low-temperature, low-pressure liquid refrigerant from the expansion device. The evaporator is a series of tubing coils that expose the cooler liquid refrigerant to the warmer air passing over them. Heat from the warmer air is transferred through the tubing to the cooler refrigerant. This causes it to boil or vaporize. It is important to realize that even though it has just boiled, it is still not considered "hot" because refrigerants boil at such low temperatures. So, it is a low-temperature, low-pressure refrigerant vapor that travels through the suction line to the compressor.

The compressor receives the low-temperature, low-pressure vapor and compresses it. It then becomes a high-temperature, high-pressure vapor. This travels to the condenser via the hot gas line.

Like the evaporator, the condenser is a series of tubing coils through which the refrigerant flows. As cooler air moves across the tubing, the hot refrigerant vapor gives up superheat and cools. As it continues to give up heat to the outside air, it cools to the condensation point, where it begins to change from a vapor into a liquid. As more cooling takes place (subcooling) all of the refrigerant becomes liquid. This high-temperature, high-pressure liquid travels through the liquid line to the input of the expansion device.

The expansion device regulates the flow of refrigerant to the evaporator. It also decreases its pressure and temperature. By the use of a built-in restriction, such as a small hole or orifice, it converts the high-temperature, high-pressure refrigerant from the condenser into the low-temperature, low-pressure refrigerant needed to absorb heat in the evaporator.

3.2.2 Refrigeration Cycle In A Typical Air Conditioning System

Figure 13 shows a typical air conditioning system. The components are divided into two sections based on pressure. The high-pressure side includes all of the components where the pressure of the refrigerant is at or above the condensing pressure. This is often referred to as the head pressure, discharge pressure, or high-side pressure. The low-pressure side includes all of the components where the pressure of the refrigerant is at or below the evaporating pressure. This is often called the suction pressure, or low-side pressure. The dividing point between the sections cuts through the compressor and the expansion device.

We will now discuss the refrigeration cycle in more detail. We will describe a typical air conditioner that uses HCFC-22 (R-22) as the refrigerant. R-22 boils at 40°F when under a pressure of 69 psig. This example will demonstrate the concepts and temperature/pressure relationships you have learned so far. For our example, assume an air temperature of 75°F for the room being cooled and an outdoor air temperature of 95°F. These values will vary due to equipment and load conditions. The numbers below correspond to the numbers shown in *Figure 13*. Follow along on the figure as we describe this system.

1. A mixture (75% liquid, 25% vapor) of R-22 is supplied from the expansion device to the evaporator. This mixture is at a pressure of 69 psig, which corresponds to the 40°F boiling point of R-22 refrigerant. (See the chart shown in *Figure 13*.) The 40°F boiling point is used here because it is typical of the temperatures normally used for evaporators in air conditioning systems.
2. Because the refrigerant flowing through the evaporator is cooler (40°F) than the warmer inside room air (75°F), it absorbs heat, causing the liquid refrigerant to boil and turn into a vapor. After traveling about 90 percent of the way through the evaporator tubing, all the refrigerant has boiled into a vapor known as the "saturated" vapor.
3. During the remaining ten percent of travel through the evaporator, the saturated vapor continues to absorb heat from the warmer air, thus raising its temperature to 50°F. In other words, the saturated vapor is superheated 10°F (50°F - 40°F). This superheated vapor flows through the suction line and is drawn into the low-pressure side of the compressor. The cooled inside room air is recirculated by the evaporator fan back into the room at a temperature of about 55°F.
4. The superheated vapor applied at the suction input of the compressor typically picks up an additional 10°F superheat (60°F - 50°F) because the vapor in the suction line absorbs more heat from the warmer surrounding air as it travels from the evaporator to the compressor.

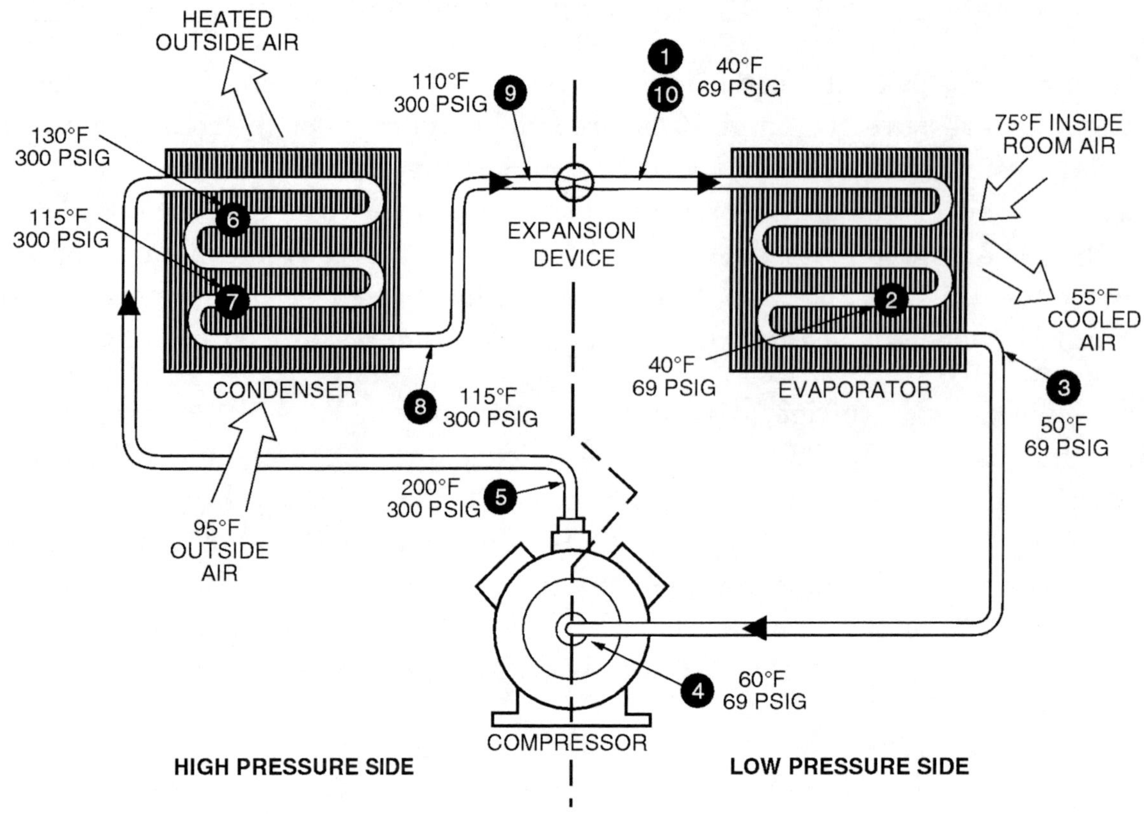

TEMPERATURE °F	REFRIGERANT R-22
27	51.2
28	52.4
29	53.6
30	54.9
31	56.2
32	57.5
33	58.8
34	60.1
35	61.5
36	62.8
37	64.2
38	65.6
39	67.1
40	68.5
41	70.0
42	71.5
43	73.0
44	74.5
45	76.0
46	77.6
47	79.2
48	80.8
49	82.4

TEMPERATURE °F	REFRIGERANT R-22
50	84.0
55	92.6
60	101.6
65	111.2
70	121.4
75	132.2
80	143.6
85	155.7
90	168.4
95	181.8
100	195.9
105	210.8
110	226.4
115	242.7
120	259.9
125	277.9
130	296.8
135	316.6
140	337.3
145	358.9
150	381.8
155	405.1

TEMPERATURE/PRESSURE CHART

H108F13.EPS

Figure 13. Typical Air Conditioning Cycle For HCFC-22 (R-22) Refrigerant

5. After compression, the highly superheated gas from the compressor flows through the hot gas line to the condenser. This hot gas may be close to 200°F at 300 psig. Since the saturated temperature corresponding to 300 psig is 130°F (see the temperature/pressure chart), the hot gas line has gained about 70°F (200°F - 130°F) superheat. This superheat must be removed before the vapor can be condensed into a liquid. The 200°F refrigerant in the hot gas line easily gives up some of its superheat to the surrounding 95°F air. The hot gas line is normally not insulated and the tubing is a good conductor of heat.

6. Because the refrigerant in the condenser is still hotter than the warmer outside air passing over the condenser, it easily gives up the remaining superheat. This drops its temperature to 130°F. As heat continues to be transferred from the vapor to the cooler outside air, the vapor begin to subcool and condense into a liquid. After the refrigerant has travelled about three quarters of the way through the condenser, all of the refrigerant has condensed into a liquid.

   The 130°F condensing temperature is set by the condenser design. A standard condenser is designed to have a condensing temperature about 35°F higher than the surrounding air. In this case, 95°F outside air is used to absorb the heat, so 95°F + 35°F = 130°F condensing temperature.

7. During the remaining one quarter of travel through the condenser, the liquid refrigerant continues to drop in temperature (subcool). This lowers its temperature about 15°F to 115°F. In other words the liquid refrigerant is subcooled 15°F (130°F - 115°F).

8. Subcooled liquid refrigerant from the condenser travels through the liquid line to the expansion device. The liquid line is usually not insulated and may be long. Thus, the 115°F liquid refrigerant is further subcooled and continues to drop in temperature as it gives up more heat to the cooler outside air. This drop could increase the subcooling by another 5°F, lowering the temperature of the liquid refrigerant to 110°F.

9. The expansion device controls the flow of liquid refrigerant to the evaporator. Subcooled liquid from the condenser enters at the high temperature of 110°F and high pressure of 300 psig. It leaves the expansion device at the low temperature of 40°F and low pressure of 69 psig, thereby lowering the boiling point of the liquid refrigerant supplied to the evaporator. In the expansion device, the subcooled liquid refrigerant at 300 psig is passed through a small opening or orifice. This changes the pressure of the liquid refrigerant from 300 psig to 69 psig, causing some of it to "flash" into a vapor. This flash gas cools the remaining liquid to produce a mixture of about 75% liquid and 25% vapor. The pressure of this mixture is 69 psig, which corresponds to the 40°F boiling point needed for correct evaporator operation. This low-temperature, low-pressure mixture from the expansion device then travels to the evaporator.

10. The refrigerant has now completed its cycle and is ready to start over again.

Refrigerants are used in cooling systems to move heat into or out of a space or substance. This section briefly describes refrigerants with the focus on their impact on the environment. You will learn more about their characteristics and specific uses later in your training. This section limits the discussion to ammonia and fluorocarbon refrigerants, since for all practical purposes they are the only ones in common use today.

4.1.0 REFRIGERANT TRADE NAMES

Traditionally, each refrigerant had a trade name or an "R" (refrigerant) name. Names like R-11, R-22, R-717, etc. were assigned by the American Society of Heating and Air Conditioning Engineers (ASHRAE). These names were substituted for the true chemical names. For example, R-22 describes the refrigerant with the chemical name of chlorodifluoromethane.

As a result of the Clean Air Act, and the concerns about refrigerants and their effect on our environment, the way refrigerants are named has changed. ASHRAE has substituted acronyms such as CFC's, HCFC's, and HFC's for the "R" in the refrigerant name. These acronyms describe the way the refrigerants are chemically structured. Their meanings will be explained in the following paragraphs. The number previously assigned to a refrigerant by ASHRAE has been retained. Both old and new names are currently being used in the trade. Some examples of changed names for commonly used refrigerants are shown in *Table 2*.

Old Name	New Name
R-11	CFC-11
R-12	CFC-12
R-22	HCFC-22
R-123	HCFC-123
R-134a	HFC-134a
R-500	CFC-500
R-502	CFC-502

Table 2. Examples Of Old And New Refrigerant Names

4.2.0   AMMONIA

Ammonia (R-717) has excellent heat transfer qualities and is used mainly in ice plants, ice skating rinks, and large food processing plants. Though not classed as poisonous, ammonia has a harsh effect on the respiratory system. Only very small quantities can be breathed safely. Exposure for 5 minutes to 50 parts per million (ppm) is the maximum exposure allowed by the **Occupational Health and Safety Administration (OSHA)**. Ammonia is hazardous to life at 5000 ppm and is flammable at 150,000 to 270,000 ppm. Ammonia has an odor that can be smelled at 3 to 5 ppm. This odor gets very irritating at 15 ppm. A respirator with a tight fitting mask should be worn when working on systems that use ammonia. As far as our environment is concerned, ammonia is considered safe.

4.3.0   FLUOROCARBON REFRIGERANTS

Man-made (synthetic) refrigerants in popular use today are all fluorocarbon or a mixture of fluorocarbon refrigerants. Included in this group are refrigerants such as CFC-11 (R-11), CFC-12 (R-12), HCFC-22 (R-22), HCFC-123 (R-123) and HFC-134a (R-134a). These refrigerants all stem from one of two base molecules, methane and ethane.

Methane and ethane are called **hydrocarbons** because they are organic compounds that contain only hydrogen and carbon atoms (*Figure 14*). When most or all of the hydrogen atoms in the methane or ethane molecule are replaced with non-metallic elements such as chlorine, fluorine, and/or bromine, the changed molecule is called a **halocarbon**, short for halogenated hydrocarbon. Chlorine, fluorine and bromine are chemically related elements called **halogens**. To halogenate means to cause some other element to combine with a halogen. When all the hydrogen atoms in a hydrocarbon molecule are replaced with chlorine or fluorine, the molecule is said to be "fully halogenated." Halocarbons in which at least one or more of the hydrogen atoms has been replaced with fluorine are called **fluorocarbons**. *Figure 14* shows the transition of a methane (R-50) molecule to a R-22 molecule. Fluorocarbon refrigerants fall into three groups, CFC's, HCFC's and HFC's, based on their chemical structure.

There is evidence that the ozone layer surrounding the earth is being destroyed by various chemicals. The ozone layer filters out harmful radiation from the sun that would otherwise reach the earth's surface and damage life. The chlorine in CFC and HCFC refrigerants is now known to contribute to this damage. Since the passage of the Clean Air Act in 1990, these refrigerants have come under increasing government regulation and control. The U.S. Environmental Protection Agency (EPA) is responsible for making and enforcing laws pertaining to the use of these refrigerants.

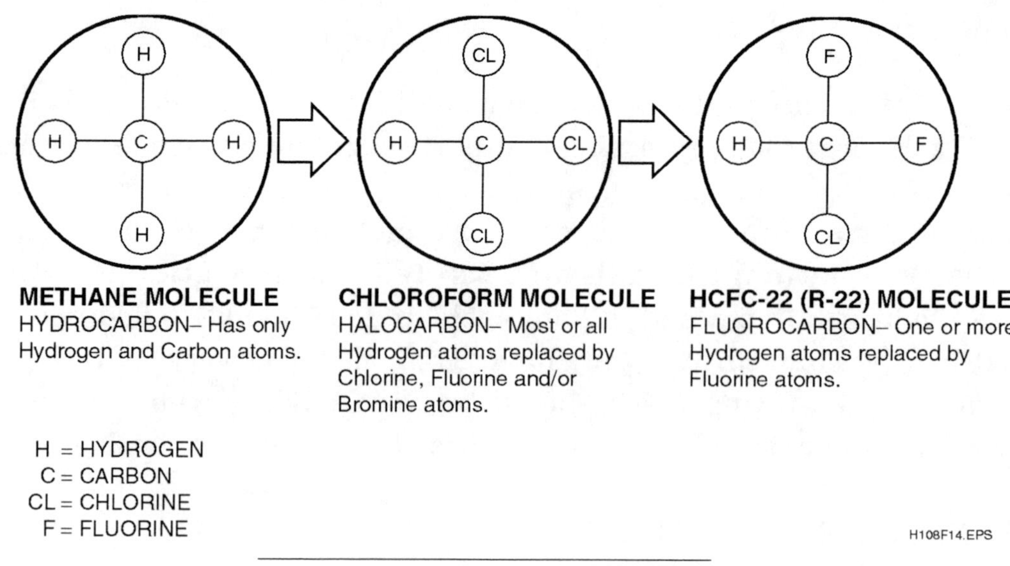

Figure 14.   Refrigerant Transition

4.3.1   Chlorofluorocarbon Refrigerants (CFC's)

Chlorofluorocarbon refrigerants, called CFC's, are made up of chlorine, fluorine, and carbon. Common CFC's include CFC-11, CFC-12 and CFC-113. CFC's are the most hazardous to the environment. All the hydrogen atoms have been removed from the molecules of CFC's. They contain only chlorine and fluorine atoms around the carbon atom. Because they contain high levels of chlorine, CFC's are under the strictest regulation. The Clean Air Act scheduled these refrigerants to phase out of production before the end of the twentieth century. Since July 1992, it has been unlawful to knowingly release CFC's into the atmosphere. If caught doing so, you can be subject to a stiff fine and possibly a prison term.

4.3.2   Hydrogenated Chlorofluorocarbon Refrigerants (HCFC's)

Hydrogenated chlorofluorocarbon refrigerants, called HCFC's, consist of hydrogen, chlorine, fluorine, and carbon. HCFC refrigerants are considered as alternates to replace the CFC refrigerants when applicable. Because they contain less chlorine, HCFC's are considered less damaging to the environment than CFC's. However, since July 1992, it has been unlawful to knowingly release HCFC refrigerants into the atmosphere. Common HCFC's include HCFC-22 and HCFC-123.

4.3.3   Hydrogenated Fluorocarbon Refrigerants (HFC's)

Hydrogenated fluorocarbon refrigerants, called HFC's, are made up of hydrogen, fluorine and carbon. Because HFC's contain no chlorine, they are considered to be least damaging to the environment. HFC-134a was developed to replace CFC-12. Its greatest use is expected to be in commercial, medium-temperature refrigeration equipment, transportation air conditioning, and refrigeration products.

4.3.4   Azeotropes

Most refrigerants are made from a single chemical compound. Others are made by combining two different refrigerants. When mixed in certain proportions, a new compound is formed that has a boiling point different than either of the base refrigerants. The mixed refrigerant is called an **azeotrope** or blend. Azeotropes have names in the 500 series. For example, CFC-500 and CFC-502 are azeotropes.

4.4.0   REFRIGERANT CONTAINERS

Refrigerants come in disposable, returnable, or refillable metal containers which vary in shape and size as shown in *Figure 15*. Low-pressure refrigerants such as CFC-11, CFC-113, and HCFC-123 come in standard steel drums or cylinders. They have boiling points close to, or slightly above, ambient (room) temperature. The pressure they exert on the container is much less than that of medium and high-pressure refrigerants, such as CFC-12, HCFC-22, HFC-134a, CFC-500, and CFC-502. These refrigerants are liquefied compressed gases. If improperly handled, the pressurized containers that contain them can burst or leak, causing damage, injury, or even death.

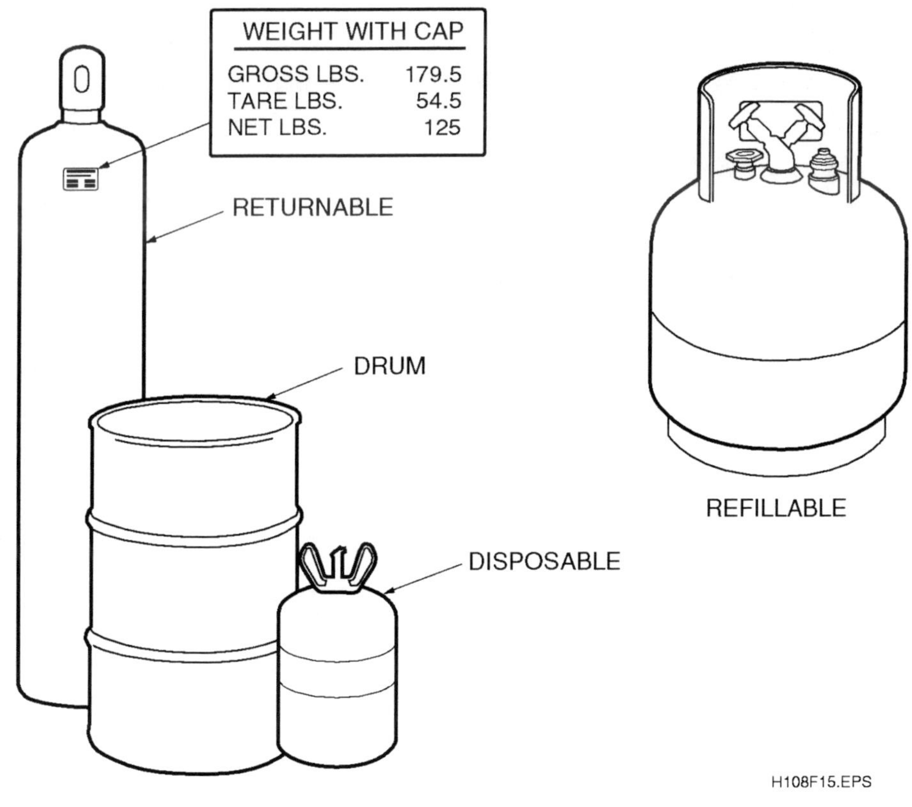

Figure 15.   Refrigerant Containers

4.4.1 Disposable Cylinders

Disposable cylinders are considered "one-way." Once emptied, their use is over. These cylinders should be stored in dry locations to prevent rusting. They should be transported carefully to prevent abrasion of their painted surfaces. As an added protection, they should be kept in their original cartons. Disposable cylinders should not be left around with quantities of refrigerant in them. Over time, rough handling or excessive heat could cause them to explode, especially if weakened by rust or corrosion.

WARNING!   Disposable cylinders must never be refilled. Not only is it dangerous, it is also against the law. Violators can be fined up to $25,000 and can face up to five years in jail.

When empty, disposable cylinders are recycled as scrap metal. Be sure that the cylinder pressure is zero pounds, then make the cylinder useless by puncturing the rupture disk or breaking off its shutoff valve.

4.4.2 Returnable Cylinders

Returnable cylinders go back to the manufacturer for reuse and refilling. They are not intended to be refilled in the field or to be used as a refrigerant recovery tank. These containers are not filled with more than 80% liquid. Excess liquid causes hydrostatic pressure that can result in an explosion. This pressure increases rapidly with even very small changes in temperature. Returnable cylinders and tanks are normally stamped with various weight values. HVAC technicians should understand the values in use. These include:

- Tare weight — The empty weight of the vessel.
- Gross weight — The combined weight of the vessel (tare weight) plus the weight of the refrigerant when the vessel is full. Be aware that the term "full" actually means a vessel that is filled to 80% of capacity. The remaining 20% of the volume must be available for expansion.
- Net weight — The weight of the contents in the vessel. For example, when ordering 50 pounds of refrigerant from a supplier, we are actually talking about the net weight. Manufacturers design vessels so that when the full net weight is reached, 20% of the volume remains for expansion.

Also stamped on the shoulder or collar of returnable cylinders is the date when the cylinder was tested. Returnable and reusable cylinders must be retested every 5 years.

4.4.3 Refillable Cylinders

Refillable cylinders are typically used and supplied with recovery/recycle units. Cylinders with a 50-pound capacity are commonly used. According to EPA regulations, all cylinders used for the recovery and storage of used refrigerant are painted gray with the top shoulder portion painted yellow. The label on the cylinder must be marked to properly identify the type of refrigerant it contains. In all cases, the protective cap that screws onto the cylinder body is painted gold. A returnable cylinder should never be substituted for a recovery cylinder.

4.5.0   IDENTIFYING REFRIGERANTS

Refrigerant containers are color-coded and marked with labels to identify the type of refrigerant they contain. These labels also include important health information about the

CFC-11	Orange	CFC-12	White
CFC-113	Dark Purple	CFC-500	Yellow
CFC-502	Purple	HCFC-22	Green
HCFC-123	Grey	HCFC-124	Dark Green
HFC-125	Medium Brown	HFC-134a	Light Blue
Refrigerant Recovery Cylinders – Gray with Yellow Top			

Table 3.   Color Codes Used For Some Common Refrigerant Containers

contents of the container. *Table 3* lists the color codes for some common refrigerant containers.

If the type of refrigerant used in a system is unknown, it can usually be identified with one of the following methods:

*   Check the manufacturer's service literature for the equipment.
*   Check the nameplate on the equipment.
*   Check the data marked on the thermostatic expansion valve (expansion device).
*   When no printed data can be found, the refrigerant type can be identified using a temperature/pressure chart. Each type of refrigerant has its own distinctive pressure/ temperature relationships. Because of these relationships, the type of refrigerant can be determined by comparing measured pressure and temperature values to the values given on a standard temperature/pressure chart. Proceed as follows:

Step 1   Take pressure and saturation temperature readings for the refrigerant when the equipment is turned off and has cooled down to the surrounding air temperature.

Step 2 Find the measured values for temperature and pressure on the temperature/pressure chart and read the corresponding R-name. For example, assume that refrigerant "X" registers 65°F and the gauge pressure measures 111 psig. Using the temperature/pressure chart shown in *Figure 16*, you would find that the refrigerant is R-22 (HCFC-22).

TEMPERATURE DEGREES °F	REFRIGERANT			
	R–12	R–22	R–500	R–502
50	46.7	84.0	57.6	97.4
55	52.0	92.6	63.9	106.6
60	57.7	101.6	70.6	116.4
65	63.8	111.2	77.8	126.7
70	70.2	121.4	85.4	137.6
75	77.0	132.2	93.5	149.1
80	84.2	143.6	102.0	161.2
85	91.8	155.7	111.0	174.0
90	99.8	168.4	120.6	187.4
95	108.3	181.8	130.6	201.4
100	117.2	195.9	141.2	216.6
105	126.6	210.8	152.4	
110	136.4			
115	146.8			

H108F16.EPS

Figure 16.   Using A Temperature/Pressure Chart To Find A Refrigerant Type

4.6.0   REFRIGERANT SAFETY PRECAUTIONS

Gloves and safety glasses must be worn to avoid getting refrigerant on your skin or in your eyes. When accidentally released to the atmosphere, refrigerant can cause frostbite or burn the skin.

Refrigerants can cause suffocation if the amount and time of exposure is great enough. Always maintain ample ventilation. Refrigerant vapor is invisible, has little or no odor, and is heavier than air. Be especially careful of low places where it might accumulate.

Equipment rooms or other areas with large machines holding large amounts of refrigerant should have alarm systems that detect small amounts of leakage and sound an alarm. When exposed to an open flame or a hot surface, refrigerant toxicity increases greatly. A self-contained respirator must be available outside equipment rooms or other areas containing large equipment. Use the respirator if leakage occurs and you must enter the contaminated area. Some equipment rooms have a mechanical ventilation system that can be used to clear contaminated air from the room.

HVAC TRAINEE TASK MODULE 03108

In addition to the above precautions, follow these rules when handling and using refrigerants:

- Always double check to be sure you are using the proper refrigerant. The containers are color-coded and labeled to identify their contents. Container labels also include product, safety, and warning information.
- Refer to technical bulletins and Material Safety Data Sheets (MSDS) available from the manufacturers for information important to your health. They describe the flammability, toxicity, reactance, and health problems that could be caused by a particular refrigerant if spilled or incorrectly used.
- Do not drop, dent or abuse refrigerant containers. Do not tamper with safety devices.
- Always use a proper valve wrench to open and close the valve.
- Replace the valve cap and hood cap to protect the cylinder valve when not in use or empty.
- Secure containers in place to prevent them from becoming damaged when moved (especially in a van or truck). Strap or chain containers in an upright position.
- Do not store containers where the temperature can exceed the cylinder relief valve settings.

5.0.0 COMPRESSORS

The compressor is the keystone of the refrigeration system. It creates the pressure difference that causes refrigerant flow around the system. In the process, it takes refrigerant vapor at a low temperature and pressure and raises the vapor to a higher temperature and pressure.

Compressors are usually driven by an electric motor. Very large compressors can be driven by internal combustion engines or steam turbines. Compressors are divided into three groups based on the way they are joined to their motors or engines (*Figure 17*).

- Open compressor — The compressor is separate from its motor. One end (the shaft) extends outside the case. A mechanical seal is used with the rotating shaft to prevent leakage of the refrigerant. The compressor motor drives the compressor using a belt (belt drive) or flexible coupling (direct drive). Belt-driven arrangements allow the motor to run at one speed, while the compressor can run at another. The proper combination of pulleys (also called drives) produces the desired speed of the compressor. Most direct-drive systems use an electric motor to drive the compressor. This means that the compressor also runs at the speed of the drive motor.
- Hermetic (welded hermetic) compressor — The compressor and motor have a common drive shaft. They are sealed in a welded steel enclosure or shell. Hermetic compressors are more compact, less noisy, and require less maintenance than open-type compressors because they have no belts or couplings to break or wear out. Because they are sealed, the entire unit must be replaced when they fail.

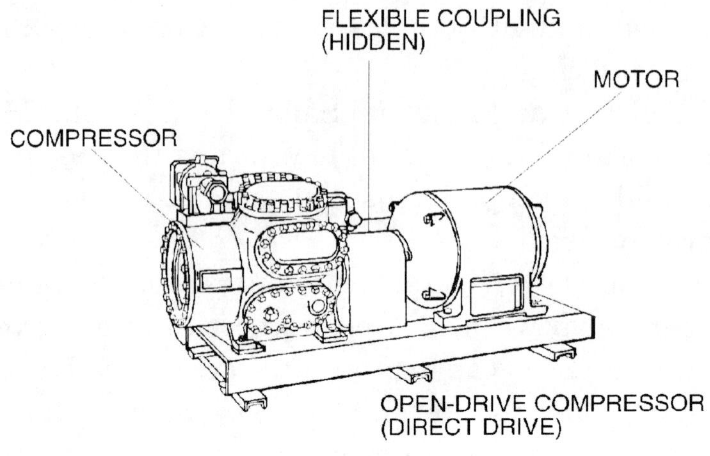

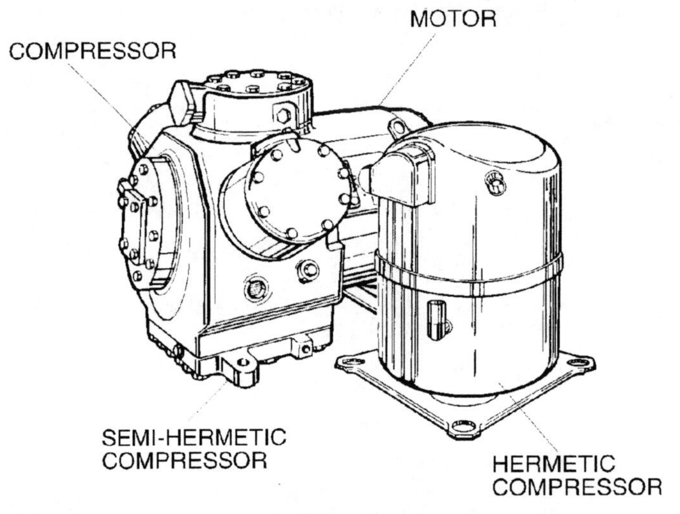

Figure 17. Compressors

- Semi-hermetic (serviceable hermetic) compressor — Similar to the hermetic compressor, the compressor and motor share the same housing and a common drive shaft. When they fail, access to the compressor or motor for repair is possible by removing the heads and/or the bottom and end plates.

Five types of compressors are commonly used in mechanical refrigeration systems:

- Reciprocating
- Rotary
- Scroll
- Screw
- Centrifugal

5.1.0 RECIPROCATING COMPRESSORS

Reciprocating compressors are the most common type. They use one or more pistons moving back and forth within a cylinder or cylinders (*Figure 18*). Piston movement is synchronized with the opening and closing of suction and discharge valves. These valves control the intake and discharge of the refrigerant. Reciprocating compressors are typically used in refrigerators, air conditioners, and commercial processing equipment. Welded hermetic compressors are most popular below 10 tons, but their use is increasing in the 10 to 20 ton range. Serviceable semi-hermetic compressors are used in commercial air conditioning and heat pumps above 10 tons. Open reciprocating compressors are used mostly for refrigeration work and on industrial and large commercial air conditioning and heat pumps anywhere in the 5 to 150 ton range.

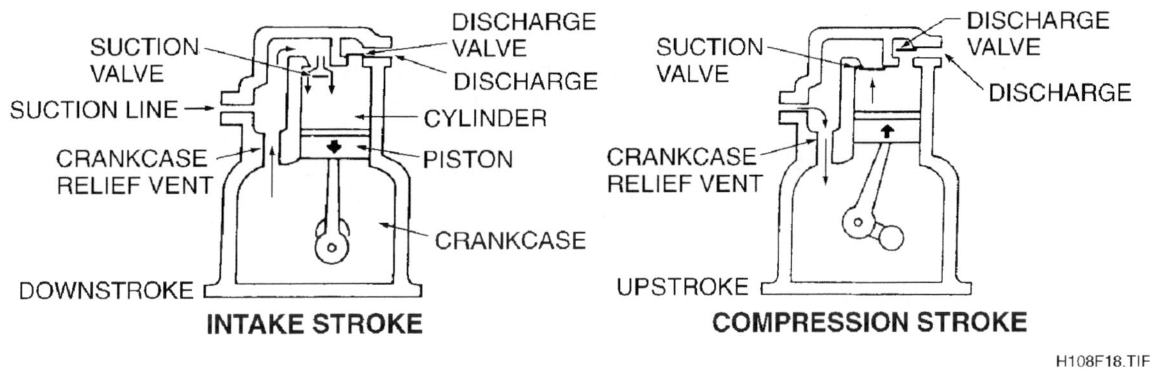

Figure 18.   Reciprocating Compressor

5.2.0 ROTARY COMPRESSORS

Rotary compressors are usually welded hermetic compressors. They are frequently used on appliances, room air conditioners, and central air conditioning below 5 tons. Rotary compressors are of two types: stationary vane and rotary vane.

In the stationary vane compressor (*Figure 19*), a shaft with an attached off-center (eccentric) rotor rotates or "rolls" around the cylinder. A stationary vane mounted in the compressor housing slides in and out and follows the rotating motion of the rotor as it moves within the cylinder. This vane also separates the suction and discharge sides of the cylinder. As the shaft turns, the rotor rolls around the cylinder, drawing suction gas in the intake opening while at the same time compressing the gas against the cylinder wall on the discharge or compression side. A valve at the discharge keeps the compressed gas from leaking back into the cylinder and into the suction side during the off cycle. This process continues as long as the compressor is running.

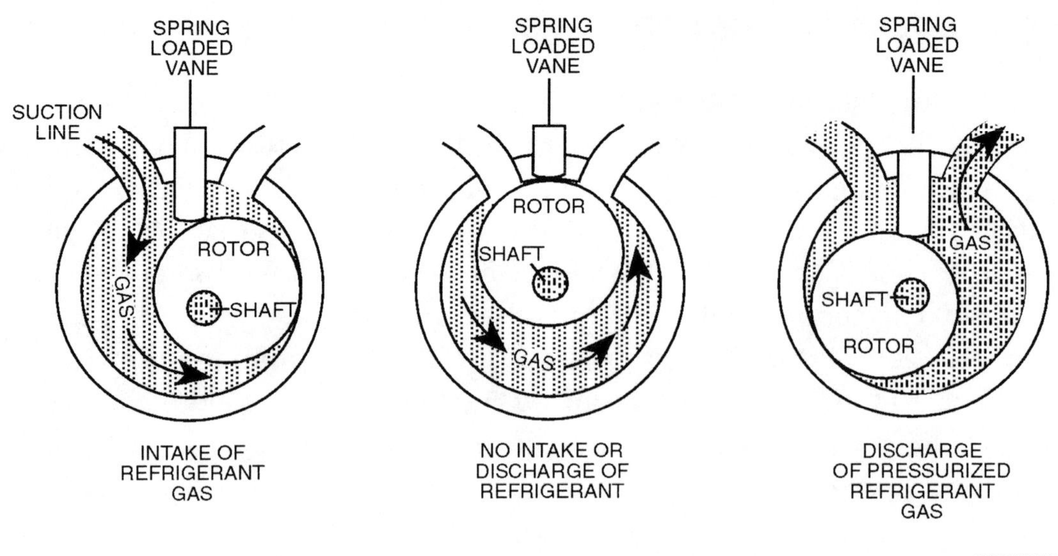

Figure 19. Stationary Vane Rotary Compressor

Rotary vane compressors have a rotor centered on the drive shaft. However, the drive shaft is positioned off-center in the cylinder. Mounted on the rotor are two or more vanes that slide in and out to follow the shape of the cylinder. As the rotor turns, these vanes trap low-pressure suction gas and compress it against the cylinder wall, then force it out the discharge opening. The vanes also keep the compressed gas from mixing with the incoming low-pressure gas.

5.3.0 SCROLL COMPRESSORS

Scroll compressors usually are welded hermetic compressors. Of all the compressor types, the scroll compressors have the fewest working parts. They operate efficiently even in applications that have large changes in refrigerant pressures such as with commercial refrigeration and heat pumps.

No suction or discharge valves are used in a scroll compressor (*Figure 20*). It achieves compression by the use of two spiral-shaped parts called scrolls. One is fixed; the other is driven and moves in an orbiting action inside the fixed one. There is contact between the two. Refrigerant gas enters the suction port at the outer edge of the scroll and after compression is squeezed out a separate discharge port at the center of the stationary scroll. The orbiting action draws gas into pockets between the two spirals. As this action continues, the gas opening is sealed off and the gas is compressed and forced into smaller pockets as it progresses toward the center.

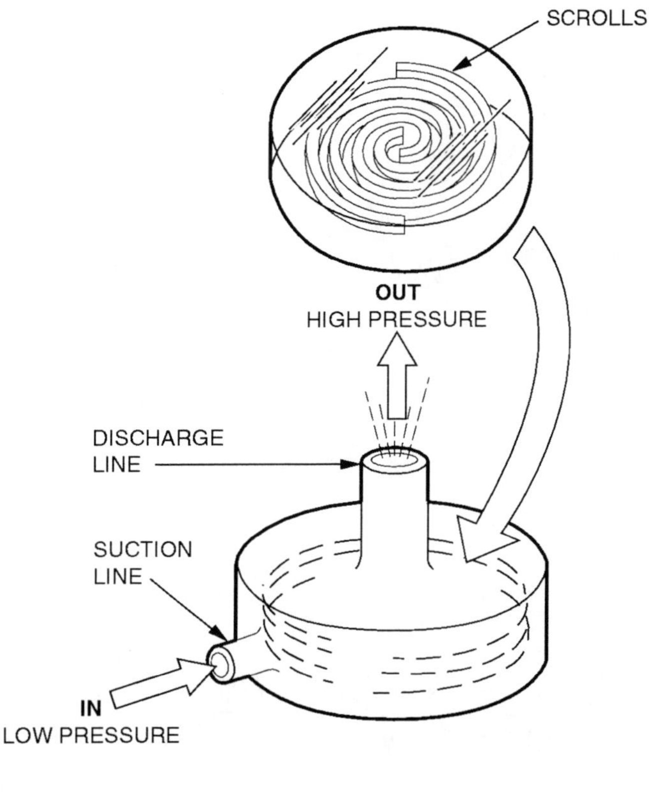

Figure 20.   Scroll Compressor

5.4.0   SCREW COMPRESSORS

Screw compressors are used in large commercial and industrial applications requiring capacities from 20 to 750 tons. They are made in both open and hermetic styles.

Screw compressors use a matched set of screw-shaped rotors, one male and one female, enclosed within a cylinder (*Figure 21*). The male rotor is driven by the compressor motor. In turn, it drives the female rotor. Normally the driven male rotor turns faster than the female rotor because it has fewer lobes than the female rotor. Typically, the male has four lobes and the female has six. As these rotors turn, they mesh with each other and compress the gas between them. The screw threads form the boundaries separating several compression chambers which move down the compressor at the same time. In this way, the gas entering the compressor is moved through a series of progressively smaller compression stages until the gas exits at the compressor discharge in its fully compressed state.

5.5.0   CENTRIFUGAL COMPRESSORS

Centrifugal compressors are made in open and hermetic designs. They are typically used on commercial and industrial refrigeration and air conditioning systems with capacities larger than 100 tons. Standard models range up to 10,000 tons of capacity with custom models exceeding 20,000 tons.

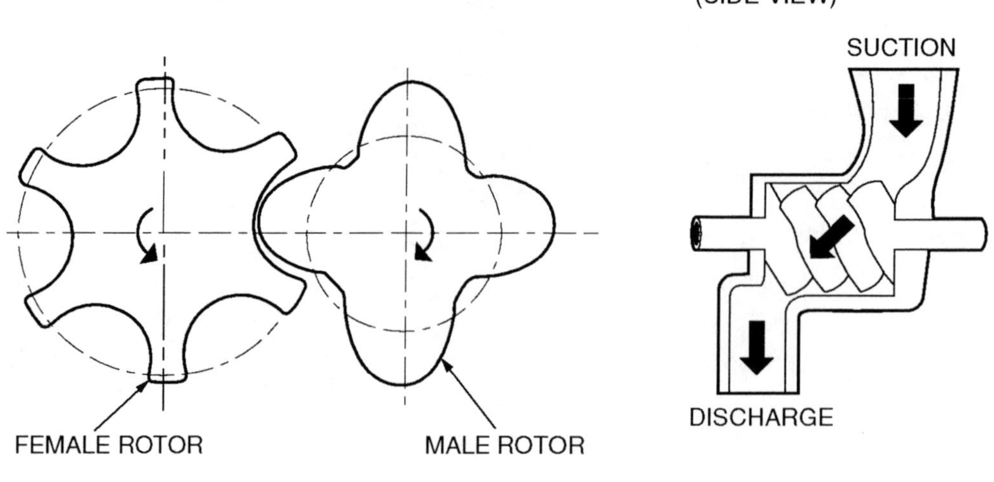

Figure 21.   Screw Compressor

Centrifugal compressors use a high-speed impeller with many blades that rotate in a spiral-shaped housing (*Figure 22*). The impeller is driven at high speeds (typically 10,000 RPM) inside the compressor housing. Refrigerant vapor is fed into the housing at the center of the impeller. The impeller throws this incoming vapor in a circular path outward from between the blades and into the compressor housing. This action, called centrifugal force, creates pressure on the high velocity gas and forces it out the discharge port. Often, several impellers are put in series to create a greater pressure difference and to pump a sufficient volume of vapor. A compressor that uses one impeller is called a single stage, one that uses two impellers is called a double stage, and so on. When more than one stage is used, the discharge from the first stage is fed into the inlet of the next stage.

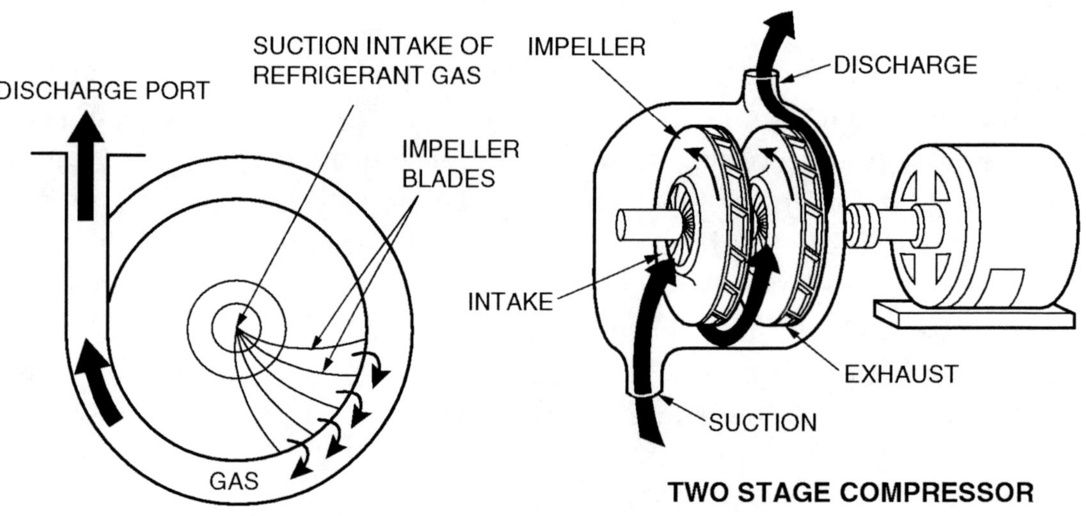

Figure 22.   Centrifugal Compressor

6.0.0 CONDENSERS

Condensers (*Figure 23*) are used for removing heat from the refrigeration system. They take in high-pressure, high-temperature refrigerant gas from the compressor and change it into a high-temperature, high-pressure liquid. They do this by transferring the heat from the refrigerant to the air, to water, or both. As the refrigerant flow progresses through the condenser, it first rejects the superheat and then fully condenses into a subcooled, high-temperature, high-pressure liquid. For the condenser to operate properly, the condensing medium of air or water must always be at a lower temperature than the refrigerant it is condensing. Condensers are grouped according to the medium used to carry the heat away from the refrigerant vapor. These groups are:

- Air-cooled
- Water-cooled
- Evaporative (a combination of air and water cooling)

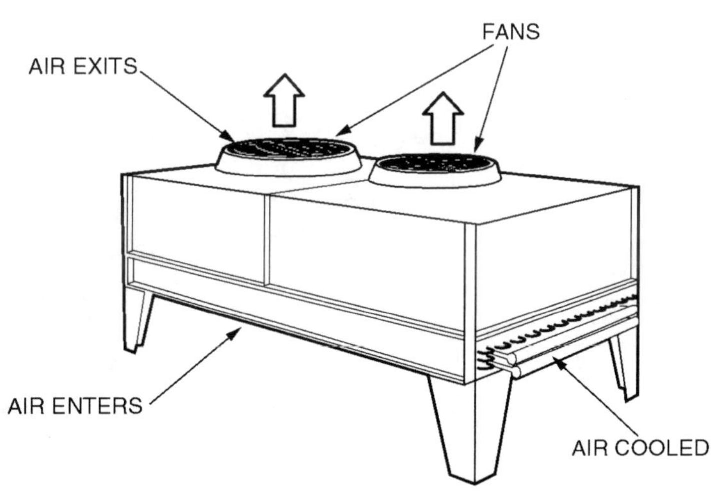

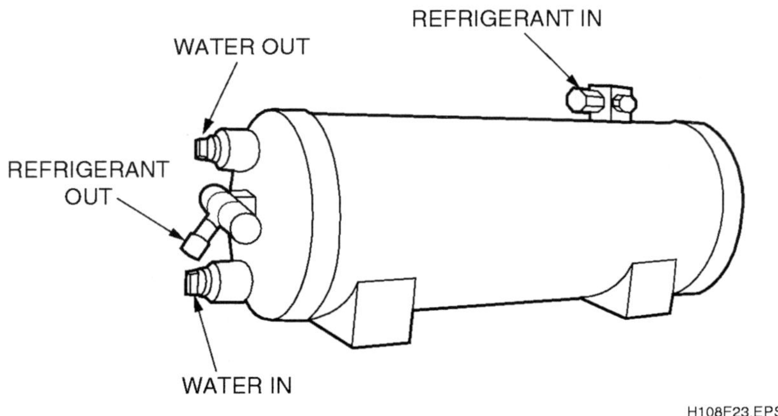

Figure 23. Condensers

6.1.0   AIR-COOLED CONDENSERS

Air-cooled condensers reject the heat absorbed by the system directly to the outdoor air. At normal design (peak load) conditions, the refrigerant flowing through the condenser is about 25°F to 35°F warmer than the outside air to which it is rejected. This means that a saturation temperature of 120°F to 130°F is typical in the condenser when the outside air is 95°F. Because the medium is outside air, this temperature tends to be greater than is needed for condensers used in water-cooled systems.

Propeller (axial) fans are used with most air-cooled condenser units to increase the amount of air being circulated across the condenser. This increases its capacity to reject heat. Because air-cooled condensers require the circulation of air over their surfaces, their location and the temperature of the surrounding air are very important to proper operation. It is important to note that the higher the temperature of the condensing air, the more work the system compressor must do to provide refrigerant vapor at an even higher temperature so heat can then be transferred from the gas out of the condenser. This causes the compressor to use more power. Air-cooled condensers are typically used in residential air conditioning up to about 5 tons and in commercial air conditioning up to about 50 tons. Air-cooled condensers are generally of two types: fin and tube and plate condensers (*Figure 24*).

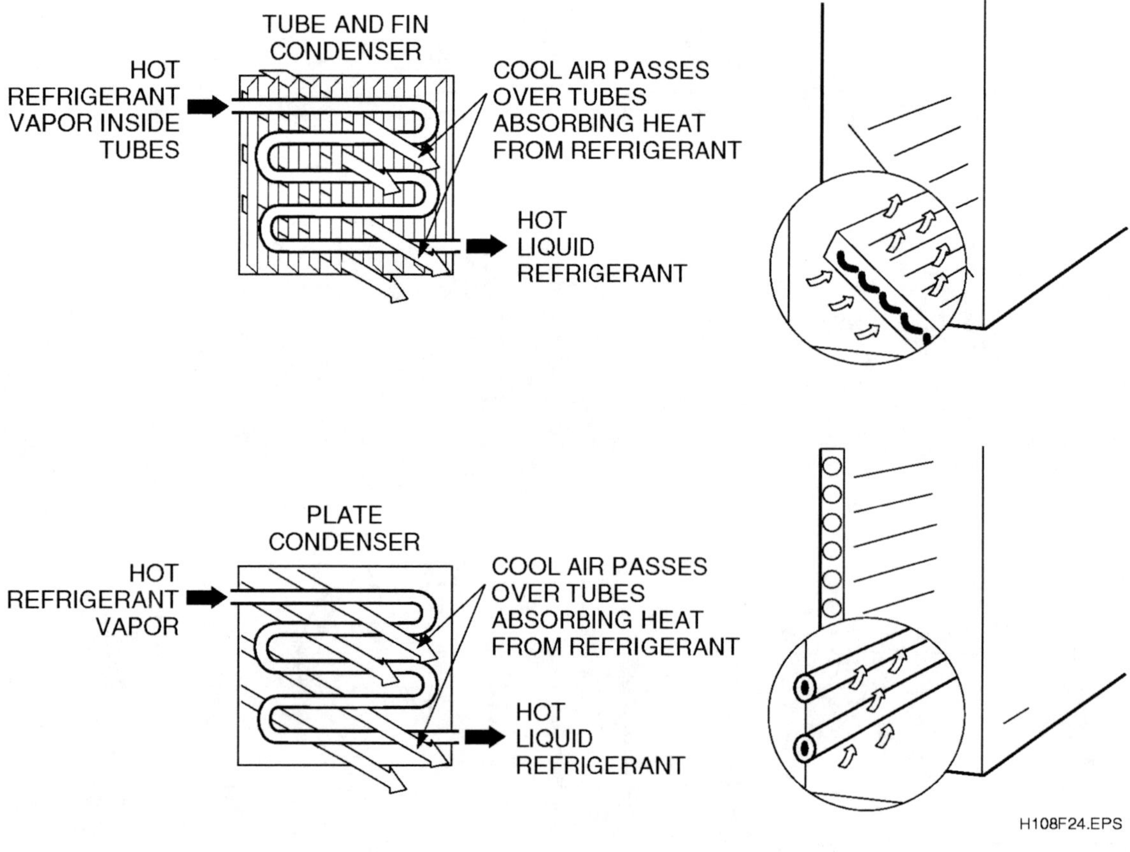

Figure 24.   Air-Cooled Condensers

6.1.1   Fin And Tube Condensers

In the fin and tube condenser, the refrigerant vapor passes through the rows of tubing. The tubing is encased in metal ribs or fins. These provide increased exposure to the surrounding air. Cooler air passing over the fins and tubing absorbs heat from the refrigerant.

6.1.2   Plate Condensers

The plate condenser operates the same as the fin and tube condenser. Basically, it is formed by two sheets of metal that have been pressed or stamped into the correct shape and then welded together. The plates provide a larger surface area for the transfer of the heat to surrounding air.

6.2.0   WATER-COOLED CONDENSERS

Water-cooled condensers are more complicated, more expensive, and require more maintenance than air-cooled condensers. However, they are more efficient and operate at much lower condensing temperatures (about 15°F lower). This allows the system compressor to run at lower head pressures, requiring the use of less power. Depending on the type, the velocity of the water flowing in a water-cooled condenser should be between three and ten feet per second. If it flows too fast, pitting of the tubing may occur. If it flows too slow, scaling will occur. In areas where water is plentiful, the water that flows through the condenser may be used once and then drained into a waste system. Most often, the water portion of a water-cooled condenser is connected via piping to a cooling tower, which is usually located on the roof of the building. In the cooling tower, the heat absorbed by the water in the condenser is rejected from the system into the atmosphere by evaporation. The cooled water is then returned to the system for reuse. There are three types of water-cooled condensers, as shown in *Figure 25*:

- Tube in tube
- Shell and tube
- Shell and coil

6.2.1   Tube In Tube Condensers

In the tube in tube condenser, the water flows through one or more curved tubes. The high-temperature, high-pressure refrigerant supplied from the compressor flows in the opposite direction. In the condenser, the hot refrigerant gives up heat to the cooler water and condenses into a subcooled liquid.

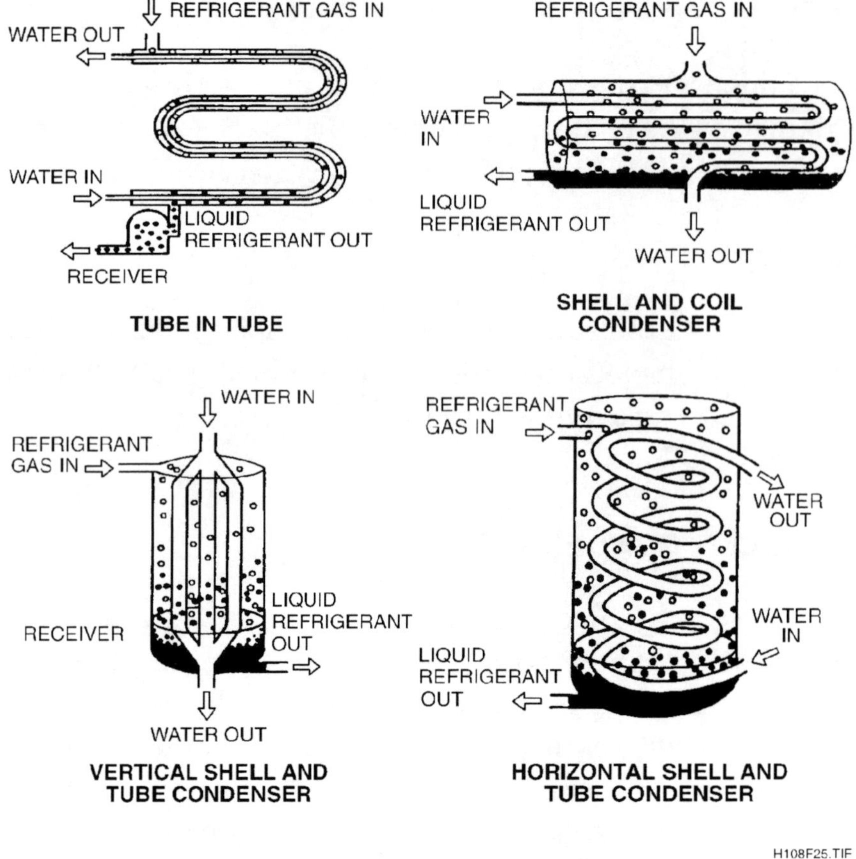

Figure 25.   Water-Cooled Condensers

6.2.2   Shell And Tube Condensers

Shell and tube condensers may be either horizontal or vertical mounted. Vertical condensers contain straight, vertical tubes encased in a metal shell. High-temperature, high-pressure refrigerant supplied from the compressor enters the top of the condenser metal shell containing the tubes. The water also enters at the top and travels down through the vertical tubes where it absorbs heat from the surrounding refrigerant, then leaves at the bottom. As the refrigerant gas condenses on the cooler water tubes, liquid refrigerant falls to the bottom of the condenser metal shell where it collects and leaves the condenser at the output. The horizontal condenser allows water flowing in the tubing to make several passes before leaving the condenser. This and the use of fins make for greater cooling efficiency since the refrigerant is exposed to more than one column of moving water.

6.2.3   Shell And Coil Condensers

Construction of the shell and coil condenser is similar to that of the shell and tube condenser, except that the tubing is wound around inside the shell rather than being a straight length. As with other water-cooled condensers, the water flowing through the tubing cools the refrigerant surrounding it. As the refrigerant gas condenses on the cooler water tubes, liquid refrigerant falls to the bottom of the condenser metal shell where it collects and leaves the condenser at the output.

   HVAC TRAINEE TASK MODULE 03108

6.2.4 Cooling Towers

In a cooling tower condenser, water containing heat from the system is exposed to the outside air, which absorbs the heat from the water. There are many types of cooling towers used with water-cooled condensers. One popular type is the natural-draft tower (*Figure 26*). Built to system capacity, these towers are mounted outdoors, usually on the roof to make use of natural air currents. They are made of a metal frame covering several layers or tiers of wooden decks. Since they use the natural air currents, no blowers are needed to move air through the tower. Water is piped up from the condenser located in the building below and is discharged in sprays over the decks. Spaces between the boards in the decks permit the water to drip or run from deck to deck, while being spread out and exposed to air breezes that enter the tower from the open sides. The cooled water is collected in a catch basin at the bottom of the tower where it is pumped back to the condenser for reuse.

Because cooling towers work partly on evaporation, any water lost due to evaporation must be replaced in order to maintain the system. This is done using a float that senses the water level in the catch basin and adds water as needed. Mechanical fans may be used to push and increase the air speed in a cooling tower. If used, the tower is called a forced-draft tower. As a result of using fans, forced-draft towers tend to be small in comparison to natural-draft towers. Another tower, called an induced-draft tower, is similar to a forced-draft tower. In it, the fans pull the air rather than push it across the wet deck surfaces.

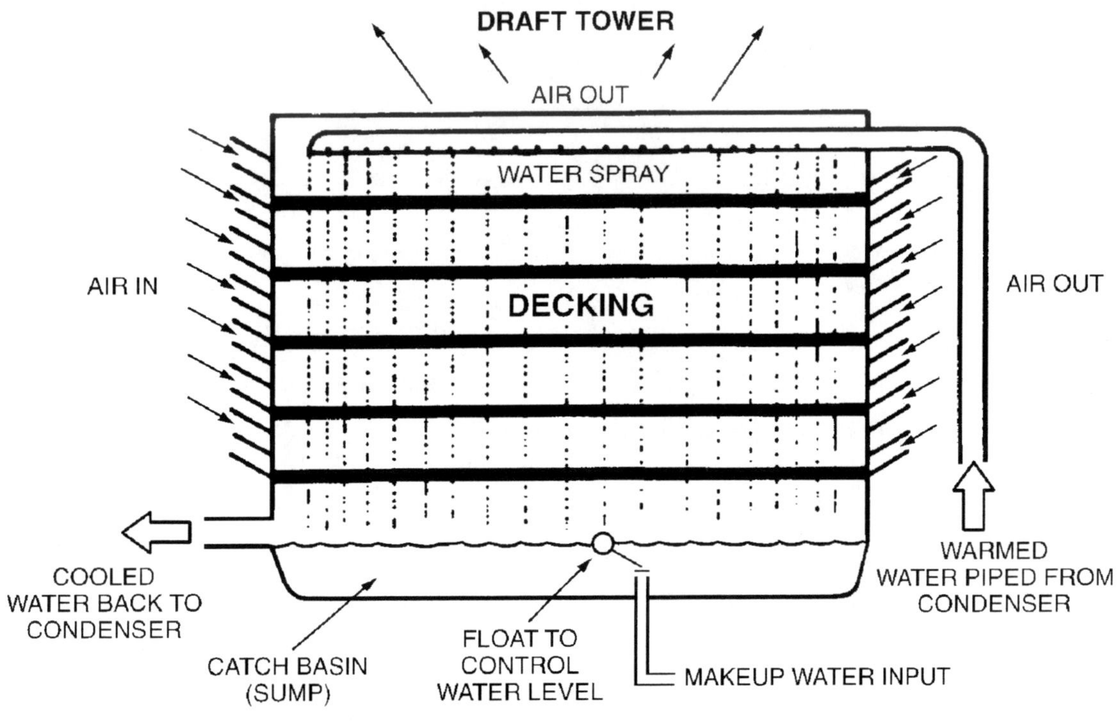

Figure 26. Natural-Draft Cooling Tower

6.3.0 EVAPORATIVE CONDENSERS

Evaporative condensers first transfer heat to water, and then from water to the outdoor air. They combine the functions of a water-cooled condenser and cooling tower in one package. The condenser water evaporates directly off the tubes of the condenser. Each pound of water that is evaporated removes about 1000 Btu's from the refrigerant flowing through the tubes.

Air enters the bottom of the unit (*Figure 27*) and flows by convection upward over the condensing coil filled with refrigerant. Simultaneously, water is sprayed over the coil. Both the air and the water absorb heat from the refrigerant in the coil. Water eliminators located above the water spray remove water from the rising air. The air is then moved out the top of the unit using one or more fans. Cooled by both air and water, the refrigerant in the coil condenses into a subcooled liquid at the output of the coil.

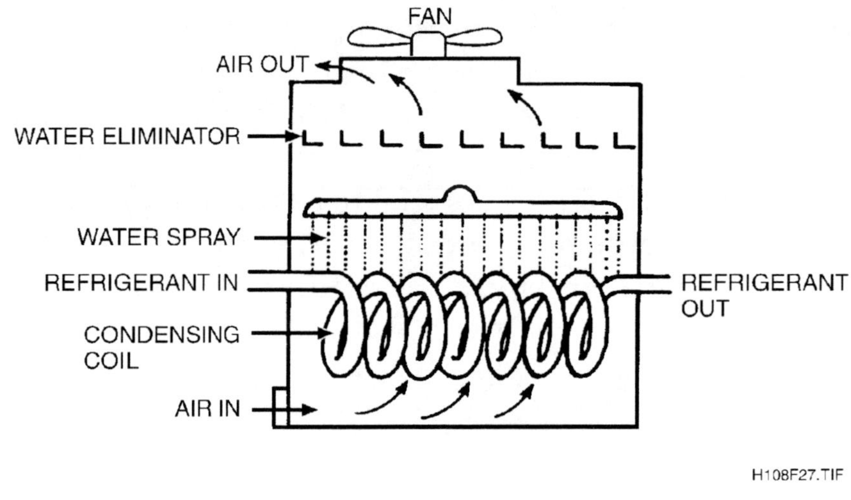

Figure 27. Evaporative Condenser

7.0.0 EVAPORATORS

Evaporators (*Figure 28*) are used to add heat to the refrigeration system. They take in low-temperature, low-pressure liquid refrigerant from the expansion device and change it into a low-temperature, low-pressure gas. This is done by transferring heat from either air or water to the refrigerant. As the refrigerant flow progresses through the evaporator, the heat in the warmer medium (air or water) causes it to boil (evaporate) and change into a vapor. When this occurs, the refrigerant absorbs heat from the medium being cooled. The amount of heat absorbed depends on how much heat is lost by the medium. The heat gain must equal the heat loss. For example, if the air passing over an evaporator gives up 800 Btu's of heat, then the refrigerant in the evaporator must gain 800 Btu's. Evaporators are of two types: the direct-expansion type and the flooded type (*Figure 29*).

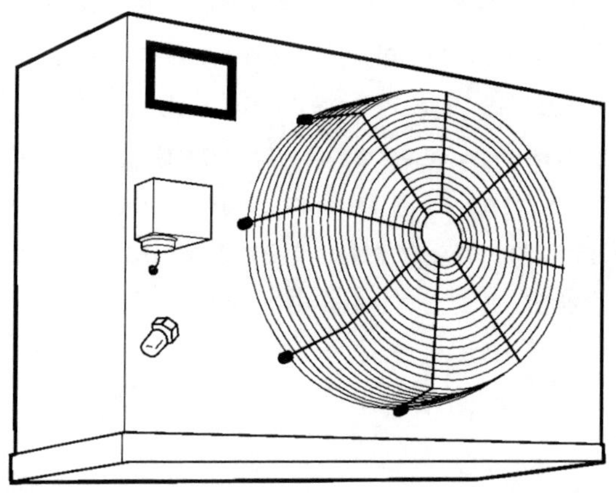

Figure 28.   Forced-Draft Evaporator

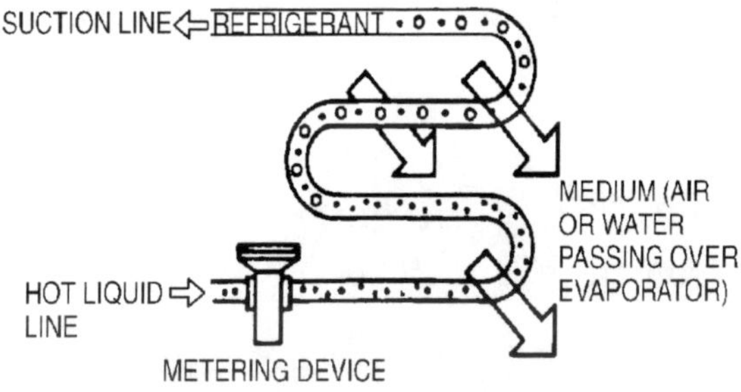

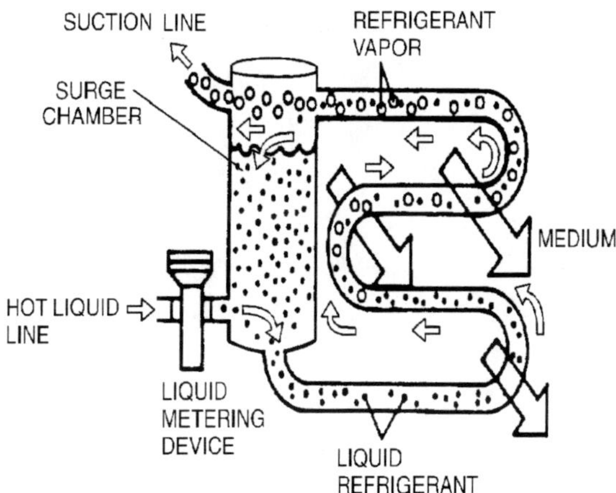

Figure 29.   Direct Expansion And Flooded Evaporators

7.1.0 DIRECT EXPANSION (DX) EVAPORATORS

Direct expansion (DX) evaporators are the most widely used. DX evaporators have one continuous tube, or coil, through which the liquid refrigerant flows. The refrigerant, with a small amount of gas mixed in, enters the evaporator and is gradually warmed by the medium until it boils and becomes a vapor near the outlet. The flow of refrigerant into the DX coil is controlled by the expansion device at its input. It supplies just the right amount of refrigerant to the evaporator so that it is all transformed into vapor by the time it reaches the evaporator output. (Actually, the metering devices used with most DX evaporators are designed to produce about a 10°F superheat of the refrigerant vapor at the evaporator output.) Movement of the medium (air or water) over the evaporator can be by natural draft, or it can be enhanced using fans or pumps (forced draft).

7.2.0 FLOODED EVAPORATORS

In a flooded evaporator, refrigerant can be circulated through the evaporator more than once (*Figure 29*). A special receptacle known as the "surge chamber" is connected between the evaporator tubing input and output. Liquid refrigerant enters the surge chamber from the metering device, then flows through the evaporator coil where it boils and then returns to the surge chamber. All of the refrigerant vapor exits through the suction line for input to the compressor. Any refrigerant not changed into a vapor collects in the surge chamber for recycling back through the evaporator. Flooded evaporators are used mainly for cold storage or with systems using ammonia. They are generally not used for air conditioning because they require much more refrigerant than the DX evaporator.

7.3.0 EVAPORATOR CONSTRUCTION

Evaporators are classified by the way they are constructed. These groups are:

- Bare-tube
- Finned-tube
- Plate-surface

7.3.1 Bare-Tube Evaporators

Bare-tube evaporators (*Figure 30*) are of two types: single circuit (path) or multiple path. Multiple-path evaporators are often used because they save space and reduce the number of metering devices needed. Each is simply a steel or copper pipe shaped in a way that best matches the job. The piping is the only surface used to transfer heat; therefore, these are often called "prime surface" coils.

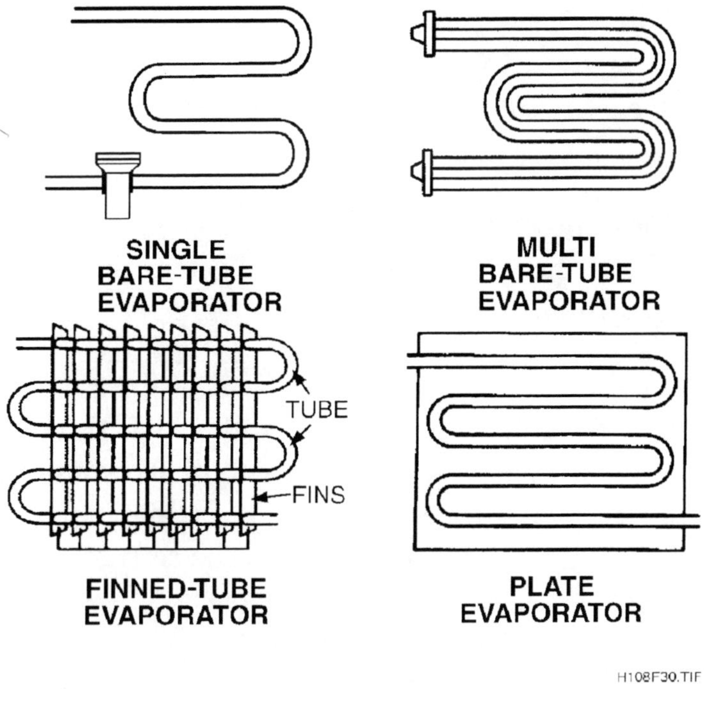

Figure 30.   Evaporator Construction Methods

7.3.2   Finned-Tube Evaporators

Finned-tube evaporators are a variation of the bare-tube evaporator. Attached to the tubing are thin spiral-wound or rectangular fins of aluminum or copper like those used with the fin and tube condenser. These fins increase the amount of surface area exposed to the heated medium. This in turn increases the amount of heat that can be transferred to the refrigerant.

7.3.3   Plate-Surface Evaporators

Plate evaporators are similar to plate condensers. They have a length of tubing weaving through a metal plate. The plate provides greater surface area for heat transfer. This type of evaporator is typically used as a shelf in older upright freezers.

7.3.4   Chilled Water System Evaporators

In many large refrigeration and air conditioning installations, cooling coils are installed at some distance apart in the building or complex. Because of the expense and the problems involved with long runs of refrigerant piping, cooling in these remote areas is done with chilled water rather than refrigerant as the medium. This chilled water is called a "secondary" refrigerant. When temperatures are below freezing, brine is used in place of water. A refrigeration system, called the "primary" system, is used to cool the secondary water or brine refrigerant. The primary system generally uses shell and tube or shell and coil evaporators called "chillers" to absorb heat from the water or brine. Like other evaporators, chillers can be of the direct expansion or flooded type. Their construction is similar to that used for shell and tube or shell and coil condensers.

In the direct expansion chiller, the colder liquid refrigerant runs through the tubing while the warmer secondary refrigerant (water or brine) circulates within the shell around the outside of the tubes. As a result, heat contained in the water or brine is transferred to the primary refrigerant flowing through the tubes. The refrigerant leaves the evaporator as a gas.

In the flooded-type chiller, the water or brine is circulated through the tubing, which is located on the bottom of the evaporator shell. This tubing is submerged below the liquid refrigerant level within the shell which is controlled by a float valve. The evaporator shell acts as a surge chamber. The top portion is left vacant so the refrigerant vapor can be properly separated from the liquid refrigerant for output to the suction line as it evaporates.

8.0.0 EXPANSION (METERING) DEVICES

The metering device is located between the condenser outlet and the evaporator inlet. High-pressure, high-temperature liquid refrigerant from the condenser enters the metering device. It leaves as a low-pressure, low-temperature mixture of liquid and vapor. Regardless of type, the metering device performs two functions:

*   It allows the liquid refrigerant to flow into the evaporator at a rate that matches the rate at which the evaporator boils liquid refrigerant into a vapor.
*   It provides a pressure drop that lowers the boiling point of the refrigerant.

There are many types of metering devices. They can be divided into two categories: fixed and adjustable. Fixed metering devices are used mainly in domestic refrigerators and freezers and residential air conditioning units. Adjustable metering devices are used most often on systems with variable load requirements. This section briefly describes the main types of fixed and adjustable metering devices. You will study these devices in greater detail later in your training.

8.1.0 FIXED METERING DEVICES

Fixed metering devices have a fixed restriction or fixed opening (orifice) size. The capillary tube (*Figure 31*) is the simplest metering device. It is a fixed length, small-diameter copper tube, usually with an inside diameter of $^1/_{16}$ to $^1/_8$ inch. Because of its high resistance to refrigerant flow, it restricts the flow of liquid refrigerant from the condenser to the evaporator. The greater its length or the smaller its diameter, the greater the pressure drop. Capillary tubes are often coiled to conserve space and protect them from damage.

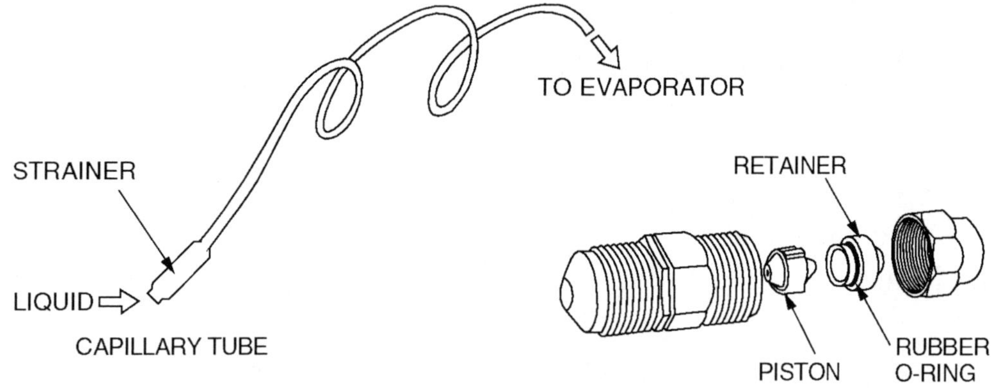

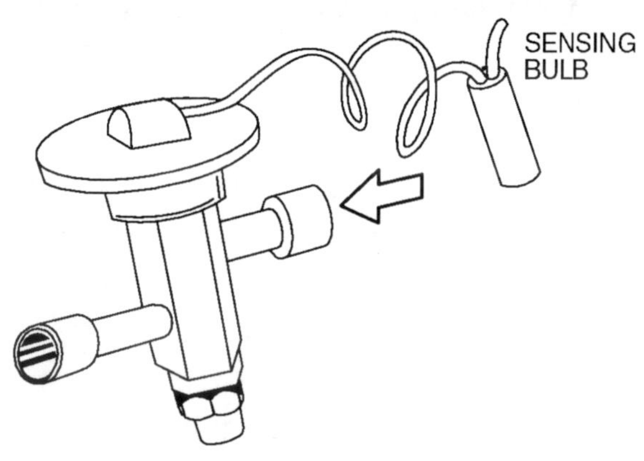

H108F31.EPS

Figure 31. Typical Metering Devices

Another type of fixed metering device similar to the capillary tube is the fixed orifice device. This device is a compact and rugged assembly that is installed at the evaporator inlet. It contains a piston. Pistons are made with different-sized orifices to match the capacities of different equipment. The smaller the orifice, the greater the pressure drop. Often, the piston is installed in this type of metering device at the time of system installation in order to match the metering device to the condensing unit.

8.2.0   ADJUSTABLE METERING DEVICES

Adjustable metering devices differ in their mechanisms and how they are controlled. They all work to regulate refrigerant flow so that the evaporator capacity matches the cooling load. There are six types of adjustable metering devices in common use:

- Hand-operated expansion valve
- Low-side float valve
- High-side float valve
- Automatic expansion valve
- Thermostatic expansion valve
- Electric and electronic expansion valves

8.2.1   Hand-Operated Expansion Valve

The hand-operated expansion valve is the simplest adjustable metering device. It is similar to a water shutoff valve and is adjusted manually to control refrigerant flow. The valve stem is built with fine threads and a needle plunger to allow for small adjustments.

8.2.2   Low-Side Float Valve

The low-side float valve (*Figure 32*) is installed in the low side of the refrigeration system. It is located inside the evaporator or just outside it in a connecting chamber. This metering device is used only with flooded evaporators. When the load on the evaporator increases and more liquid refrigerant boils away, the level of refrigerant in the evaporator and the float chamber drops. This causes the float to drop and the float needle to open so more liquid refrigerant from the condenser can enter the metering device. If the load on the evaporator decreases, less liquid is boiled away. This causes the float to rise and close the float needle, reducing the refrigerant flow.

8.2.3   High-Side Float Valve

The high-side float valve (*Figure 32*) is installed in the high side of the refrigeration system. It is a separate chamber located in the liquid line after the condenser. Within the float valve chamber, a float ball rests on the surface of the liquid refrigerant. The amount of liquid refrigerant leaving the chamber depends on, and is equal to, the condenser condensation rate. When the load on the evaporator increases, more vapor enters the condenser. As a result, more vapor is condensed into liquid by the condenser, thus raising the level of the liquid refrigerant in the chamber. As the liquid level climbs in the float chamber, the float ball rises. This causes the high-side valve to open, increasing the amount of liquid refrigerant supplied from the chamber to the evaporator. If the evaporator load decreases, the condenser in turn produces less liquid refrigerant. This causes the float ball to drop, thereby shutting off the valve and decreasing the amount of liquid refrigerant supplied to the evaporator.

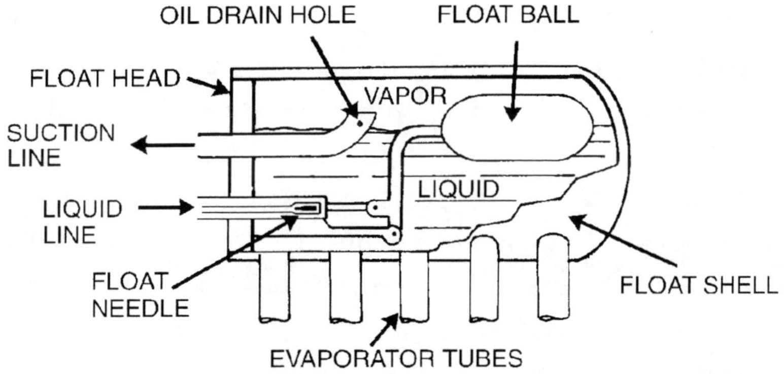

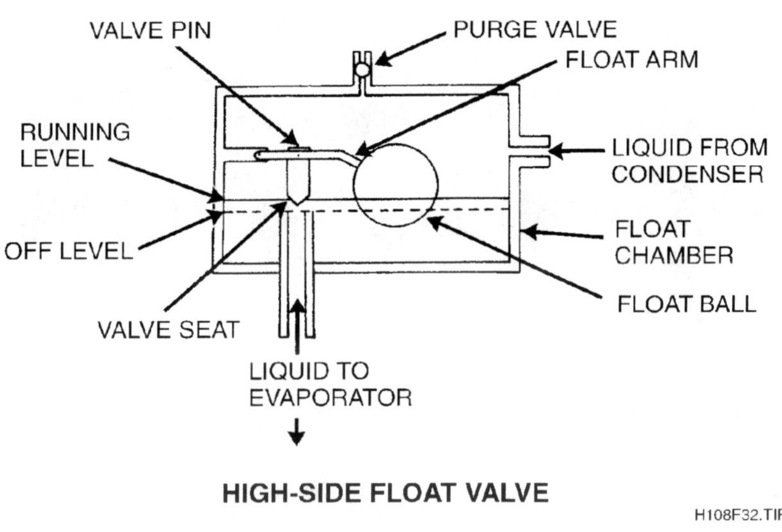

Figure 32. Low-Side And High-Side Float Valves

8.2.4 Automatic Expansion Valve

The automatic expansion valve (*Figure 33*) is a metering device that maintains a constant pressure in the evaporator. It uses evaporator pressure to close or open the valve. It consists of a diaphragm, which has a valve and seat below it and an adjustable pressure spring above it. Pressure from the refrigerant in the evaporator pushes up against the bottom of the diaphragm while the pressure from the adjustable spring pushes down against the top. Any pressure difference between the two causes the diaphragm to bend; this in turn moves the valve into or out of the seat. This increases or decreases the flow of refrigerant applied to the evaporator.

8.2.5 Thermostatic Expansion Valve

The thermostatic expansion valve or TXV controls the amount of refrigerant flowing through the evaporator by sensing the level of superheat in the suction line at the evaporator output. Like the automatic expansion valve, a TXV has a diaphragm with a valve and seat below it (*Figure 33*). The difference between the two is that the adjustment spring is below the diaphragm rather than above it. The pressure exerted on the underside of the diaphragm is a combination of evaporator and adjustment spring pressures. The spring is adjusted to exert a pressure that represents the desired level of evaporator superheat. Pressure on top of the diaphragm is applied via tubing from a remote sensing bulb used to monitor the superheat at the evaporator outlet. This bulb contains a refrigerant charge, or other suitable chemical charge, as the sensing fluid. If the superheat increases or decreases as a result of changing load, a corresponding pressure change is transmitted from the bulb to the valve. This pressure change counteracts the combined evaporator and adjustment spring pressures to increase or decrease the valve opening, allowing more or less refrigerant flow through the valve.

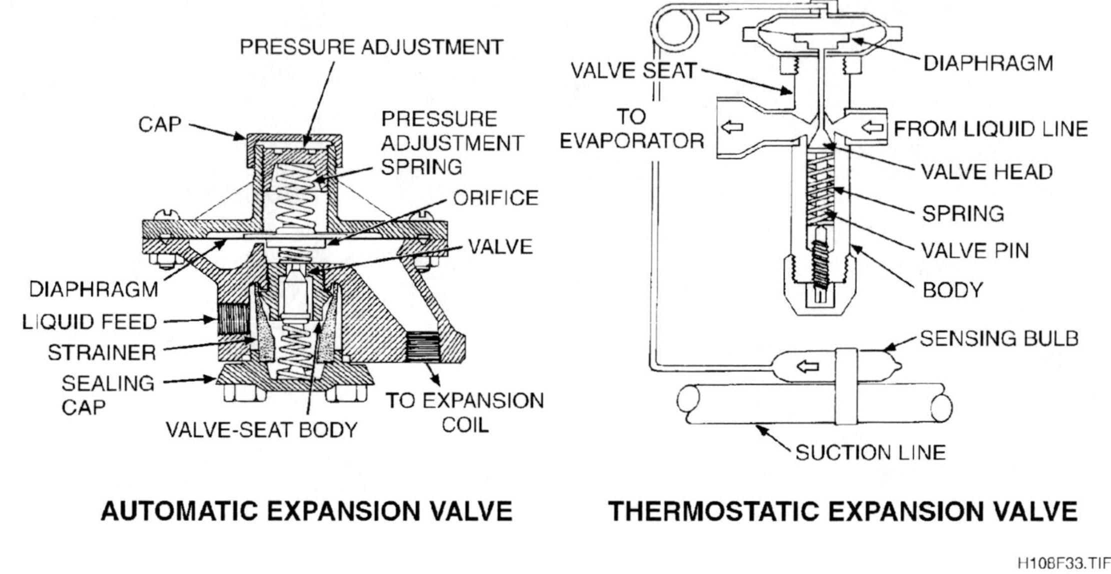

Figure 33.   Automatic And Thermostatic Expansion Valves

8.2.6   Electric and Electronic Expansion Valves

Electric and electronic alternatives to the thermostatic expansion valve are available and in common use for larger systems. Electronic expansion valves typically use microprocessor-controlled stepper motors to increase or decrease the flow of liquid refrigerant from the condenser to the evaporator. The stepper motor gets its control signal from an electronic control panel that is attached to a sensor that measures superheat. In response to the control signal, the stepper motor shaft moves an attached sleeve in and out of the valve body in tiny steps. This movement of the sleeve causes more or less refrigerant metering slots in the valve to be exposed, allowing more or less refrigerant to flow. Another electronically-controlled valve in use is the pulsating solenoid expansion valve. Using an electronic input signal, the solenoid valve pulses rapidly, opening and closing in response to changes in the cooling load.

Additional components (*Figure 34*) are usually added to the basic refrigeration system in order to improve safety, endurance, efficiency or servicing. Some of these components are factory installed, while others may be installed in the field. This section briefly describes the most commonly used components. They include:

- Filter-drier
- Sight glass/moisture liquid indicator
- Suction line accumulator
- Crankcase heater
- Oil separator
- Heat exchangers
- Receiver
- Service valves
- Compressor muffler

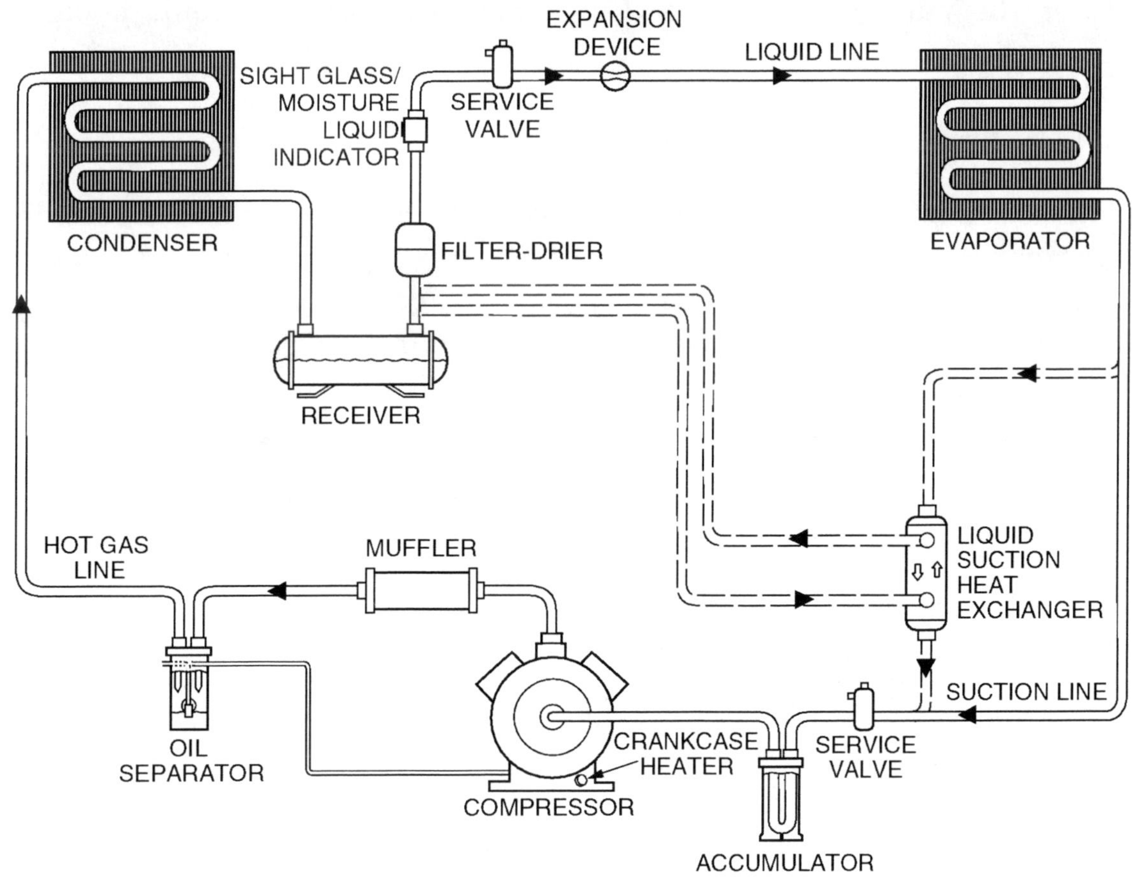

Figure 34.   Other Refrigeration System Components

9.1.0   FILTER-DRIER

The filter-drier or filter-strainer combines the functions of a refrigerant filter and a refrigerant drier in one device. The filter protects metering devices and the compressor from foreign matter such as dirt, scale or rust. The drier removes moisture from the system and traps it where it can do no harm. Filter-driers are normally installed in the liquid line in front of all metering devices. Filter-driers are replaced periodically during system maintenance or immediately after a system repair, such as a compressor burnout.

9.2.0   SIGHT GLASS AND MOISTURE LIQUID INDICATOR

The sight glass is like a window that allows the technician to view the condition of the system refrigerant. It is typically used when checking the refrigerant charge. The normal location for the sight glass is in the liquid line, as close to the condenser outlet or receiver as possible. A moisture liquid indicator is a sight glass with a small moisture-indicating device installed in it. This moisture indicator is exposed to the refrigerant and changes color depending on the amount of moisture in the refrigerant. When the moisture is within the limits set by the manufacturer, the indicator is one color. If too much moisture is present, the device will change color.

9.3.0   SUCTION LINE ACCUMULATOR

The suction line accumulator is a trap used to prevent the compressor from taking in slugs of liquid refrigerant or compressor oil. If liquid refrigerant is allowed to enter the compressor, noisy operation, high consumption, and compressor damage may result. Accumulators are installed in the suction line as near the compressor suction inlet as possible. At this location, any liquid refrigerant or oil will be trapped temporarily in the accumulator. The trapped refrigerant remains in the accumulator until it is evaporated back into a vapor. Some accumulators have heaters that help to vaporize refrigerant liquid. Trapped oil is usually piped back to the compressor.

9.4.0   CRANKCASE HEATER

Crankcase heaters are installed on compressors to prevent liquid refrigerant from migrating to the compressor and causing damage. These heaters work by evaporating refrigerant from the oil. They are usually fastened to the bottom of the crankcase or inserted directly into the compressor crankcase (immersion type). Wrap-around or "bellyband" heaters that encircle the outside shell of welded hermetic compressors are also used.

9.5.0   OIL SEPARATOR

Oil is used in a refrigeration system for three purposes:

1. It helps seal the system.

2. It dampens compressor noise.

3. It acts as a coolant for the compressor and compressor motor.

Because there is oil in the compressor, it mixes with the refrigerant and travels with it to other areas of the system. Oil separators minimize the amount of oil that circulates through the system. Oil coats the inside of every component through which it passes. It reduces the heat transfer ability and efficiency of the evaporator and condenser. Another reason for the oil separator is to slow down the accumulation of oil in places from which oil return is difficult.

Oil separators are seldom used on residential or commercial air conditioning systems. Their use is mainly in refrigeration and industrial systems. Typically, they are installed in the hot gas line as close to the compressor discharge as practical. Separators usually have a reservoir (sump) to collect the trapped oil. A float valve in the sump keeps a seal between the high-pressure and low-pressure sides of the system. This valve automatically returns the oil to the compressor through an orifice.

9.6.0   HEAT EXCHANGERS

Two types of heat exchangers can be used with refrigeration systems: the liquid-to-suction type and the refrigerant water pre-heater. The liquid-to-suction heat exchanger transfers some of the heat from the warm liquid refrigerant leaving the condenser to the cool suction gas leaving the evaporator. (See path shown as dashed lines on *Figure 34.*) This increases efficiency and helps sub-cool the liquid refrigerant. In some applications it is used to evaporate the small amount of liquid refrigerant expected to return from the evaporator in the suction line to the compressor. Operation of the heat exchanger is similar to that of a water-cooled condenser. The liquid refrigerant leaving the condenser and the cool suction gas leaving the evaporator flow in opposite directions through the heat exchanger. The amount of heat that can be exchanged between the gas and liquid is determined by the temperature difference between the two, the amount of surface area, and how much time there is for heat exchange to occur.

The refrigerant water pre-heater is used to pre-heat the water supplied at the input of a hot water heater. In this heat exchanger, heat is transferred from the compressor hot gas line to the water. This reduces energy consumption whenever heat needs to be rejected from the system and helps to de-superheat the discharge gas leaving the compressor. Instead of rejecting heat outdoors, the heat is transferred to the hot water system. Shell and coil and tube in tube type heat exchangers are typically used as water pre-heaters.

9.7.0   RECEIVER

The receiver is a tank or container used to store liquid refrigerant in the system. This storage is needed on some systems to accommodate changes during operation, to freely drain the condenser of refrigerant, and to provide a place to store the system charge during system service procedures or prolonged shutdowns. The receiver is installed in the liquid line between the condenser and the metering device. Receivers are used to store the excess refrigerant created by varying cooling loads in many systems that use self-adjusting metering devices. Note that some residential and commercial air conditioning equipment store this excess in the condenser. In these cases, the condenser is large enough to hold the excess while delivering less than peak capacity and subcooling.

9.8.0   SERVICE VALVES

Service valves enable the service technician to seal off parts of the system while installing gauges and to provide access to the system for servicing. These valves are manually operated and come in several types. Service valves can be installed in any line that may have to be valved off, usually at a place that is easily accessible to the service technician.

9.9.0   COMPRESSOR MUFFLER

Mufflers are used most often in systems with open or semi-hermetic reciprocating compressors. Reciprocating compressors generate sound that can be transmitted along the piping. A muffler installed in the discharge line, as near the compressor as practical, is used to remove or dampen these pulsations. The muffler lowers the system noise and prevents possible damage from vibration.

10.0.0   CONTROLS

Controls are the devices used to start, stop, regulate and/or protect the components of the mechanical refrigeration system. They can be divided into two groups: primary and secondary. Primary controls start or stop the refrigeration cycle either directly or indirectly as a result of sensing temperature, humidity, pressure, or time. Secondary controls regulate and/or protect the cycle and its components. You have already been introduced to many controls in *Basic Electricity*, Module 03107. This section will introduce you to other types of control devices. You will study these devices again in greater detail later in your training.

10.1.0   PRIMARY CONTROLS

Primary controls start or stop the refrigeration cycle either directly or indirectly as a result of sensing temperature, humidity, pressure, or time. Primary controls include:

*   Thermostat
*   Pressurestat
*   Humidistat
*   Time clock

10.1.1 Thermostats

All thermostats sense and respond to the temperature in a conditioned space. They switch the system on or off at a preset temperature called the "setpoint" by opening a set of contacts in the system control circuit. This may be done in several different ways. Some are activated by the warping of a bimetal strip (*Figure 35*). A bimetal device operates on the principle that different metals expand or contract at different rates when heated or cooled. A familiar form of the bimetal thermostat is the mercury bulb room thermostat. Its contacts are enclosed in an airtight glass bulb containing a small amount of mercury. Expansion and contraction of the bimetal element tilts the bulb in different directions. In one position, the mercury rolls to the low end of the bulb and closes or "makes" the electrical circuit; in the other, it rolls to the high end of the bulb and opens or "breaks" the electrical circuit. If the thermostat makes on a temperature rise and breaks on a temperature drop, it is a cooling thermostat. If it makes on a temperature drop and breaks on a temperature rise, it is a heating thermostat.

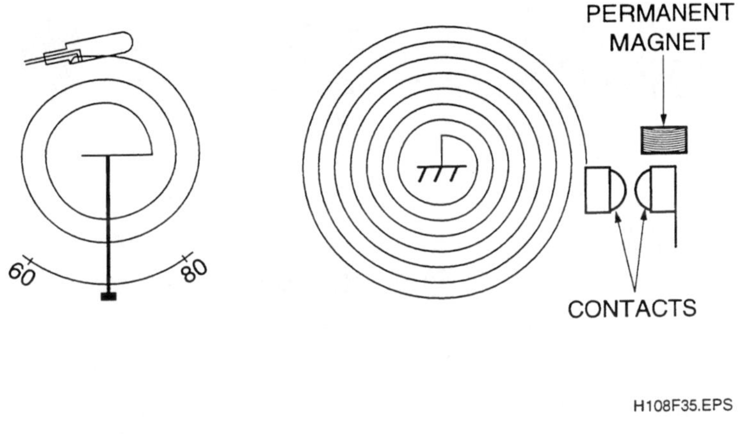

Figure 35. Bimetal Strip Used As A Thermostat

Another type of thermostat is activated by pressure applied from a bellows attached to a chemical-filled sensing bulb with a capillary tube (*Figure 36*). When filled with a refrigerant, the bulb pressure increases or decreases as the temperature rises or drops. An increase in bulb pressure causes the bellows to expand, mechanically closing or opening the electrical contacts, depending on whether it is a heating or cooling thermostat. Cooling thermostats close on a pressure rise while heating thermostats open on a pressure rise. The action is the opposite for a pressure drop. This type of thermostat is sometimes called a "remote bulb thermostat" because the bulb can be located at a different location than the rest of the thermostat. Another pressure-actuated type of thermostat is a diaphragm thermostat. In this type of thermostat, the bulb pressure moves a diaphragm rather than a bellows to open and close the electrical contacts.

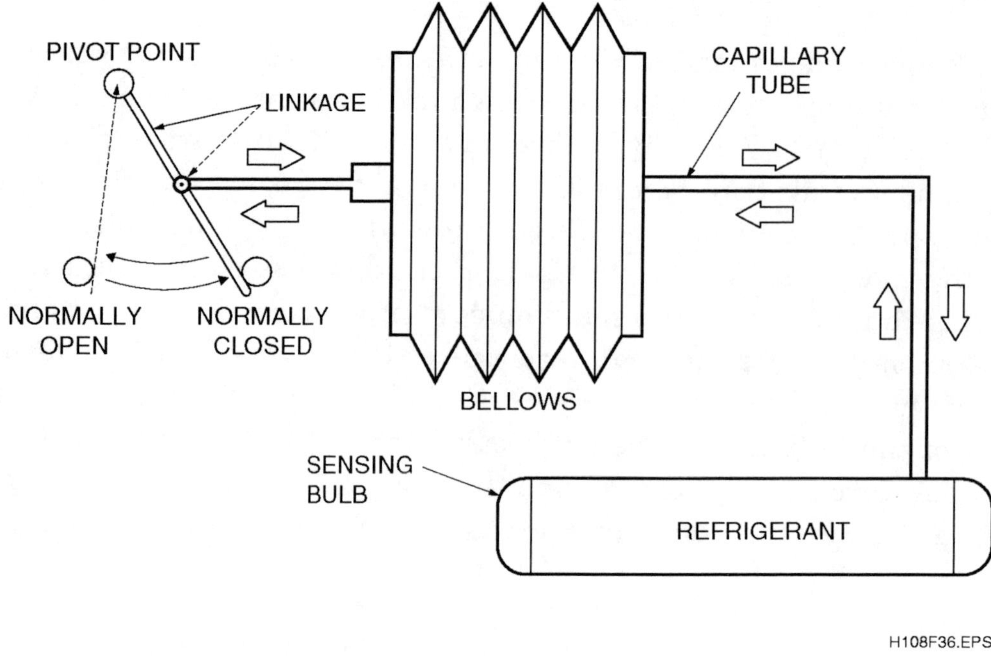

Figure 36. Remote Bulb Thermostat

Electronic thermostats use electronic components to sense temperature changes and perform switching functions. These thermostats generally use either a thermocouple or thermistor to sense the temperature. A thermocouple sensor is made of two dissimilar wires welded together at one end called a "junction." When the junction is heated, it generates a low level DC voltage that is applied to the switch circuits in the thermostat. A thermistor sensor is a semiconductor device in which the resistance changes with a change in temperature. As the temperature being measured varies, the thermistor resistance varies, causing a change in the current applied to the thermostat switch circuits. When the setpoint is reached, the switches open or close to control the related components. Electronic thermostats are more accurate and reliable than other types. Many contain microprocessor chips that allow them to be programmed for automatic start-up, shutdown, or setpoint changes.

10.1.2 Pressurestat

Pressurestats control systems using variations in pressure. They work on both the high-pressure and low-pressure sides of the compressor. Pressurestats allow operation of the compressor when low-side pressure has reached preset conditions. They also act as a cutout to open the circuit should the high-side pressure exceed a safe level. Two versions of pressurestats are the bellows type and the Bourdon type. The bellows type is directly connected to the refrigeration system through a capillary tube. As the pressure within the system changes, the pressure in the bellows also changes. The bellows expand with a pressure increase and contract with a pressure decrease. The movement in and out of the bellows makes or breaks an electrical circuit as contacts mechanically linked to the bellows open or close.

The Bourdon pressurestat uses a thin C-shaped tube that is closed at one end and connected through a capillary tube to the refrigeration system. A mercury bulb switch like that used in a room thermostat is linked to the Bourdon tube through a spring-loaded lever. As the pressure inside the Bourdon tube increases, the tube straightens and, by the action of the spring-loaded lever, opens or closes the contacts in the mercury switch. If a pressure rise opens the switch contacts and a drop closes the contacts, the device is a high-pressure switch. If a pressure rise closes the contacts and a drop opens the contacts, the device is a low-pressure switch.

10.1.3   Humidistat

Humidistats sense moisture or humidity. Electro-mechanical humidistats use materials such as human hair or nylon that expand when they absorb moisture from the air (hygroscopic materials). As the humidity increases, the hygroscopic material becomes moist and expands. This movement, applied through a spring-loaded lever, opens or closes the humidistat electrical contacts. A reduction in humidity causes the hygroscopic material to dry and contract. If a rise in humidity closes the switch contacts and a drop opens the contacts, the device being controlled turns on with a rise in humidity and cycles off with a drop in humidity. If a rise in humidity opens the switch contacts and a drop closes the contacts, the device being controlled turns on with a humidity drop and cycles off with a humidity rise.

Electronic humidistats use electronic sensing elements like lithium salts with hygroscopic properties. Another method uses carbon particles embedded in a hygroscopic material. These sensing elements work like a thermistor. Changes in humidity affect the resistance of the material and thereby change the current in an electronic circuit.

10.1.4   Time Clock

Time clocks are often used to start and stop a refrigeration system or selected components within a system. Time clocks typically are built into thermostats. When a time clock is used, the system thermostat, pressurestat or humidistat act as the primary controls during normal operation. However, when a timed event occurs, such as start-up or shutdown, the time clock circuit overrides the other controls. Timers are used to activate defrost circuits in heat pumps.

10.2.0   SECONDARY CONTROLS

Secondary controls regulate and/or protect the refrigeration system while it is in operation. They are often referred to as "operating controls" because they keep the cycle adjusted and running properly. Operating controls include:

*   Metering devices
*   Relays, contactors, and starters
*   Condenser water valve

- Refrigerant solenoid valve
- Back pressure valve
- Check valve
- Timed devices

Controls that protect the system and its components from damage are called safety controls. Safety controls include:

- Electrical overloads
- Current/temperature devices
- Thermostats
- Pressurestats
- Fusible plug
- Rupture disc
- Oil safety switches
- Flow switches

You have already been introduced to the electrical controls listed above in *Basic Electricity*, Module 03107. Most of the other ones listed have been covered previously in this module. The following paragraphs introduce you to those devices not discussed previously.

10.2.1 Condenser Water Valve

The condenser water valve is used in systems with water-cooled condensers. It regulates the head pressure at a constant level or at a preset minimum level. It is usually a self-contained, pressure-actuated bellows valve.

10.2.2 Back Pressure Valve

The back pressure valve maintains a constant pressure and, therefore, a constant saturation temperature in the evaporator. It does this regardless of changes in pressure elsewhere in the system. It is typically a self-contained, pressure-actuated bellows valve. Its operating pressure is adjusted using a spring-tension screw.

10.2.3 Check Valve

Check valves are used to make sure of single-direction flow. They prevent or "check" reverse flow. They use either a movable flapper or ball that moves away from its seat to allow flow in the desired direction and seals on the seat to stop the flow in the wrong direction.

10.2.4 Fusible Plug and Rupture Disc

Fusible plugs and rupture discs protect the refrigeration system or specific components from damage caused by an over-pressure condition. A fusible plug has a soft metal core with a low melting point. In the case of extreme pressure or fire, the high temperature melts the core. This allows the gas to escape before hazardous pressures build up. The rupture disc is a thin metal disc that ruptures at pressures above the maximum desired pressure.

10.2.5 Oil Safety Switches

An oil safety switch is a pressure-actuated, electrical safety control used to protect the compressor against damage caused by a loss of oil pressure.

10.2.6 Flow Switches

Flow switches are used to shut down the system when either air or water flow to the evaporator or condenser is inadequate. They also prevent the system from starting with inadequate flow.

11.0.0 PIPING

Most piping used in refrigeration systems is ACR copper tubing. Aluminum, steel, stainless steel, and plastic tubing may also be used for certain applications. The piping layout is usually made by the system designer. The HVAC technician is interested in piping mainly from a servicing viewpoint.

There are four requirements for a good piping layout:

1. It provides refrigerant paths.
2. It avoids excessive pressure drops.
3. It returns oil.
4. It protects compressors.

The purpose of the piping system is to provide a path for the flow of refrigerant from one component to another. Refrigerant flow must be accomplished without excessive pressure drops, such as those caused by friction, long risers, restrictions, and other piping conditions.

Some oil circulates in all refrigeration systems. The piping layout, therefore, must return the oil to the compressor crankcase. Liquid refrigerant or oil entering the compressor in the suction line can seriously damage the compressor. Good piping practices minimize the potential for damage.

In *Figure 37*, the major pipe lines of the refrigeration cycle are identified. The suction line carries cold low-pressure gas from the evaporator to the compressor. The hot gas line carries hot high-pressure gas from the compressor to the condenser. Where separate receivers are used, a condensate line is installed to drain refrigerant from the condenser to the receiver. The liquid line carries the liquid refrigerant from the receiver to the metering device.

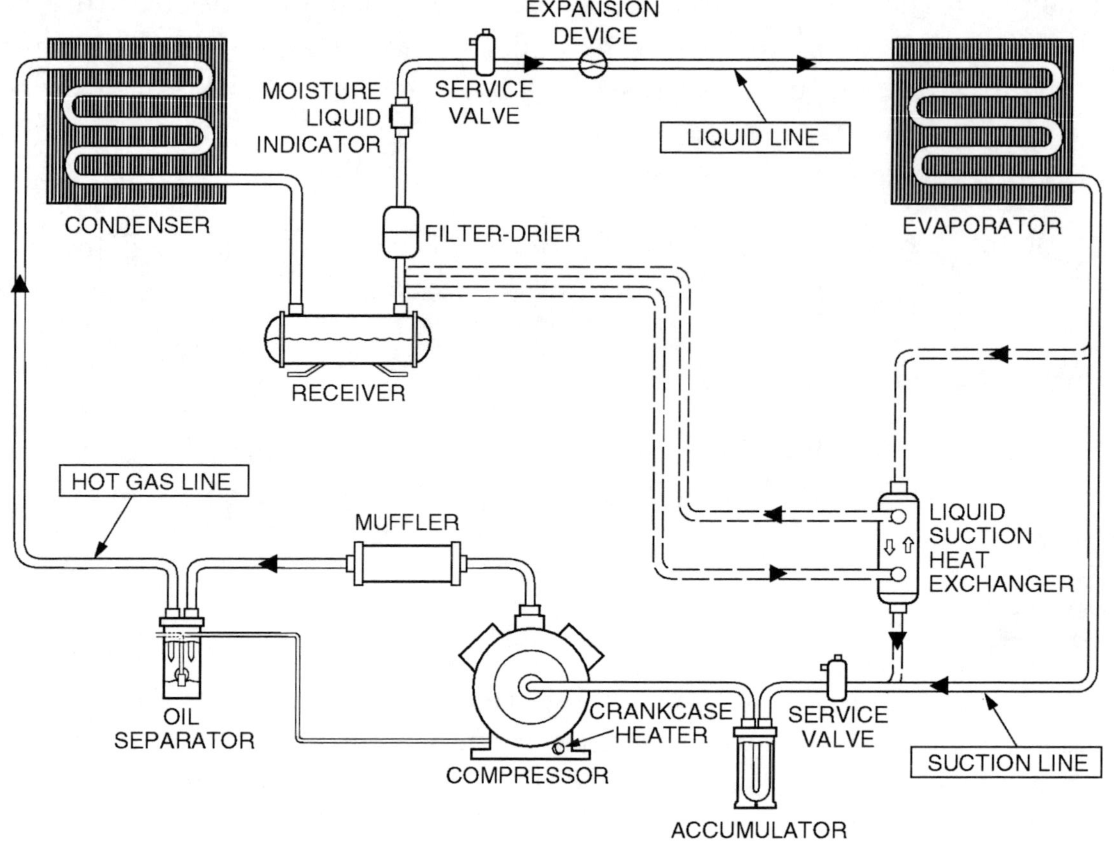

Figure 37. Refrigeration System Major Pipe Lines

11.1.0 BASIC PRINCIPLES

Certain basic principles apply to all refrigerant lines: keep them simple and pitch horizontal lines in the direction of flow. This helps to maintain the flow of oil in the right direction, prevent oil traps, and avoid backward flow during shutdown.

Unnecessary oil pockets (*Figure 38*) should be avoided. Poor layout or complicated routing can result in pockets in which oil can collect. Also, allow room for expansion and vibration. All lines expand and contract with changes in temperature.

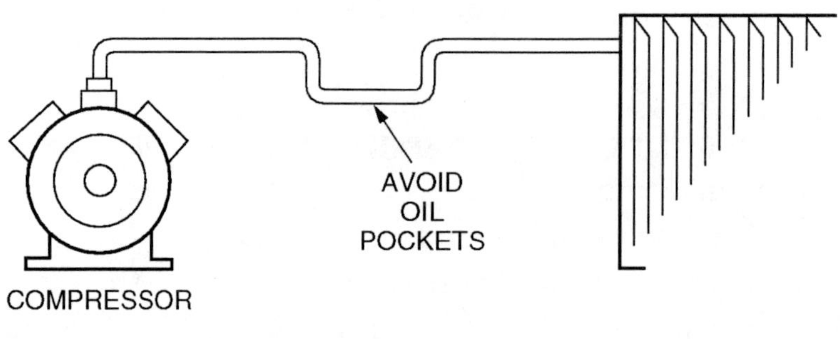

Figure 38. Avoid Oil Pockets

HVAC TRAINEE TASK MODULE 03108

11.2.0 SUCTION LINE

The suction line is the first line to be considered. It is the most critical from a layout standpoint. The pressure drop at full load must be within practical limits and oil return must be maintained under minimum load conditions. Its design must prevent liquid refrigerant from draining to the compressor during shutdown. It must also prevent oil and refrigerant from returning to the compressor in slugs during operation.

In suction risers (*Figure 39*), oil is carried upward by the refrigerant gas. A minimum gas velocity must be maintained to keep the oil moving. The sump at the bottom of the riser promotes free drainage of liquid refrigerant away from the TXV bulb, so that the bulb senses suction gas superheat instead of evaporating liquid temperature.

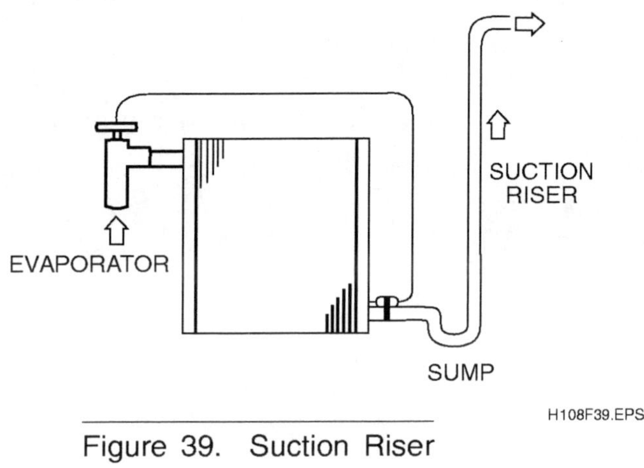

Figure 39. Suction Riser

Where system capacity is varied through compressor capacity control or some other arrangement, a short riser will usually be sized smaller than the remainder of the suction line to insure oil return up the riser (*Figure 40*). Where system capacity is variable over a wide range, it may not be possible to find a pipe size for a single suction riser that will insure oil return and still have a reasonable pressure during maximum conditions. A double suction riser (*Figure 41*) should then be used.

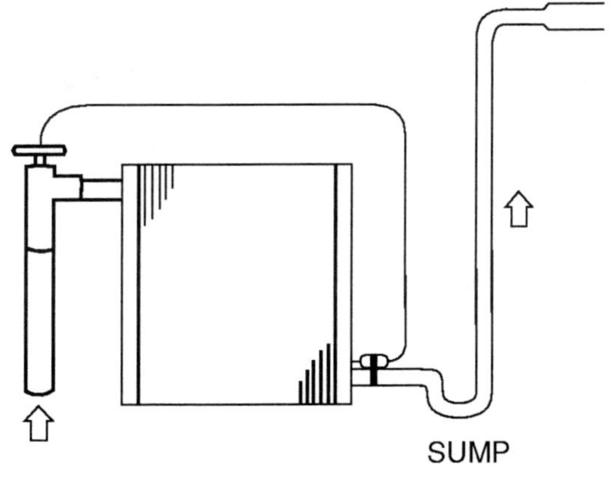

Figure 40. Reduced Riser

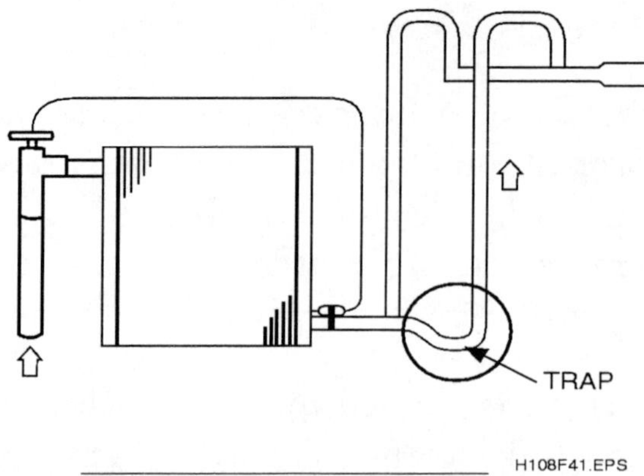

Figure 41.   Double Riser

When the compressor is on the same level or below the evaporator (*Figure 42*), a rise to at least the top of the evaporator must be placed in the suction line. This is to prevent liquid draining from the evaporator into the compressor during shutdown. The suction line loop can be left out and the piping simplified if the system is on pump-down control.

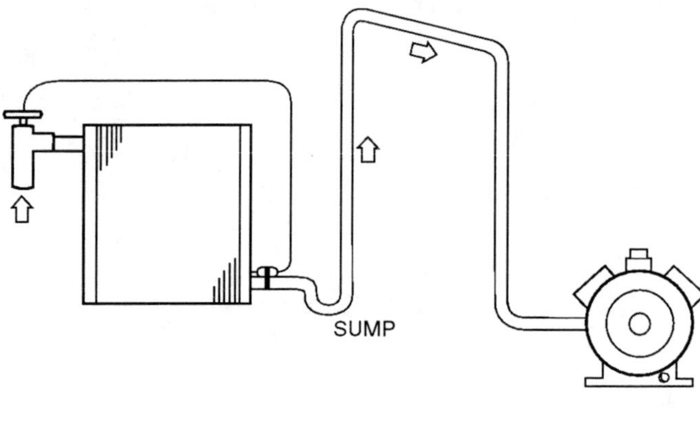

Figure 42.   Inverted Loop

Pump-down control is accomplished by placing a solenoid valve in the liquid line ahead of the metering device. A simple suction line, draining by gravity directly to the compressor and without traps, is allowed in this situation.

It is important to prevent liquid refrigerant from draining from the evaporator to the compressor during shutdown, but it is equally important to avoid unnecessary traps in the suction line near the compressor. Such traps would collect oil, which could be carried to the compressor in the form of slugs during start-up, thereby causing serious damage. The part of the suction line near the compressor should be free-draining into the compressor.

11.3.0   HOT GAS LINE

Considerations for the hot gas lines are similar to those for suction lines. They must also insure that the pressure drop at full load is maintained within limits; that oil return or circulation is kept under minimum load conditions; and that refrigerant and oil are prevented from draining back to the head of the compressor during shutdown.

Figure 43 shows a hot gas line in its simplest form. It should pitch down from the compressor to the condenser. If a riser must be placed in the line, smaller pipe is used to assure oil return under light load conditions. Where the system load varies over a wide range, a double riser is required.

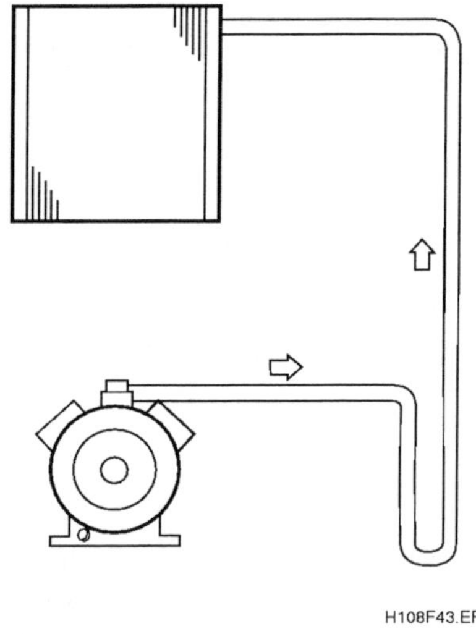

Figure 43.   Hot Gas Line

Since the hot gas line connects to the head of the compressor, provisions must be made to prevent oil or condensed refrigerant from flowing back through the line and into the compressor during shutdown. A loop to the floor between the compressor and the riser will normally provide an adequate reservoir to trap and hold the oil or condensed refrigerant.

If the compressor is located where its temperature can be lower than the temperature at the condenser or receiver, a check valve should be used. The preferred location for a muffler is in the downflow side of the hot gas loop as close to the compressor as possible. If the muffler is placed in a horizontal section of the hot gas line, position it vertically so that the outlet connection comes off at the bottom to avoid trapping oil.

The piping between the condenser and the receiver must provide for the drainage of condensed refrigerant to the receiver and the venting of the gas generated in the receiver back to the condenser. This line should be large enough to allow gas formed in the receiver to flow back to the condenser without restricting the drainage of liquid refrigerant from the condenser. When the horizontal distance between the condenser and the receiver is more than six feet, a separate gas equalizer line is required.

11.4.0   LIQUID LINE LAYOUT

The layout of the liquid line is the least critical portion of the piping system. Oil mixes freely with most refrigerants when they are in liquid form. Therefore, it is not necessary to provide high velocities in liquid lines to insure oil return. Traps in the liquid do not create oil return problems.

It is desirable to have a slightly sub-cooled liquid reach the metering device at a sufficiently high pressure for proper operation. Excessive pressure drop can result in the loss of capacity at the metering device; pressure drop without sufficient sub-cooling will cause some of the refrigerant to flash into gas. Pressure drop in liquid lines can be due to pipe friction, vertical rise and accessories. Liquid pipe sizes are normally chosen for pressure drops; vertical rise in the liquid line is normally dictated by the job conditions. Total pressure drop through accessories should not exceed 4 psi.

11.5.0   PIPE SUPPORTS

Refrigerant line pipe supports should be made of a piece of insulation and a strap. The insulation serves as a vibration isolator and prevents transmission to the structure.

11.6.0   INSULATION

Liquid lines are not normally insulated except where the surrounding temperature is higher than the liquid refrigerant. Hot gas lines are generally above the surrounding temperature and need only be insulated for personnel protection. Suction lines should be insulated to prevent condensation. Some heat absorption is desirable to evaporate any slop-over, but excessive heat gain by the suction gas must be avoided. Suction line insulation must be covered with a vapor barrier and weatherproofed when outdoors.

The three major considerations in refrigerant piping layout are compressor protection, oil return, and pressure drop. *Figure 44* illustrates the methods by which these piping objectives are achieved.

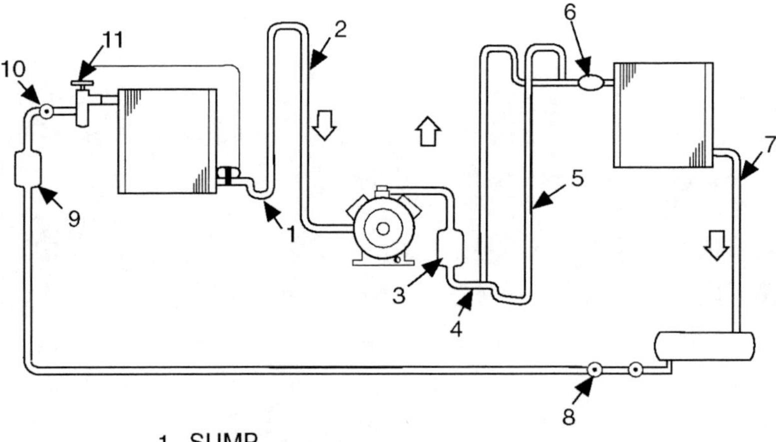

1. SUMP
2. SUCTION LOOP
3. MUFFLER
4. HOT GAS LOOP
5. DOUBLE HOT GAS RISER
6. CHECK VALVE
7. CONDENSATE LINE (AMPLY SIZED)
8. RECEIVER SIGHT LINE
9. FILTER-DRIER
10. SIGHT GLASS
11. METERING DEVICE

H108F44.EPS

Figure 44.   Piping Layout

SUMMARY

Air conditioning makes it possible to change the condition of the air in an enclosed area. It is a process that heats, cools, cleans, and circulates air and controls its moisture content. The cooling portion of air conditioning depends on refrigeration. Refrigeration is also used for the preservation of food and in many industrial processes. The refrigeration cycle is based on the following concepts:

- Heat always flows from a warmer substance or location to a cooler substance or location.
- Heat must be added to or removed from a substance before a change in state can occur.
- The flow of a gas or liquid is always from a higher pressure area to a lower pressure area.
- The temperature at which a liquid or gas changes state is dependent on pressure.
- Cold is merely the absence of heat.
- Heat is always present.

References

For more advanced study of topics in this task module, the following books are suggested:

Basic Refrigeration (Slides and Student Handbook), York International Corporation, Publications Distribution Center, York, Pennsylvania.

General Training Air Conditioning (Fundamentals) – GTAC-I, Carrier Corporation, Literature Services, Syracuse, New York.

Modern Refrigeration and Air Conditioning, The Goodheart-Willcox Company, Inc., South Holland, Illinois.

Refrigeration & Air Conditioning Technology, Second Edition, Delmar Publishers, Inc., Albany, New York.

1. The Fahrenheit scale is based on boiling water having a sea level temperature of:
   a. 459°.
   b. 212°.
   c. 180°.
   d. 100°.

2. Five pounds of water heated to raise the temperature two degrees requires:
   a. 25 Btu's.
   b. 5 Btu's.
   c. 10 Btu's.
   d. 100 Btu's.

3. Sensible heat is:
   a. heat that produces a change in state without a change in temperature.
   b. heat that changes a liquid to a vapor.
   c. heat that can be sensed by a thermometer.
   d. none of the above.

4. Superheat is added:
   a. after all of a solid changes to a liquid.
   b. in changing vapor to liquid.
   c. after all liquid has been changed to vapor.
   d. in changing liquid to a vapor.

5. The transfer of heat from one object to another by direct contact is called:
   a. radiation.
   b. change of state.
   c. convection.
   d. conduction.

6. A ton of refrigeration is equal to:
   a. 2,880,000 Btu's per day.
   b. 144 Btu's per hour.
   c. 12,000 Btu's per hour.
   d. 180 Btu's per hour.

7. Zero gauge pressure corresponds on the absolute scale to:
   a. 44.7 psi.
   b. 27.4 psi.
   c. 14.7 psi.
   d. none of the above.

8. The boiling temperature of a liquid will be lower as:
    a. the pressure is increased.
    b. the pressure is decreased.
    c. it changes state.
    d. none of the above.

9. A gas or liquid always flows:
    a. from a lower pressure to a higher pressure.
    b. from a higher temperature to a lower temperature.
    c. from a higher pressure to a lower pressure.
    d. in a straight line.

10. In a refrigeration system, the high-pressure, high-temperature gas is converted into a high-pressure, high-temperature liquid by the:
    a. compressor.
    b. evaporator.
    c. expansion (metering) device.
    d. condenser.

11. In a refrigeration system, the low-pressure, low-temperature liquid is converted into a low-pressure, low-temperature vapor by the:
    a. compressor.
    b. evaporator.
    c. expansion (metering) device.
    d. condenser.

12. In a refrigeration system, the low-pressure, low-temperature vapor is converted into a high-pressure, high-temperature vapor by the:
    a. compressor.
    b. evaporator.
    c. expansion (metering) device.
    d. condenser.

13. The low side of the refrigeration system includes the:
    a. suction side (input) of the compressor.
    b. muffler.
    c. discharge side of the compressor.
    d. condenser.

14. Head pressure refers to the pressure in the:
    a. low side of the system.
    b. evaporator.
    c. expansion (metering) device.
    d. high side of the system.

15. If a gas is superheated 15 degrees to a temperature of 87°F, we can determine that its:
    a. boiling point is 72°F.
    b. freezing point is 32°F.
    c. saturation point is 72°F.
    d. both a and c.

16. The four most commonly used refrigerants listed in order, from the most to the least damaging to the ozone layer, are:
    a. R-11, R-12, R-502 and R-22.
    b. R-22, R-12, R-11 and R-502.
    c. R-502, R-12, R-22 and R-11.
    d. R-11, R-12, R-22 and R-502.

17. The difference between a halocarbon and a fluorocarbon refrigerant is that:
    a. there is no difference.
    b. the fluorocarbon has fluorine in it while the halocarbon still has carbon.
    c. a halocarbon never has chlorine in it while the fluorocarbon does.
    d. the fluorocarbon always has fluorine in it but not all halocarbons do.

18. An unknown refrigerant registers 80°F and has a pressure of 102 psig at room temperature. Using a temperature/pressure chart (*Figure 16*), you determine that the refrigerant is:
    a. R-22.
    b. R-502.
    c. R-500.
    d. R-12.

19. A compressor with a piston that travels back and forth in a cylinder is a _____ compressor.
    a. reciprocating
    b. rotary
    c. centrifugal
    d. none of the above

20. The main purpose of a condenser is to:
    a. store liquid refrigerant.
    b. remove heat from the refrigerant.
    c. remove water from the refrigerant.
    d. add heat to the refrigerant.

21. The main purpose of an evaporator is to:
    a. store liquid refrigerant.
    b. remove heat from the refrigerant.
    c. remove water from the refrigerant.
    d. add heat to the refrigerant.

22. An automatic expansion valve is controlled by:
    a. suction line superheat.
    b. evaporator pressure.
    c. spring pressure.
    d. b and c.

23. Components (accessories) most often found in the refrigeration system liquid line are:
    a. receiver, sightglass/moisture indicator, filter-drier.
    b. liquid-suction heat exchanger, suction line accumulator.
    c. muffler, oil separator.
    d. water pre-heater.

24. A secondary control is used to:
    a. regulate the cycle.
    b. protect the cycle.
    c. regulate and/or protect the cycle.
    d. control conditions within the cycle.

25. Horizontal piping runs used in refrigeration systems should:
    a. pitch toward the compressor.
    b. pitch away from the compressor.
    c. pitch in the direction of flow.
    d. be level.

ANSWERS TO SELF CHECK REVIEW / PRACTICE QUESTIONS

<u>Answer</u>	<u>Section Reference</u>
1. b	2.1.1
2. c	2.1.2
3. c	2.1.3
4. c	2.1.3
5. d	2.2.1
6. c	2.2.5
7. c	2.3.2
8. b	2.3.3
9. c	2.3.4
10. d	3.2.1
11. b	3.2.1
12. a	3.2.1
13. a	3.2.2
14. d	3.2.2
15. d	3.2.2
16. a	4.1.0 and 4.3.0
17. d	4.3.0
18. c	4.5.0
19. a	5.1.0
20. b	6.0.0
21. d	7.0.0
22. d	8.2.4
23. a	9.0.0
24. c	10.2.0
25. c	11.1.0

The NCCER makes every effort to keep these manuals up-to-date and free of technical errors. We appreciate your help in this process. If you have an idea for improving this manual, or if you find an error, a typographical mistake, or an inaccuracy in the NCCER's Craft Training Manuals, please write us, using this form or a photocopy. Be sure to include the exact module number, page number, a description of the problem, and the correction, if possible. Your input will be brought to the attention of the Technical Review Committee. Thank you for your assistance.

Instructors – If you found that additional materials were necessary in order to teach this module effectively, please let us know so that we may include them in the Equipment/Materials list in the Instructor's Guide.

Write:	Curriculum and Revision Department
	National Center for Construction Education and Research
	P.O. Box 141104
	Gainesville, FL 32614-1104
Fax:	352-334-0932

Craft _______________________ Module Name _______________________

Module Number _______________________ Page Number(s) _______________________

Description of Problem _______________________

____

____

____

(Optional) Correction of Problem _______________________

____

____

(Optional) Your Name and Address _______________________

____

Introduction to Heating

Module 03109

INTRODUCTION TO HEATING

Objectives

Upon completion of this module, the trainee will be able to:

1. Explain the three methods by which heat is transferred and give an example of each.
2. Describe how combustion occurs and identify the byproducts of combustion.
3. Identify the various types of fuels used in heating.
4. Recognize the major components and accessories of a forced-air furnace and explain the function of each component.
5. State the factors that must be considered when installing a furnace.
6. Identify the major components of a gas furnace and describe how each works.
7. With supervision, use a manometer to measure and adjust manifold pressure on a gas furnace.
8. Identify the major components of an oil furnace and describe how each works.
9. Describe how an electric furnace works.
10. With supervision, perform basic furnace preventive maintenance procedures such as cleaning and filter replacement.

Prerequisites

Successful completion of the following Task Modules is required before beginning study of this Task Module: Common Core Curricula, HVAC Modules 03101 through 03108.

Required Student Material

1. Student Module
2. Appropriate Personal Protective Equipment

Course Map Information

This course map shows all of the *Wheels of Learning* task modules in the first level of the HVAC curricula. The suggested training order begins at the bottom and proceeds up. Skill levels increase as a trainee advances on the course map. The training order may be adjusted by the local Training Program Sponsor.

Course Map: HVAC, Level 1

TABLE OF CONTENTS

Trade Terms Introduced In This Module

Annual Fuel Utilization Efficiency (AFUE): HVAC industry standard for defining furnace efficiency.

Atomize: The process by which a liquid is converted into a fine spray.

British thermal unit (Btu): The amount of heat needed to raise the temperature of one pound of water one degree Fahrenheit.

Celsius (Centigrade): The temperature scale used in the metric system, in which the freezing point of water is 0° and the boiling point is 100°.

Combustion: The process by which a fuel is ignited in the presence of oxygen.

Condensing furnace: A furnace that contains a secondary heat exchanger that extracts latent heat by condensing flue gases.

Conduction: A means of heat transfer in which heat is moved from one substance to another by means of direct contact.

Convection: The transfer of heat by the flow of liquid or gas.

Electrode: An electrical terminal that will conduct a current.

Fahrenheit: The temperature scale commonly used in the U.S. in which the freezing point of water is 32° and the boiling point is 212°.

Flame rectification: The process by which a flame produces a sensible electrical current.

Heat exchanger: A device, usually metal, that is used to transfer heat from a warm surface or substance to a cooler surface or substance.

Hot surface igniter: A ceramic device that glows when an electrical current flows through it. Used to ignite gas in a gas furnace.

Induced-draft furnace: A furnace in which a motor-driven fan draws air from the surrounding area or from outdoors to support combustion.

Infiltration: Air that enters a building through doors, windows, and cracks in the construction.

Manometer: An instrument that measures air or gas pressure by the displacement of a column of liquid.

Natural-draft furnace: A furnace in which the natural flow of air from around the furnace provides the air to support combustion.

Oil burner: The main component of an oil-fired furnace. It combines oil and air and sprays the combination into the combustion chamber.

Orifice: A precisely drilled hole that controls the flow of gas to the burners.

Piezoelectric: The property of a quartz crystal that causes it to vibrate when a high frequency voltage is applied to it.

Primary air: Air that is pulled or propelled into the combustion process along with the fuel.

Radiation: The movement of heat in the form of invisible rays or waves, similar to light.

Redundant gas valve: A gas control containing two gas valves in series. If one fails, the other is available to shut off the gas when needed.

Relative humidity: The amount of moisture in the air in relation to the capacity of the air to hold moisture.

Safety pilot: A pilot light with a flame sensing element.

Secondary air: Air that is added to the mix of fuel and primary air during combustion.

Standing pilot: A gas pilot that is on continuously.

Spud: A threaded metal device that screws into the gas manifold. It contains the orifice that meters gas to the burners.

Thermocouple: A device made up of two unlike metals that generates electricity when there is a difference in temperature from one end to the other.

1.0.0   INTRODUCTION

We rely on heating systems in our homes, schools, and places of business to keep the temperature in our comfort range. There are many types of heating systems. Most of them burn oil or natural gas; some use electricity as an energy source. In this module, you will learn the basic principles of heating. You will also learn about the various types of heating systems. This module focuses on gas-fired and oil-fired warm air furnaces. In later lessons you will learn about boilers and heat pumps.

2.0.0   HEATING FUNDAMENTALS

2.1.0   HEAT TRANSFER

For heat transfer to occur, there must be a difference in temperature between two objects. The larger the difference, the greater the amount of heat transfer. Heat always transfers from a warmer region to a cooler region. Any object, including the human body, gives off heat if the air around it is cooler than the object. The reason you feel cold in the winter is that your body is losing heat to the cooler air around it. To keep the heat from escaping you have two choices: wear layers of insulation (clothing) or warm the surrounding air so that the body stops giving away heat. In hot weather, your body feels hot because it cannot transfer heat to the warmer surrounding air.

In the last module, you learned that there are three methods by which heat is transferred: conduction, convection, and radiation. Heating systems use one or more of these methods to warm the air in the conditioned space. (See *Figure 1*.)

2.1.1   Conduction

Conduction is the flow of heat from one part of a material to another part or to a substance in direct contact with it. The rate at which a material transmits heat is known as its conductivity. The amount of heat transmitted by conduction between two surfaces is determined by the surface area, thickness, and conductivity of the materials, and the temperature difference between the two.

An example of conduction is the transfer of heat from the gas burners to the heat exchangers in a furnace (*Figure 1A*). Another is the transfer of heat from a stove burner to a frying pan.

2.1.2   Convection

Convection is air motion due to the warmer portions rising and the denser, cooler portions falling. For convection to occur, there must be a difference in temperature between the source of heat and the surrounding air. The greater the difference in temperature, the greater the movement of air by convection. The greater the movement of air, the greater the transfer of heat.

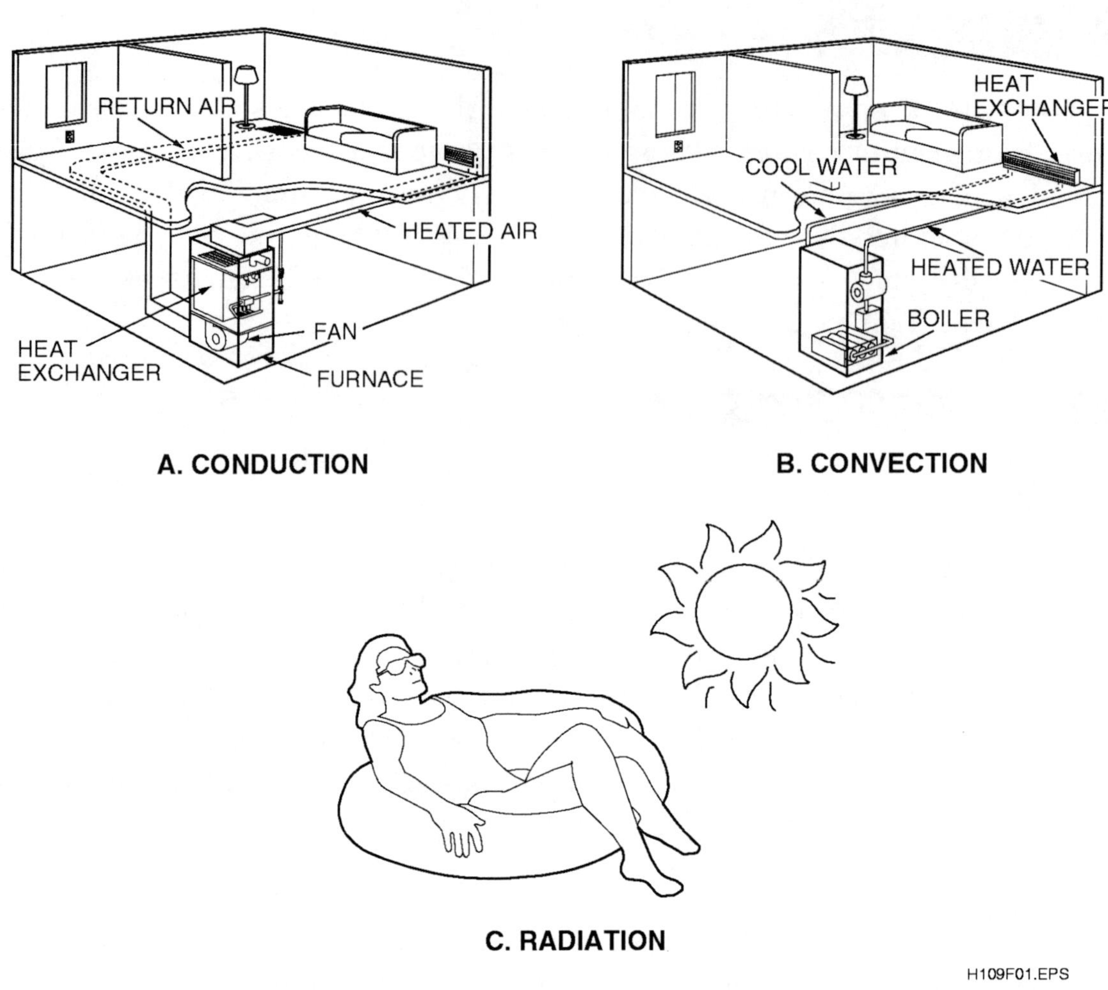

Figure 1.   Heat Transfer Methods

Hot water heating elements such as the one shown in Figure 1B rely on convection. The room air at the baseboard heater is heated, causing it to rise. As it rises, it gives up its heat to the cooler room air and becomes colder. The cold air then falls to the floor and the process repeats.

2.1.3   Radiation

Radiation is the transfer of heat through space by wave motion. Heat passes from one object to another without warming the space between them. The amount of heat transferred by radiation depends upon the area of the radiating body, the temperature difference between the two objects, and the distance between the heat source and the object being heated.

The heat from the sun is radiated heat (*Figure 1C*). Another example of radiated heat is a plug-in electric space heater. The heat from radiation heat sources tends to be intense near the source and to heat only surfaces that it can "see." When you are sitting at a campfire, for example, the front of your body can be hot while your back is cold.

2.1.4　Humidity

Heat transfer is affected by the moisture content of the air, which is known as humidity. You will most often hear the term **relative humidity.** This term describes the capacity of air to hold water. A relative humidity of 50% means that the air contains half the moisture it is capable of holding. The body loses heat more rapidly when the humidity is low. Therefore, homes in many climates are equipped with humidifiers that add moisture to the air during winter months when the air is dry. Humidifiers are frequently added to forced-air furnaces.

2.2.0　TEMPERATURE

Temperature is defined as the degree of hotness or coldness measured on a numerical scale. It can be accurately measured using a thermometer. There are two thermometer scales in frequent use. One type is the English system known as the **Fahrenheit** scale; the other is the metric system and is called the **Celsius** scale. It is important to understand both of these scales, because some temperature information is given in Fahrenheit degrees and some in Celsius.

The Celsius thermometer, though not generally familiar to most Americans, is the easier of the two to understand. Zero degrees on the Celsius scale refers to the freezing point of water and 100 degrees represents the boiling point of water. The freezing point of water on the Fahrenheit scale is 32 degrees and the boiling point of water is 212 degrees. To reduce confusion about which scale is involved, a capital C is placed after a Celsius degree indication and a capital F is placed after the number of Fahrenheit degrees.

There are times when it may be necessary to change from one scale to the other. These conversions can be done using the following formulas:

Degrees Celsius = 5/9 x (degrees Fahrenheit - 32 degrees)

Degrees Fahrenheit = (9/5 x degrees Celsius) + 32 degrees

For example, 32°F = 0°C [5/9 x (32 - 32) = 5/9 x 0 = 0].

2.3.0　HEAT MEASUREMENT

The unit of measurement of heat in the British system is the **British thermal unit (Btu).** One Btu is the amount of heat required to raise one pound of water one degree Fahrenheit.

It is also important to be aware of the rate of heat change. That is, how fast or how slowly a piece of heating equipment produces heat. The term for this is Btu's per hour (Btuh). For example, the capacity of a furnace may be stated as 110,000 Btuh.

2.4.0 COMBUSTION

Combustion is the burning of fuel to create heat. During combustion, oxygen is combined rapidly with a fuel to release the stored energy in the form of heat. There are three conditions necessary for combustion to take place.

- First, there must be fuel. The fuel can be gas, such as natural gas; liquid, such as fuel oil; or solid, such as coal. Two elements that all fuels have in common are carbon and hydrogen.
- Second, fuel must be ignited in order to burn. A pilot burner, electronic ignition, or hot surface igniter can be used to ignite a gas burner, while an electric spark is used to ignite fuel oil.
- Third, oxygen must be present for the burning of fuel to take place.

2.4.1 Complete Combustion

Complete combustion takes place when carbon combines with oxygen to form carbon dioxide (CO_2). Carbon dioxide is nontoxic and can be exhausted to the atmosphere. The hydrogen combines with oxygen to form water vapor (H_2O) which is also harmless and can be exhausted to the atmosphere.

2.4.2 Incomplete Combustion

Incomplete combustion results from a lack of oxygen and causes undesirable products to form. These undesirable products are carbon monoxide, pure carbon (soot), and aldehyde. Both carbon monoxide and aldehyde are toxic. Soot causes coating of the heating surface of the furnace and reduces heat transfer.

Fuel-burning devices must be adjusted so that complete combustion of the fuel always takes place. Sufficient air must be provided for proper combustion to take place and to eliminate the hazards of incomplete combustion. Furnaces are normally adjusted to provide from 5 to 50% excess air in order to guard against the possibility of incomplete combustion.

2.4.3 Combustion Efficiency

When fuel is burned in a furnace, a certain amount of heat is lost in the hot gases that are vented through the chimney. This function is necessary for disposal of the products of combustion, but the loss must be kept down to allow the furnace to operate at its peak efficiency. Air entering the furnace at room temperature or lower is heated to fuel gas temperatures that range from 350°F to 600°F depending upon the design and adjustments of the furnace.

If the amount of heat lost is 20%, the furnace efficiency would be 80%. Charts are available for determining the efficiency of a furnace based on the temperature and the carbon dioxide content of the flue gases. Knowing the efficiency of the furnace makes it possible for a technician to calculate the furnace output.

The current industry standard for defining furnace efficiency is **Annual Fuel Utilization Efficiency (AFUE)**. AFUE takes into account operating efficiency as well as combustion efficiency. The National Appliance Energy Conservation Act of 1987 requires that all furnaces built after 1991 have an AFUE of no less than 78%. Most existing **natural-draft furnaces** that rely on convection for combustion air have an AFUE of less than 78%. With the addition of special features such as electronic ignition and vent dampers, it is possible to bring a natural-draft furnace up to an AFUE of 80%. They have been replaced by **induced-draft furnaces**, which use a fan to draw in and exhaust combustion air. Induced-draft furnaces have an AFUE that ranges from 78% to 89%, depending on the kinds of efficiency options installed. **Condensing furnaces** have the highest efficiency — greater than 90%. The condensing furnace uses a condensing coil as a secondary heat exchanger to extract heat from the combustion byproducts before they are vented outdoors.

2.4.4 Flames

The type of flame and the intensity with which it burns have a direct relationship to the efficiency of the heating unit. Pressure-type oil burners burn with a yellow flame, while gas burners burn with a blue flame that has an orange tip. The difference is mainly due to the manner in which air is mixed with the fuel.

A yellow flame in a gas burner denotes incomplete combustion and an obstructed burner. In this case, the burner should be checked. A blue flame is produced when approximately 50% of the air (**primary air)** is mixed with the gas before ignition. The balance of air, called **secondary air**, is supplied during combustion to the exterior of the flame. Improper gas flames are the result of inefficient or incomplete combustion. They can be caused by too much primary air, a lack of secondary air, or by contact between the flame and a cool surface.

2.5.0 FUELS

Fuels are available in three forms: gases, liquids, and solids. Gases include natural gas, manufactured gas, and liquefied petroleum (LP). Liquids include fuel oils. There are six grades of fuel oil, with No. 2 being the most common. Solids include coal and wood. This program focuses on gas and oil, which are the types most likely to be encountered in the HVAC trade. *Table 1* shows the heating values of common fuels.

Fuel	Heat Released
COAL	
Bituminous	12,000 to 15,000 Btu/lb.
Anthracite	13,000 to 14,000 Btu/lb.
OIL*	
Grade 1	137,000 Btu/gallon
Grade 2	140,000 Btu/gallon
Grade 4**	141,000 Btu/gallon
Grade 5	148,000 Btu/gallon
Grade 6	152,000 Btu/gallon
GAS	
Natural ***	900 to 1200 Btu/cubic ft.
Manufactured	500 to 600 Btu/cubic ft.
Liquefied Petroleum (LP)	2500 to 3200 Btu/cubic ft.
WOOD	6200 Btu/lb. (avg.)

* Grades are determined by the American Society for Testing and Materials (ASTM).
** This grade is not commonly used.
*** Check with local gas company for specific values.

Table 1. Heating Values Of Common Fuels

2.5.1 Gaseous Fuels

There are three types of gaseous fuels: natural gas, manufactured gas, and liquefied petroleum.

- Natural gas — Natural gas comes from the earth and usually accumulates in the upper part of oil wells. It is colorless and nearly odorless. An odorant, such as a sulfur compound, is added so that leaks can be detected. The content of this gas varies by locality and has a bearing on the Btu content, which varies from 900 to 1200 depending on the locality, but is usually in the range of 1000 to 1050. The chief component of natural gas is methane. It also includes other hydrocarbons.

- Manufactured gas — Manufactured gases are combustible gases that are normally produced from solid or liquid fuel and are used mostly in industrial processes. They are produced mainly from coal, oil, and other hydrocarbons, and are comparatively low in Btu's per cubic foot (500 to 600). They are not considered an economical space-heating fuel.

- Liquefied petroleum — LP is a byproduct of the oil refining process. It is stored in liquid form, but is vaporized when used. There are two types of LP gas: propane and butane. Propane is more useful as a space heating fuel because it boils at -40°F and can be readily vaporized for heating in a northern climate. Butane vaporizes at about 32°F. Propane has a heating value of about 2500 Btu's per cubic foot, whereas butane has a heating value of approximately 3200. LP gas is usually propane with a small amount of butane added. When LP gas is used as a heating fuel, the equipment must be designed or modified for its use.

WARNING!   Propane and butane vapors are considered to be more dangerous than those of natural gas because they have a higher specific gravity. The vapor is heavier and tends to accumulate near the floor, increasing the danger of an explosion at ignition.

LP gas (also known as bottled gas) is an alternative in areas where natural gas is not available. Manufacturers of gas appliances make LP conversion kits to adapt natural gas furnaces for LP. However, the use of LP gas is heavily regulated by state and local laws in some locales. It is thus not always a good alternative to natural gas.

2.5.2   Fuel Oils

Fuel oils are rated according to their Btu/gallon content and the American Petroleum Institute (API) gravity. The API gravity is an index related to the heating values for standard grades of fuel oil. There are six common grades of oil: No's. 1, 2, 4, 5 (light), 5 (heavy) and 6. The lighter-weight oils have a higher API gravity.

Grade 1: A light-grade distillate for use in vaporizing-type oil burners.

Grade 2: A heavier distillate for use in domestic pressure-type oil burners.

Grade 4: A light residue or heavy distillate used for higher-pressure commercial oil burners.

Grade 5 (light): A medium weight, residual-type fuel used in commercial oil burners that are specifically designed for it.

Grade 5 (heavy): A residual-type fuel for commercial oil burners. Usually requires preheating.

Grade 6: Also termed Bunker C; a heavy residue used for commercial burners. Preheating is necessary in the tank to permit pumping; additional preheating is needed at the burner to convert the fuel into a fine spray or **atomize** it.

2.5.3　Coal

Coal is made principally of carbon (as much as 80%). There are four different types of coal: anthracite, bituminous, sub-bituminous and lignite. The types vary in their Btu content and burning and handling characteristics.

1. Anthracite — Anthracite is a clean, hard coal that burns with an almost smokeless short flame. It is difficult to ignite, but burns freely when started. It is noncaking and leaves a fine ash that does not clog grates and ash removal equipment.
2. Bituminous — Bituminous coals consist of a wide range of coals varying from high grade in the east to low grade in the western U.S. These coals are more brittle than anthracite and break up readily into small pieces for grading and screening. The flame is long, but varies for different grades. Much smoke and soot can result unless burning is carefully controlled.
3. Sub-bituminous — Sub-bituminous coal tends to break up when dry. It ignites easily and can ignite spontaneously when stored. It burns with a medium flame. It forms little soot and smoke and is desirable for its noncaking qualities.
4. Lignite — Lignite has a woody consistency. It is clean to handle, but is high in moisture content and low in heating value. It has a strong tendency to break up. Because of its high moisture content, it is difficult to ignite. It forms little smoke and soot and is noncaking.

2.5.4　Wood

The amount of heat produced by wood varies with the weight of the wood. Wood weight can vary from a low of about 25 pounds per cubic foot to more than 70.

Wood provides about 6200 Btu/lb. of heat on average. The lightest woods are generally the easiest to burn. Heavier, denser woods provide more heat, but can be difficult to ignite. Resinous woods also provide a lot of heat, but create other problems, such as creosote buildup in chimneys. Lighter woods, such as pines with lots of resins, are useful in starting fires.

There are about 700 species of trees in the United States. Some are too small for fuel use, some are too localized for general use, and some are too rare. The poor efficiency of wood as a fuel, combined with the destruction of trees that might otherwise be used for lumber, are two reasons why the burning of wood for fuel on a large scale makes poor economic sense. Because of the amount of smoke and ash it produces, wood also makes poor environmental sense. Its use is strictly regulated in some areas.

3.0.0 FORCED-AIR FURNACES

See *Figure 2*. Fuel-burning furnaces provide heat by the combustion of fuel within a **heat exchanger**. Air is passed over the outside surfaces of the heat exchanger, transferring the heat from the fuel to the air.

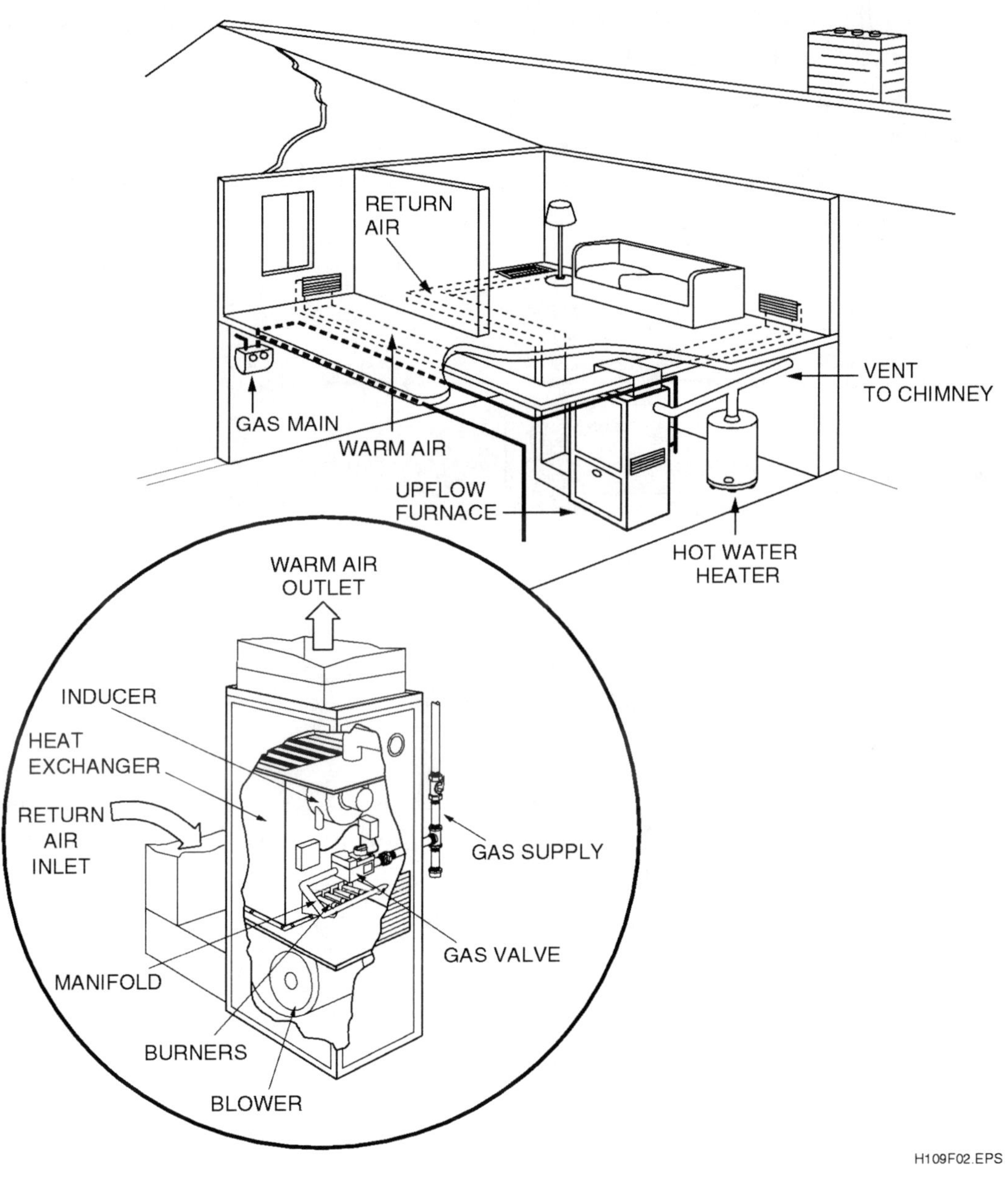

Figure 2. Forced-Air Heating

In the fuel-burning furnace, the products of combustion are exhausted to the atmosphere through the flue passages and the chimney. In an electric furnace, air passes directly over the heated elements without the use of a heat exchanger, since no products of combustion are formed. A forced-air furnace uses a fan to move the air over the heat exchanger and to circulate the air through the distribution system (ductwork).

3.1.0 TYPES OF FORCED-AIR FURNACES

There are four different designs of forced-air furnaces in common use. Each requires a different arrangement of the basic components.

1. The upflow design (*Figure 2*) is used in the basement or in a first-floor equipment room. The blower is located below the heat exchanger. Air enters at the bottom or lower sides of the unit and exits through the top into the duct plenum.
2. The horizontal furnace (*Figure 3*) is used in attics or crawlspaces where the height of the furnace must be kept at a minimum. Air enters at one end of the unit through the blower and filter compartment and is forced horizontally over the heat exchanger, exiting at the opposite end.

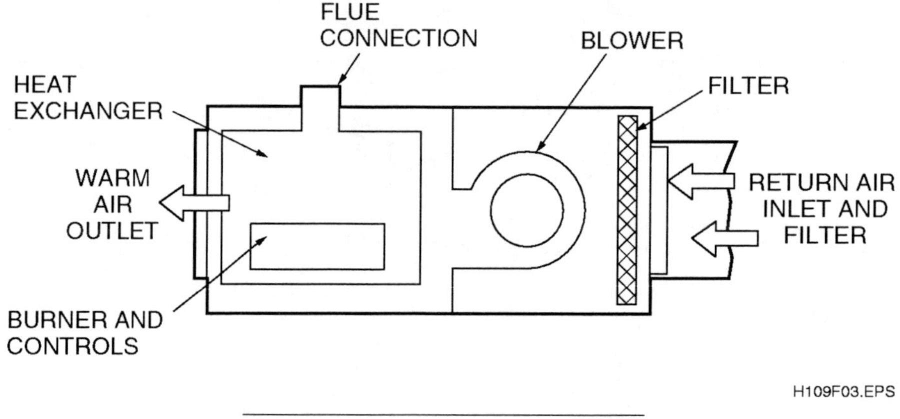

Figure 3.   Horizontal Furnace

3. The low-boy furnace (*Figure 4*) occupies more floor space and is lower in height than the upflow design. It is best suited for basement installation. The blower is placed alongside the heat exchanger, the return air plenum is built above the blower compartment, and the warm air outlet or heat supply plenum is built above the heat exchanger.

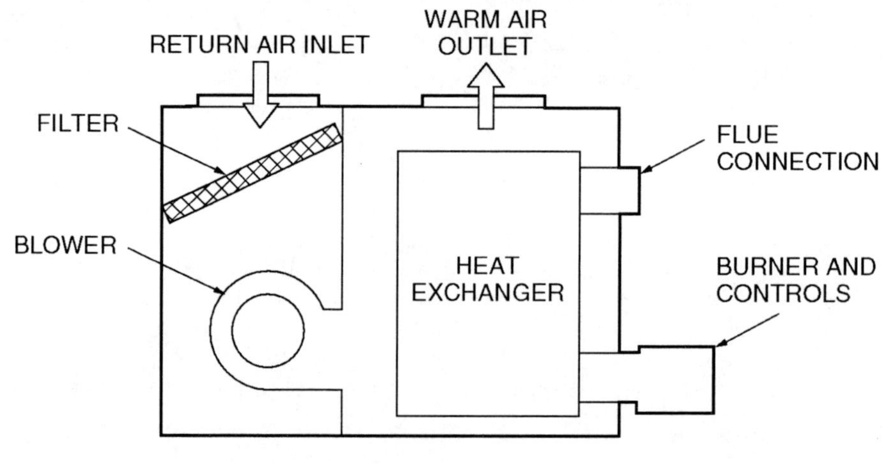

Figure 4.   Low-Boy Furnace

4. The downflow or counterflow furnace (*Figure 5*) is used in houses with an under-the-floor distribution system. The blower is located above the heat exchanger and the return air plenum is connected to the top of the blower area. The warm air supply plenum is connected to the bottom of the enclosure cabinet.

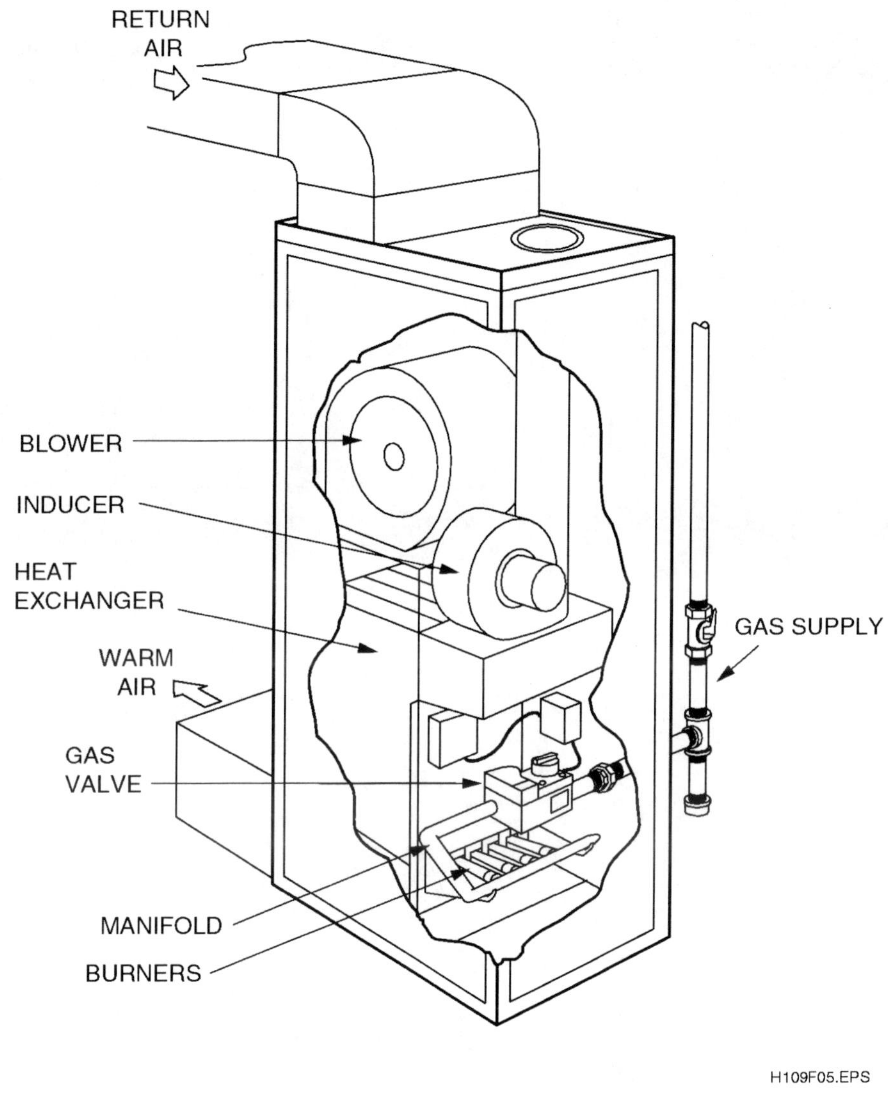

Figure 5.   Downflow Furnace

3.2.0   HEAT EXCHANGERS

The heat exchanger is the part of the furnace where combustion takes place. The heat exchanger is usually made of cold-rolled, low-carbon steel, fabricated with welded seams. There are several types of heat exchangers. The types shown in A, B, and C of *Figure 6* are commonly found in gas-fired furnaces. In these three types, the burners fire directly into the heat exchanger and the hot flue gases flow through the heat exchanger and out the other end into a collector box that directs the gases to the vent system.

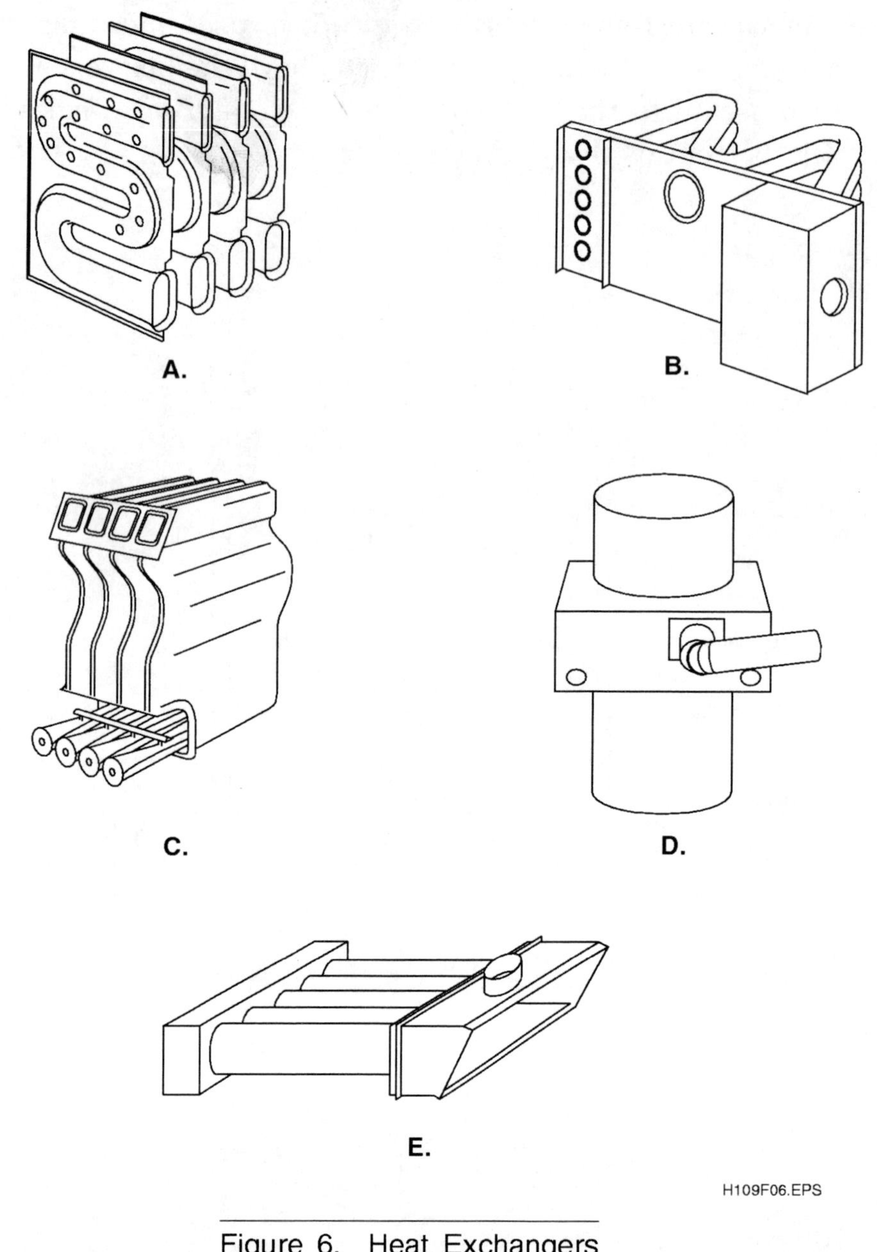

Figure 6.   Heat Exchangers

The heat exchangers shown in D and E are typical of those found in oil-fired furnaces. The cylindrical heat exchanger (D) has a primary and secondary surface. The primary surface is in direct contact with the flame; the secondary surface extracts heat from the hot flue gases. Oil does not burn as cleanly as gas. Therefore, the long, thin heat exchangers used for gas heating would get plugged up if used in oil-fired furnaces.

The number of heat exchanger sections depends on the furnace capacity. In a gas furnace, each section has its own burner. A low capacity furnace might have two sections fed by two burners; a high capacity furnace might have four sections and four burners. The heat exchanger sections terminate in a collector box that directs the flue gases into the flue pipe.

3.3.0   CONDENSING FURNACES

Condensing furnaces are equipped with a secondary heat exchanger that often looks like a refrigeration condensing coil. A pulse-type condensing furnace is shown in *Figure 7*. Instead of continuous burning, gas is burned in a series of pulsed combustions that occur inside the special heat exchanger. The condensing coil is in the path of the conditioned air flowing through the furnace. The flue gas from the combustion process is piped through the condensing coil, which condenses the flue gas into a liquid. The latent heat created by the condensing process is transferred to the conditioned air. Condensing furnaces can reach efficiencies of 95% and more. Even though they may be more expensive initially, in the long term they save money.

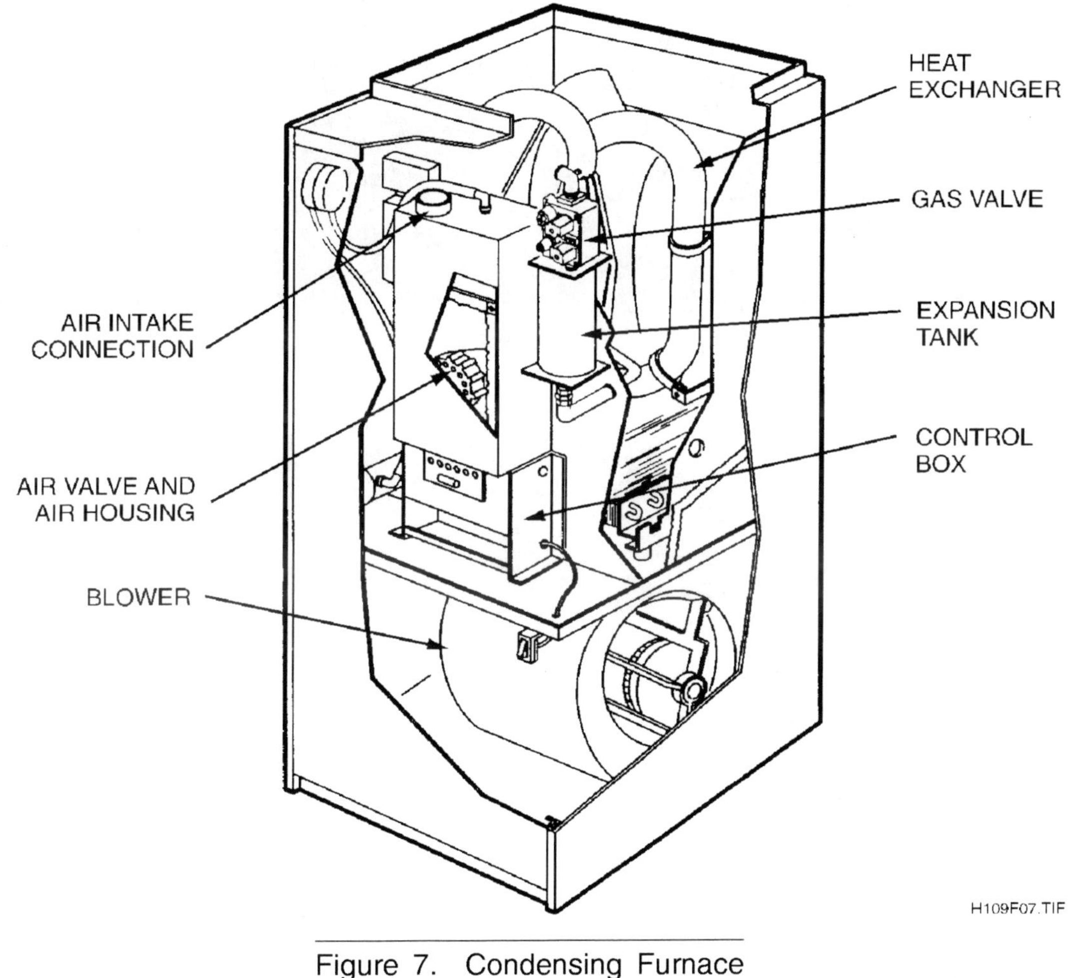

Figure 7.   Condensing Furnace

3.4.0   FANS AND MOTORS

3.4.1   Induced-Draft Fan

As previously mentioned, the induced-draft fan is used on products in the mid-efficiency and high-efficiency classes. This fan draws combustion air into the heat exchangers to support combustion. It also forces the combustion products out through the vent system. *Figure 8* shows the flow of combustion air and flue gases.

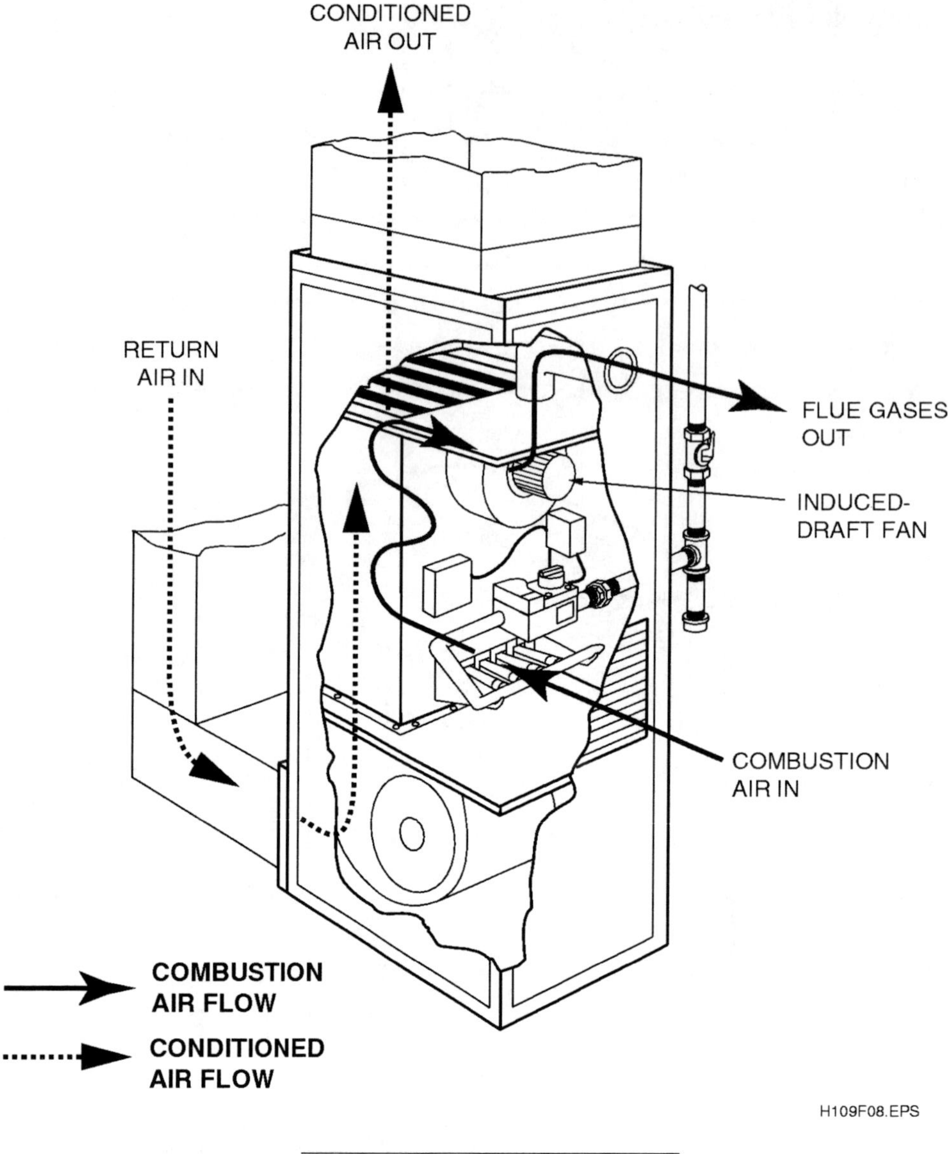

Figure 8. Furnace Air Flow

The induced-draft fan usually draws air from the space around the furnace. In condensing furnaces, or in tightly sealed buildings, combustion air may have to be piped in from the outside.

3.4.2 Blower

Much of the successful performance of a heating system depends upon the proper operation of the fan and fan motor called the blower. The fan and motor circulate the air through the system, obtaining it from return and outside air and forcing it over the heat exchanger and through the supply distribution system to the space to be heated. Centrifugal fans with forward curved blades are generally used. Air enters through both ends of the wheel and is compressed at the outlet by centrifugal force.

The speed of the air (velocity) is usually stated in cubic feet per minute (CFM).

Fans and motors must be matched to each other and to the system in order to deliver the required amount of air against the system resistance. It is usually necessary to make some adjustment of the air quantities at the time of installation because the resistance of the system cannot be accurately determined prior to installation.

The two types of fan motor arrangements in use are the belt drive and the direct drive. The belt-drive arrangement uses a fixed pulley on the fan shaft and a variable pulley on the motor shaft. The speed of the fan is in direct proportion to the ratio of the pulley diameters. The variable-pitch pulley diameter can be changed by adjusting the position of the outer flange of the pulley. On the belt-drive arrangement, the belt tension is usually sufficient to prevent slippage. The motor is mounted on rubber isolators to prevent sound transmission and reduce vibration.

The direct-drive fan arrangement has the fan mounted on an extension of the motor shaft. Fan speeds are changed only by altering the motor speed. The motor speed is altered by the use of extra windings on the motor. This is sometimes called a tap-wound or multi-speed motor. This type of motor has a series of electrical "taps," each of which provides a different speed. Often, one speed is selected for heating and another for cooling.

3.5.0   AIR FILTERS

Air filters remove dust, pollen, molds and other particles from the air circulating through the building. They are most commonly located at the return air entrance to the furnace. Filters come in different efficiency levels and their capacity to capture particles is measured in microns. A micron is 1/1000 of a millimeter. If you placed 25,000 microns side by side, they would equal an inch. Standard-efficiency filters will remove particles of dust, dirt, pollen, and molds as small as 10 microns. That sounds pretty good, but it really means that 50% or more of the particles in the air are getting through the filter. The replaceable fiberglass filters you can buy at a hardware store are in this class. So are most of the permanent washable filters, which are made of metal or polyurethane foam.

High-efficiency filters will capture particles down to about half a micron. That covers mold, pollen and much of the dust in the air. This class of filters includes self-magnetizing electrostatic filters and bag filters.

An electronic air cleaner is needed to capture really fine particles, including smoke and vapors. Electronic air cleaners can be obtained as stand-alone units or can be added to a furnace as an accessory.

Furnace filters should be replaced or cleaned (depending on the type) at least twice during the heating season. As dirt and other captured particles accumulate on the filter, they restrict airflow and thus reduce the efficiency of the furnace.

3.6.0 AUTOMATIC VENT DAMPER

The automatic vent damper is an energy-saving device. It is optional equipment on some heating units. One type of vent damper can be connected to the spark ignition control of the gas furnace (*Figure 9*). This type of system prevents the heated air from going up the flue when the furnace is not operating. It also provides immediate venting for products of combustion. Automatic vent dampers must be installed only on heating units for which they have been designed. If installed on other furnaces, they would not only be inefficient, but could be dangerous.

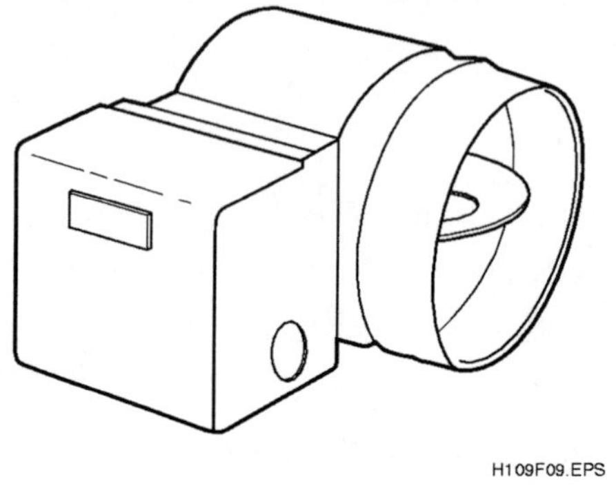

H109F09.EPS

Figure 9.   Automatic Vent Damper

The vent damper installation package consists of a damper assembly, a damper operator, a flue adaptor, and a damper cable. The damper can be installed in a vertical flue or in a flue that is not inclined more than 20 degrees from vertical. The damper operator opens and closes the vent damper upon demand of the space thermostat. When the thermostat calls for heat, the damper motor or operator is energized and the damper will open. The vent damper will remain open while the main burner is operating.

Another type of vent damper is the thermally-actuated vent damper (*Figure 10*). One type of thermally-actuated automatic vent damper uses a bimetal strip that transforms heat energy into mechanical energy. As the bimetal strip twists due to the heat from the furnace, the damper plate opens.

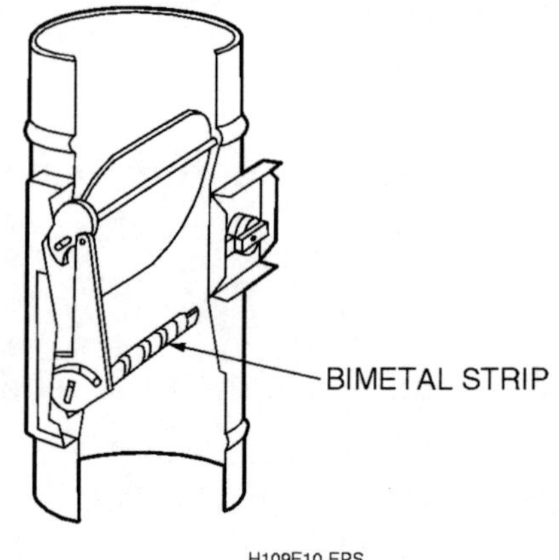

H109F10.EPS

Figure 10.   Thermal Vent Damper

3.7.0   HUMIDIFIERS

Humidifiers are used to add moisture to indoor air. The amount of humidity depends upon the outside temperatures, the building construction, and the relative humidity that the interior of the house will withstand without a condensation problem. It is commonly held that a relative humidity of 30 to 50 percent is desirable. Too much humidity can cause condensation problems and too little humidity can cause static electricity problems and may cause furniture to crack or come apart at glued joints. There are several types of humidifiers. In this discussion, we are mainly concerned with humidifiers that can be added to heating systems. Most of the humidifier types use some medium to pick up water from a reservoir and expose it to the airstream in the ductwork.

3.7.1   Plate-Type Humidifier

The plate-type evaporative humidifier has a series of porous plates mounted in a rack. The lower section of the plates extend down into water that is contained in a pan. A float device regulates the supply of water to maintain a constant level in the pan. The pan and plates are mounted in the warm air (supply) plenum.

3.7.2   Rotating-Drum Humidifier

The rotating-drum evaporative humidifier has a slowly revolving drum that is covered with a polyurethane pad partially submerged in a supply of water. The water supply is controlled by a float system. As the drum rotates, it absorbs water. The humidifier is mounted on the side of the return air plenum and the air from the supply plenum is ducted into the side of the humidifier. The air passes over the wet surface, picks up moisture, and then goes into the return air plenum.

3.7.3   Rotating-Disk Humidifier

The rotating-disk evaporative humidifier (*Figure 11*) is similar to the drum type, in that the water-absorbing material revolves. It is normally mounted on the underside of the main warm air supply duct.

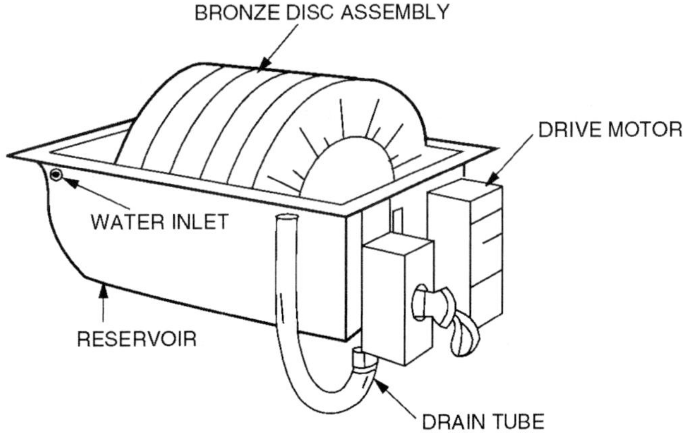

Figure 11.   Rotating Disk Humidifier

3.7.4 Fan-Powered Humidifier

The fan-powered evaporative humidifier (*Figure 12*) is mounted on the warm air plenum. Air is drawn in by the fan and forced over the wet core and is then delivered back into the supply air plenum. The water flow over the core is controlled by a valve. A humidistat senses the humidity level and is used to turn the humidifier on and off. It controls both the fan and the water supply valves. The humidifier operates only when the furnace fan is running.

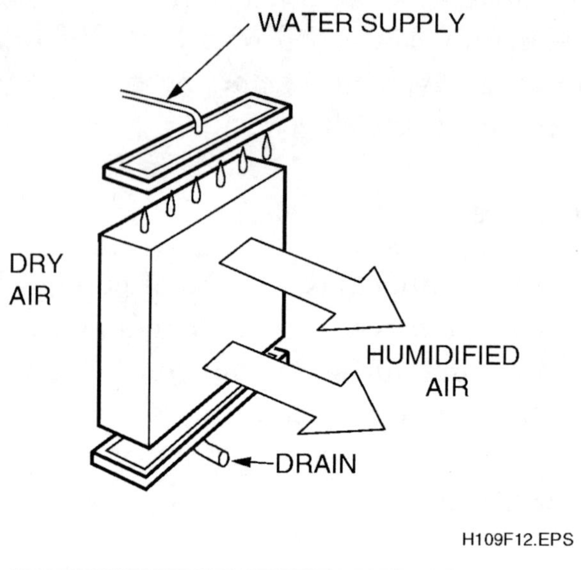

Figure 12. Fan-Powered Humidifier

3.7.5 Atomizing Humidifier

The atomizing humidifier consists of a metal enclosure containing a stainless steel water atomizing nozzle. A finely dispersed water mist is produced that is capable of instant evaporation in the furnace ductwork. A minimum of 40 psi normal household water supply is required for efficient atomization. The humidifier operates only when the furnace fan is running.

3.7.6 Vaporizing Humidifier

The vaporizing humidifier uses an electrical heating element immersed in a water reservoir to evaporate moisture into the furnace supply air plenum. A timed flush cycle is included so that the accumulated solids from the water will not remain in the reservoir. A humidistat is connected into the system so that it not only starts the water heater but also turns on the furnace fan, if it is not already running.

3.7.7 Bypass Humidifier

Bypass humidifiers (*Figure 13*) are usually installed on either plenum of any type of forced-air furnace. They operate on the bypass principle. That is, air movement is accomplished by the static pressure differential between the supply and return plenums. Water is supplied to the water panel evaporator. Dry air that is forced through the wet panel picks up the available moisture and distributes the humidity through the conditioned space. Minerals and solid residues not trapped by the panel evaporator are flushed down the drain. For situations where water hardness is not a problem, some bypass humidifiers incorporate a water circulating system.

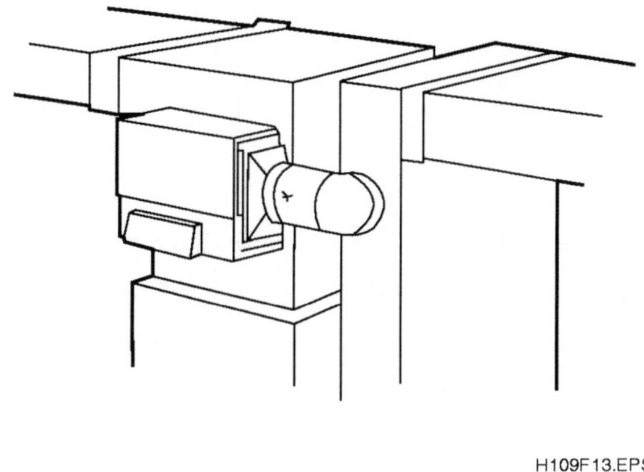

H109F13.EPS

Figure 13. Bypass Humidifier

3.7.8 Ultrasonic Humidifier

The ultrasonic humidifier contains a crystal known as a **piezoelectric** crystal. The crystal vibrates at a high frequency when an electric current is applied to it. Water dripping onto the vibrating crystal is atomized and injected into the airstream.

3.7.9 Steam Humidifier

In this type of humidifier, water is heated and converted into steam. Steam humidifiers (*Figure 14*) have been in use for many years. Due to their light weight, one worker can usually install them.

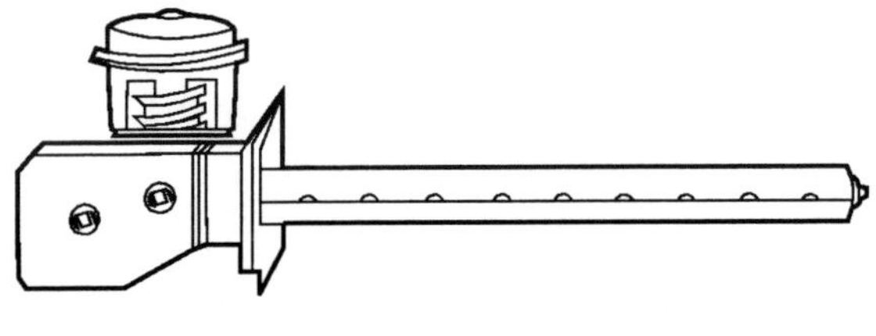

H109F14.EPS

Figure 14. Steam Humidifier

The unjacketed discharge manifold cools down completely when not in operation, but instant start-up is also considered to be one of the positive factors of steam humidifiers. In large installations where steam may be available, steam humidification may be injected directly into the air distribution system. Factories and health care facilities are examples.

Problems can arise in the use of humidifiers when the mineral content (hardness) of the water is too great. When the water hardness is more than 10 grains of dissolved mineral particles per gallon, a water treatment unit should be added to the humidifying system.

3.8.0 INSTALLATION

Proper installation of a furnace is important for the safety of building occupants and efficient operation of the furnace. Factors that must be considered during installation of any furnace are: location, safety controls, clearances, venting, and combustion air.

Applicable national, state, and local codes should be followed when installing a furnace. You will find, however, that manufacturer's instructions are often more stringent than the codes. This is done to make sure that codes are not inadvertently violated.

Proper handling of furnaces is especially important. They are made of sheet metal, which can be damaged if the furnace is dropped or mishandled. Also, some components, such as ceramic igniters, can be easily damaged.

3.8.1 Location

A furnace should be installed in a central location to reduce the length of duct runs. There should be plenty of space around the furnace to permit easy access for servicing and repair. Furnaces should never be located near a source of aerosol sprays, bleaches, detergents, air fresheners, and cleaning solvents. Even small concentrations of such materials can corrode a furnace. In such environments, it is necessary to bring combustion air in from the outdoors. Manufacturers may refuse to honor warranties if a furnace is exposed to corrosive materials.

3.8.2 Safety Controls

Most furnaces come equipped with the required safety controls. It is important to make sure that all controls are installed and working properly. Some of the furnace safety controls are:

- Flame rollout switch — Shuts off the furnace if the flame escapes the fire box.
- Heat limit switch — Shuts off the furnace if the flue is blocked or a fan stops working.
- Control to shut off gas flow if ignition does not occur.
- Control to shut off the furnace if fuel flow is interrupted.

3.8.3   Clearances

Local codes usually specify the required distance between the furnace and combustible materials. Flammable materials must not come into contact with the heat exchangers, burners, or any other hot surfaces, such as the flue vent.

3.8.4   Furnace Venting

The flue gases leaving the furnace contain carbon monoxide, carbon dioxide, or both. Carbon monoxide is highly poisonous; it's the same kind of gas that comes from an automobile exhaust pipe. Carbon monoxide is caused by incomplete combustion.

Carbon dioxide results if complete combustion occurs. It is not poisonous, but it replaces oxygen. A large concentration of carbon dioxide can cause the body to suffer from lack of oxygen. In extreme cases, it can cause asphyxiation. For these reasons, flue gases must be vented in accordance with local and national codes.

Standard-efficiency furnaces and mid-efficiency furnaces both produce hot flue gases. These gases must be vented through a special metal vent or a masonry chimney. (See *Figure 15*.) It is often necessary to use a double wall metal vent for the vent connector. If a masonry chimney is used, it must be lined and have the correct dimensions. Even if the chimney has a tile liner, it is sometimes necessary to add a metal liner to meet code requirements. A lot depends on such factors as the distance from the furnace to the vent, the amount of heat in the flue gases, and whether the furnace is common-vented with another gas appliance such as a hot water heater.

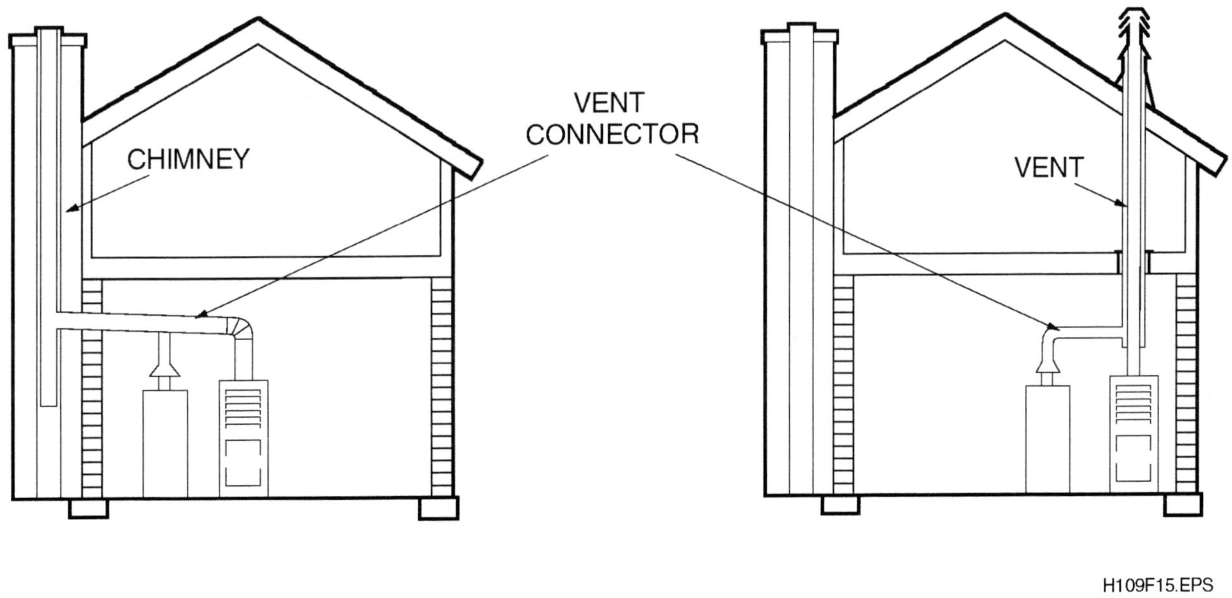

Figure 15.   Furnace Venting

The Gas Appliance Manufacturers Association (GAMA) provide tables and instructions that will help installers determine how to vent a particular furnace. Some manufacturers have simplified the tables and instructions for their product lines and provide the information in their installation instructions.

High-efficiency furnaces such as condensing furnaces can usually be direct-vented through an outside wall using PVC pipe.

3.8.5 Combustion Air

Many furnaces obtain the air for combustion from the air supply around the furnace. This approach is common with natural-draft and mid-efficiency furnaces because the air used for combustion can usually be made up by **infiltration**. If the furnace is installed in a confined space such as an equipment closet, the room must be vented to allow for sufficient airflow.

High-efficiency furnaces need a lot of combustion air. For that reason, they must have combustion air piped in from the outdoors. The same is true for mid-efficiency furnaces in tightly sealed buildings where there is not enough infiltration to provide makeup air. If these furnaces try to use indoor air for combustion, the pressure differential it creates can cause flue gases to be drawn into the building. It can also cause problems like difficulty in closing doors.

4.0.0 GAS FURNACES

In the preceding section, we discussed forced-air furnaces in general terms. Much of the discussion applied to both gas and oil furnaces. In this section and the next, we will discuss specific components used in gas and oil furnaces.

In a gas furnace, gas fuel is supplied at low pressure into a burner head, where it is mixed with the air required for burning. The combustion forms hot gases which pass through the furnace heat exchanger or boiler into the vent pipe and chimney. The rate at which gas is supplied to the burner is controlled by a gas valve. The pressure is controlled by an automatic pressure regulator, which is usually built into the gas valve assembly.

The function of a gas-burning unit is to produce a proper fire at the base of the heat exchanger. In order to do this, the equipment must control and regulate the flow of gas, assure the proper mixture of gas with air, and ignite the gas under safe conditions. To accomplish these functions, the gas burner assembly (*Figure 16*) needs four major parts or sections: gas valve, ignition device, manifold and orifice, and burners.

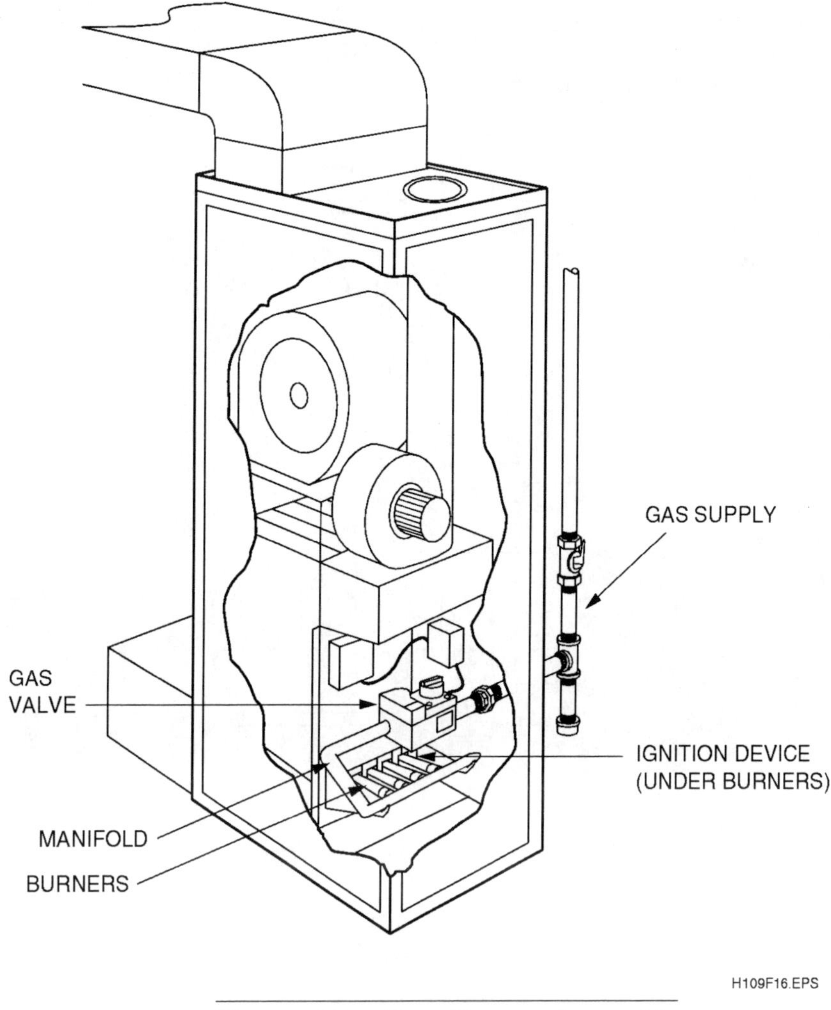

Figure 16. Gas Burner Assembly

4.1.0 FLAME IGNITION

Combustion begins when the gas is ignited. There are several ways that gas ignition can be accomplished. (See *Figure 17.*)

4.1.1 Standing Pilot

Older furnaces use **standing pilots**, which produce a small gas flame that remains on all the time. When gas begins to flow, it is immediately ignited by the pilot light. The **safety pilot** igniter contains a **thermocouple** that converts the heat from the pilot flame into a small electrical current. The thermocouple is connected to the gas valve control circuit. If the pilot flame stops burning, the current from the thermocouple stops flowing and the gas valve is shut off.

Some safety pilots use a bimetal switching device instead of a thermocouple. It reacts to the heat from the pilot and keeps the switch closed as long as the pilot is lit.

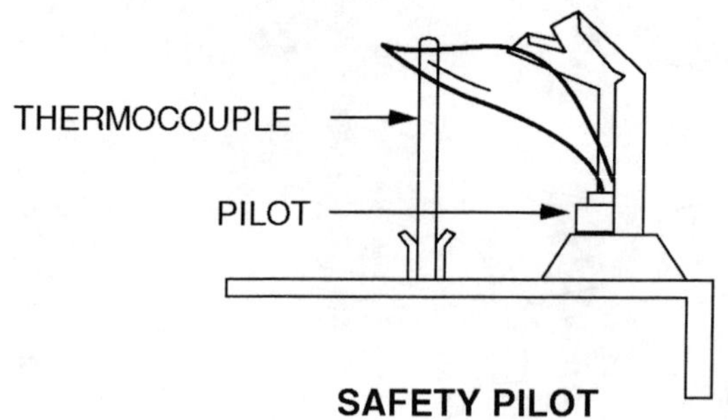

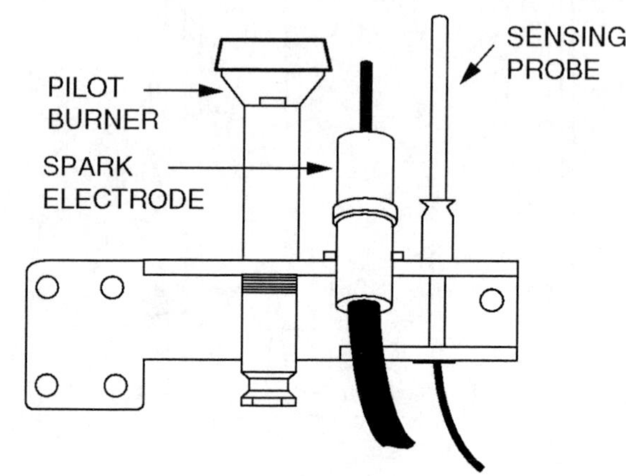

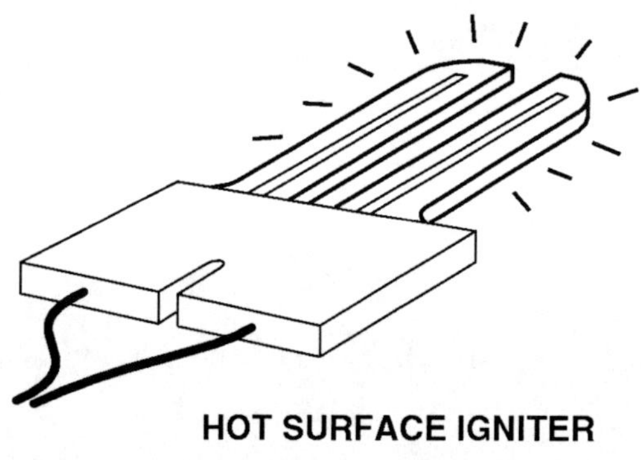

Figure 17. Gas Furnace Ignition Devices

4.1.2 Re-Ignition Pilot

The re-ignition pilot, or intermittent igniter, saves energy because, unlike the standing pilot, it does not require continuous gas flow to the pilot. When the thermostat calls for heat, a special transformer in the control circuit produces a high voltage spark (10,000 volts or more) that ignites the pilot gas. These pilots often use a process known as **flame rectification** for pilot safety. This process is based on the fact that the pilot flame will produce a tiny current in the microampere range. As long as this current is sensed, it means that the pilot is on. If the current stops, the control circuit shuts off the gas supply to the furnace.

4.1.3   Direct Ignition

Many modern furnaces use a device known as a **hot surface igniter**. It is placed next to a burner in place of the pilot. When the thermostat calls for heat, a current flows through the igniter, causing it to become extremely hot. The heat ignites the gas. When direct ignition is used, a separate flame sensor must be placed near the burners as a safety shutoff device. Hot surface igniters are made of ceramic and are thus very fragile. They must be handled carefully.

4.2.0   GAS VALVE ASSEMBLIES

Older systems used a gas supply arrangement in which the gas valve and pressure regulator were separate components (*Figure 18*). On systems manufactured since the 1970's, however, these functions, along with pilot safety control (on pilot-type furnaces) have been combined into a single assembly. The modern gas valve is a **redundant gas valve**; that is, it contains two independent gas valves in series. If one fails, the other will shut off the gas flow when required.

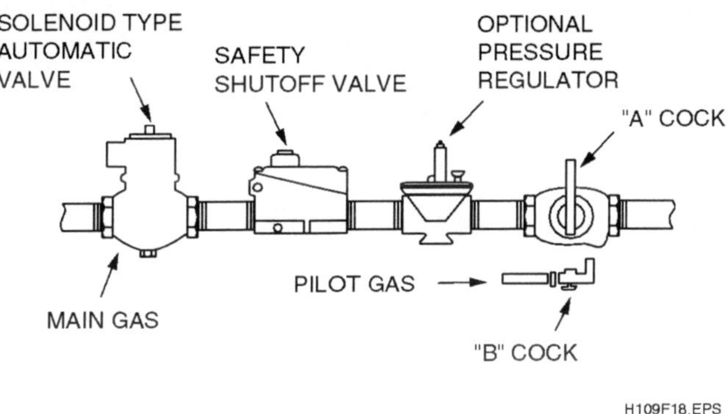

Figure 18.   Older Gas Supply

The gas valve assembly shown in *Figure 19* is just one of the many types available. It has a gas outlet for the pilot, as well as a pilot control which would not be found on the gas control for a direct-ignition furnace. Gas valves on some high-efficiency furnaces are designed to control two levels, or stages, of heat. These gas valves supply full or partial gas pressure, depending on the heat demand.

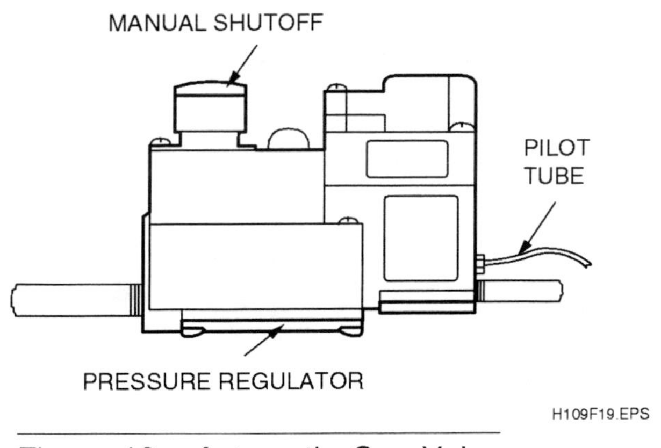

Figure 19.   Automatic Gas Valve

4.2.1 Automatic Gas Valve

The principle function of the automatic gas valve is to control the gas flow. Some of these valves control the gas flow directly; others regulate the pressure on a diaphragm, which in turn regulates the flow of gas. There are five principal types of automatic gas valves:

- The solenoid-operated valve (*Figure 20*) uses electromagnetic force to operate the valve plunger. When the thermostat calls for heat, the plunger is raised, opening the gas valve. These valves are often filled with oil to eliminate noise and lubricate the unit.

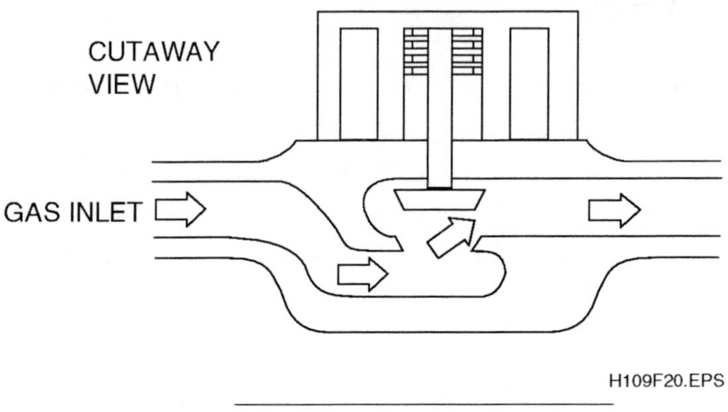

Figure 20.   Solenoid Valve

- The diaphragm-operated valve (*Figure 21*) uses gas pressure above and below the diaphragm to control the device. When the coil is energized, the gas supply to the upper section (above the diaphragm) is cut off and the pressure in the upper section is then bled off to the atmosphere. The pressure of the gas in the lower section then bends the diaphragm up, allowing gas to flow through the valve, as shown in *Figure 21*. When the coil deenergizes, the upper section is pressurized. With equal pressure above and below the diaphragm, the weight of the diaphragm forces it down to shut off the gas flow.

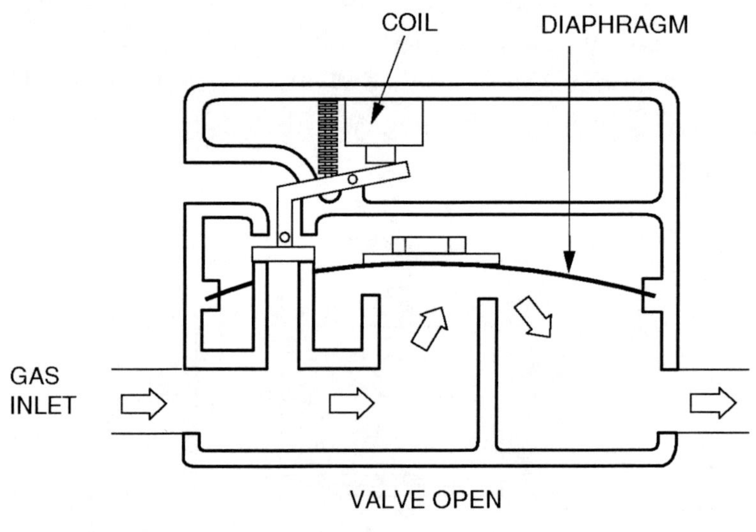

Figure 21.   Diaphragm Valve

- Bimetal valve operators have a high resistance wire wrapped around the diaphragm. When the thermostat calls for heat, current is supplied to the wire, causing it to heat and "warp" the diaphragm. The warping action opens the valve. The action of this valve is slow and the unit is sometimes referred to as a delayed action valve.
- Bulb-type valve operators depend upon the expansion of a liquid-filled bulb to provide the operating force. The sensing bulb is attached to a bellows that expands and contracts, activating the gas flow valve.
- The heat motor valve (*Figure 22*) uses an electric heating coil to give off heat, which moves an expandable rod. The force necessary to operate the valve is provided by the movement of the expandable rod. When the rod is contracted and the valve is in the de-energized mode, the spring below the valve disc keeps the valve closed. The rod expands to open the valve when the heater produces enough heat.

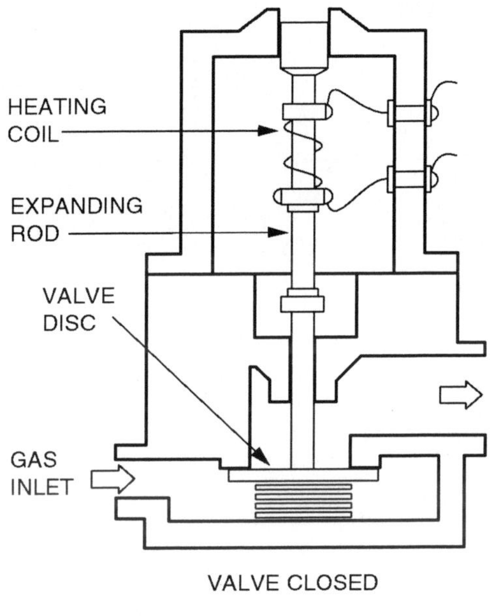

Figure 22. Heat Motor Valve

In the energized mode, the expansion of the rod moves the valve stem downward, forcing the disc off the seat and allowing the gas to flow through the valve. Due to the time needed to heat the rod, a delay of about 20 seconds takes place. There is a delay of about 40 seconds when the valve is de-energized.

4.3.0 MANIFOLD AND ORIFICES

The manifold (*Figure 23*) delivers gas equally to all the burners in a furnace. It is usually made of $1/2$ to 1 inch black iron pipe. A **spud** is screwed into the manifold. The **orifice**, which is the size of the opening in the spud, determines how much gas is delivered to the burner. The size of the orifice selected depends upon the type of fuel gas, the pressure in the manifold, and the gas input required for each burner. A number on the spud shows the orifice size. If the number is not visible, a numbered drill can be used to determine the size.

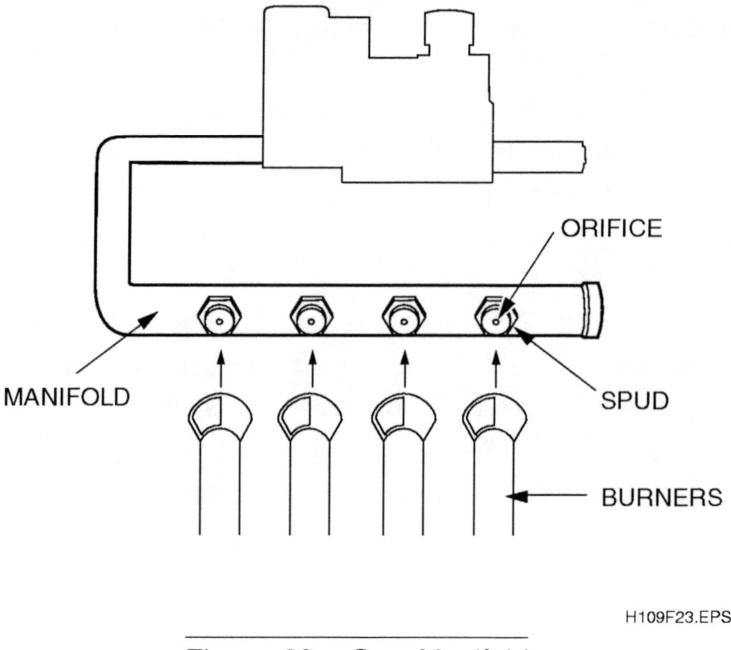

Figure 23. Gas Manifold

As discussed earlier, most gas burners require that some air be mixed with the gas before combustion; this air is called "primary" air. Primary air (*Figure 24*) amounts to approximately one-half of the total air required for proper combustion. Too much primary air causes the flame to lift off the burner surface. Too little primary air causes a yellow flame.

Air supplied to the burner at the time of combustion is called "secondary" air. If too little secondary air is present, carbon monoxide forms. As discussed earlier, most gas burner units are designed to operate on approximately 50% excess secondary air supply. For correct flame generation the proper ratio between primary and secondary air must be maintained.

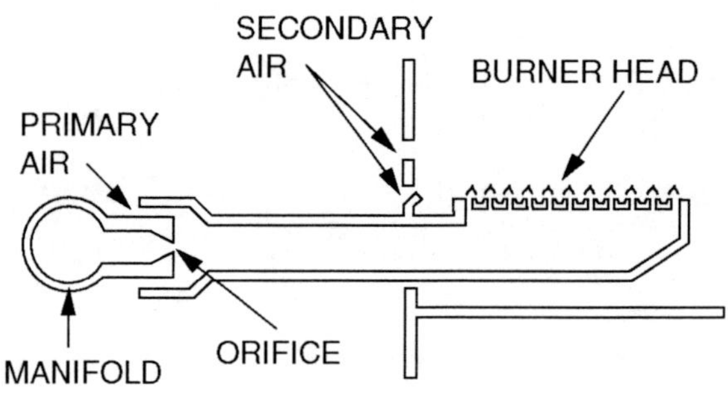

Figure 24. Combustion Air Supply

4.4.0 GAS BURNERS

Gas burners can be either single-port or multi-port (*Figure 25*). Single-port burners direct a flame into an opening in the heat exchanger. Multi-port burners distribute flame through slots or holes along the length of the burner. Unlike single-port burners, multi-port burners are located inside the heat exchangers.

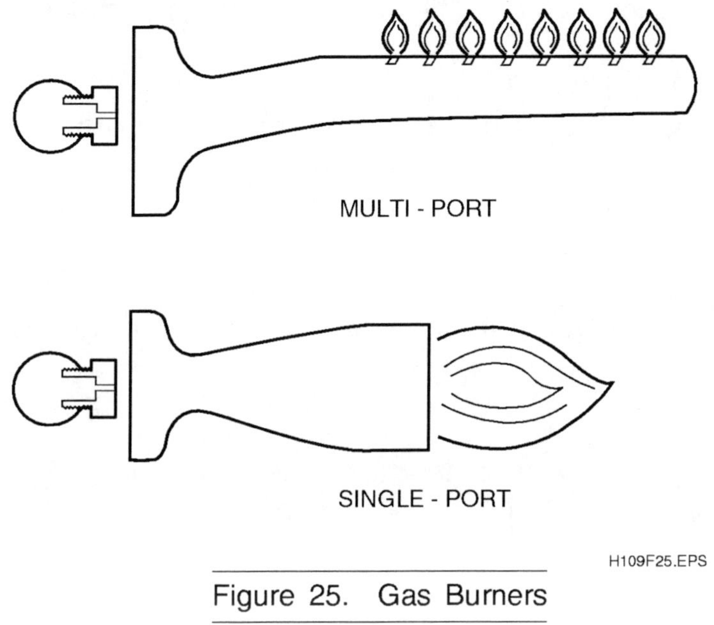

Figure 25. Gas Burners

4.5.0 MAINTENANCE

Gas furnaces require periodic cleaning as well as maintenance and service on the gas burner assembly, the pilot assembly, and the automatic gas valve.

4.5.1 Maintenance Checks

At the beginning of each heating season, certain maintenance tasks should be performed. Filters must be cleaned or replaced. The blower motor and heat exchangers must also be cleaned. The heat exchangers are cleaned with a flexible wire brush attached to a drill.

Operation of the burners and pilot (if any) should also be checked. The burners should produce a steady blue flame with an orange tip. A wavering blue flame indicates too much draft. A yellow flame indicates a lack of primary air and shows that carbon monoxide may be forming. Lack of primary air to the burners may also produce other symptoms:

- Flashback to the spud (also caused by low manifold pressure)
- Soot on the burners and heat exchangers

Any excess of primary air will cause the flame to lift away from the spud.

4.5.2 Manifold Pressure

If the manifold is not receiving the correct pressure, the heat exchangers may not warm up enough to prevent condensation and incomplete combustion may occur. The manufacturer's installation instructions supplied with the furnace will specify the correct manifold pressure. You need to know the heat value and specific gravity of the gas, which you can get from the local utility. You also need to know the size of the orifice, which you can get by looking at the spud.

One way to check the manifold pressure is to connect a **manometer** to the pressure port on the gas valve (*Figure 26*). A manometer measures pressure by the amount of water that is pushed up the column. It is calibrated in inches of water column (in. w.c.). A typical manifold pressure might be 3.5 in. w.c.

If the pressure needs to be adjusted, the pressure adjustment screw on the gas valve is used.

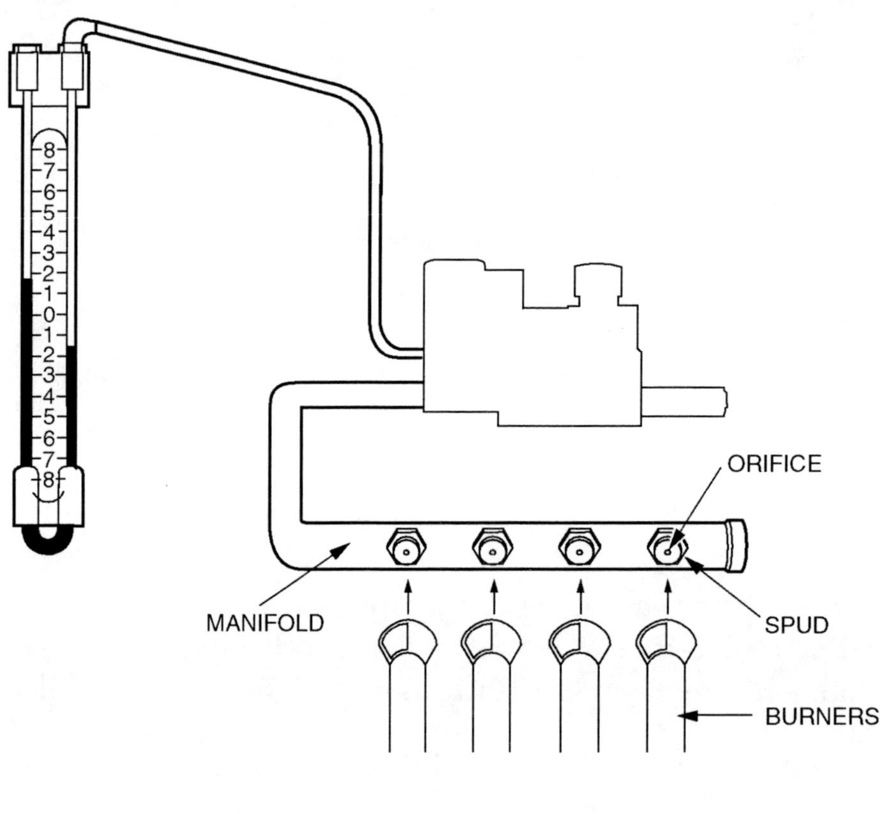

Figure 26.   Adjusting Manifold Pressure

5.0.0   OIL FURNACES

In a typical oil-fired furnace, the **oil burner** receives oil from a nearby storage tank. The oil burner shoots a spray of oil into the combustion chamber, where it is mixed with air, which is supplied by a blower fan that is part of the oil burner. **Electrodes** located at the point where the oil spray enters the combustion chamber provide a high voltage spark that ignites the oil.

5.1.0 PRESSURE BURNER

The pressure-type burner (*Figure 27*) consists of a pump, an air tube and nozzle assembly, and a fan. Oil is pumped to the nozzle, which converts it to a fine mist. The oil mist is mixed with air in the air tube (also known as a "blast tube") and ignited by an electric spark. The burner is not operated continuously, but is turned on or off in accordance with the demands of a room thermostat. The refractory firepot makes the unit more efficient. Most of the burner mechanism is outside the furnace or boiler, making it accessible for servicing. Grade 2 fuel oil is commonly used with pressure-type oil burners.

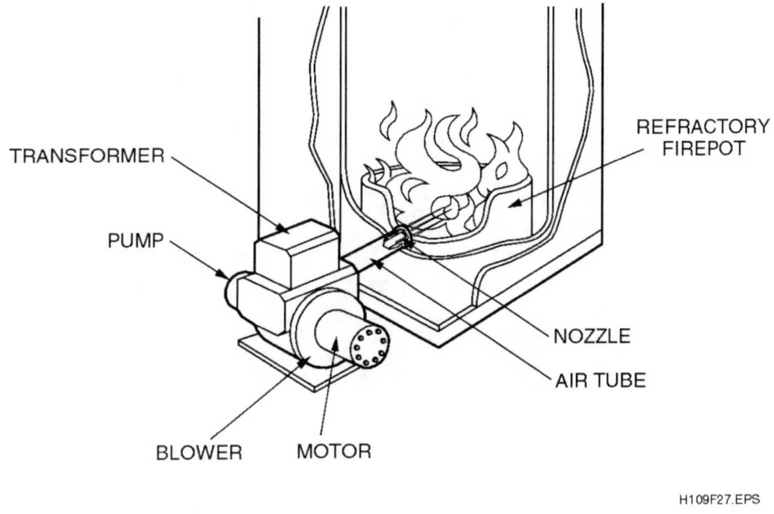

Figure 27. Pressure Burner

5.2.0 ROTARY BURNER

The rotary burner (*Figure 28*) consists of a horizontally spinning cup or set of tubes which spray the fuel by centrifugal force. The oil spray is further broken up and mixed with air by a fan attached to the underside of the spinner. This mixture is ignited by an electric spark. The burner is not operated continuously, but is turned on or off in accordance with the demands of a room thermostat. The entire unit is located inside the furnace or boiler. It occupies a small amount of floor space, but is not as accessible for servicing as the pressure burner.

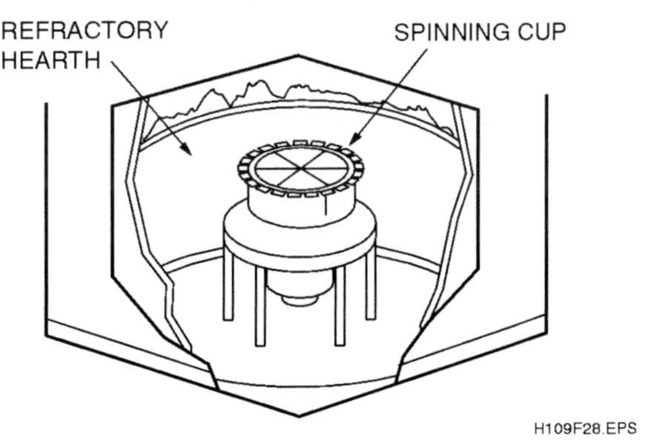

Figure 28. Rotary Burner

5.3.0 POT-TYPE BURNER

The vaporizing or pot-type burner (*Figure 29*) consists of a pot containing a pool of oil and a control that regulates the oil flow. Heat from the burning process vaporizes the oil. The air necessary for burning is admitted into the vapor above the oil pool by natural draft or by a small fan. There are few moving parts and the burner operates quietly.

The fire is manually started with oily waste, after which a low pilot flame keeps the fire alive. When heat is called for, more oil is fed into the pool and a hot flame is created. Careful adjustment of both fuel and air is necessary in order to avoid a smoky, sooty flame. The initial cost of this type of burner is low, but No. 1 grade fuel oil is required.

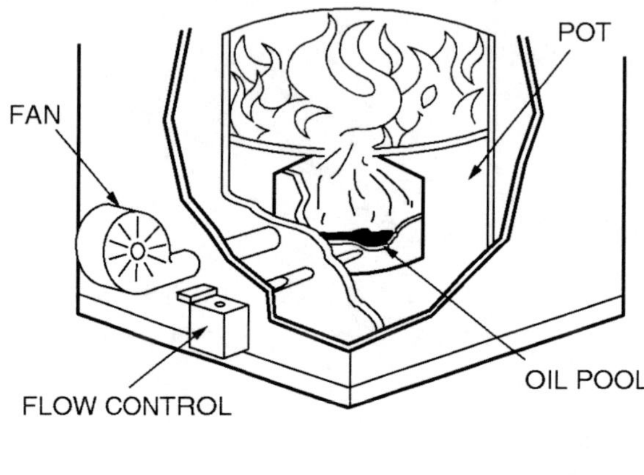

Figure 29.   Pot-Type Burner

5.4.0 OIL BURNER OPERATION

The pressure burner is the most common type. Its operation is discussed here in some detail to acquaint you with the operation of oil-fired furnaces.

There are two types of pressure-type oil burners: the low-pressure gun-type and the high-pressure gun-type.

In a low-pressure gun-type burner, oil and primary air are mixed before going through a nozzle. A pressure of 1 to 15 psi on the mixture, plus the action of the nozzle orifice, causes the oil to atomize. Secondary air is drawn into the spray mixture after it is released from the nozzle. An electric spark is used to light the combustible mixture.

A high-pressure gun-type burner forces oil through the nozzle at a pressure of 100 psi. This breaks the oil into fine, mist-like droplets. The atomized oil spray creates a low-pressure area into which the combustion air flows. Combustion air is supplied by a vane fan, creating turbulence and complete mixing action. The high-pressure gun-type burner is the most popular domestic burner. It is simple in construction and efficient in operation. The parts are mass-produced, readily available, and relatively low in cost.

5.4.1 Power Assembly

The power assembly of the high-pressure burner unit consists of the motor, fan, and oil pump (*Figure 30*).

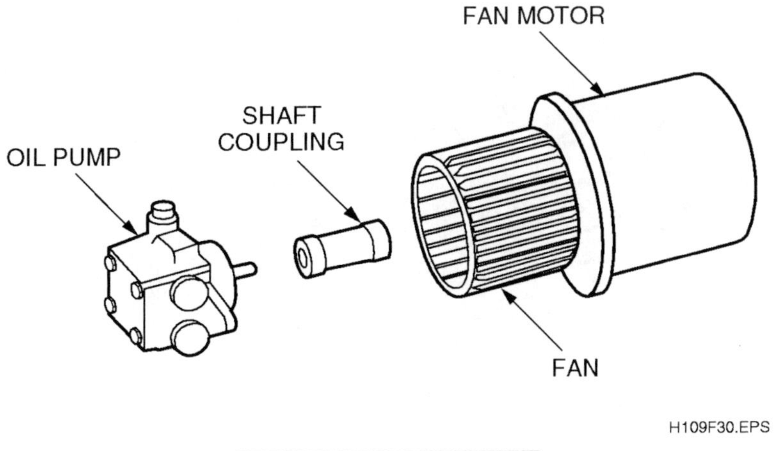

Figure 30.   Fan Motor

The motor drives the fan and oil pump. The fan forces air through a blast tube to provide combustion air for the atomized oil. The oil pump draws oil from the storage tank and delivers it to the nozzle. The fuel/air mixture is ignited by an electric spark that is formed between two electrodes in the nozzle assembly. The motor supplies power to operate the oil pump and fan. The fan delivers air for combustion. The inlet to the fan has an adjustable opening so that the air volume can be controlled manually. The fan outlet delivers the combustion air through the blast tube of the burner.

A pressure-regulating valve on the oil pump has an adjustable spring that permits adjustment of the oil pressure. The pump delivers more oil than the system can use. The excess oil is returned to the tank or is dumped back into the supply line.

An automatic cutoff valve stops the flow of oil as soon as the pressure drops, thereby preventing oil from dripping into the combustion chamber. Oil pumps are designed for single-stage or two-stage operation. The single-stage pump is used where the oil supply is above the burner and the oil is fed to the pump by gravity. The two-stage unit is used where the storage tank is below the burner. The first stage draws the oil to the pump and the second stage provides the pressure required by the nozzle. Pump suction should not exceed 15 inches of vacuum.

5.4.2 Piping

Piping connections to the oil pump are of two types: single-pipe and two-pipe. In the single-pipe system there is only one pipe from the storage tank to the burner. In the two-pipe system, two pipes are run from the tank to the burner. One carries supply oil; the other carries return oil. Compression fittings should never be used on oil system piping. The lines must be purged of air in order to avoid problems. A two-pipe system is easier to keep free of air.

5.4.3 Nozzle

The nozzle assembly (*Figure 31*) consists of the oil feed line, the nozzle, ignition electrodes, and the transformer connections.

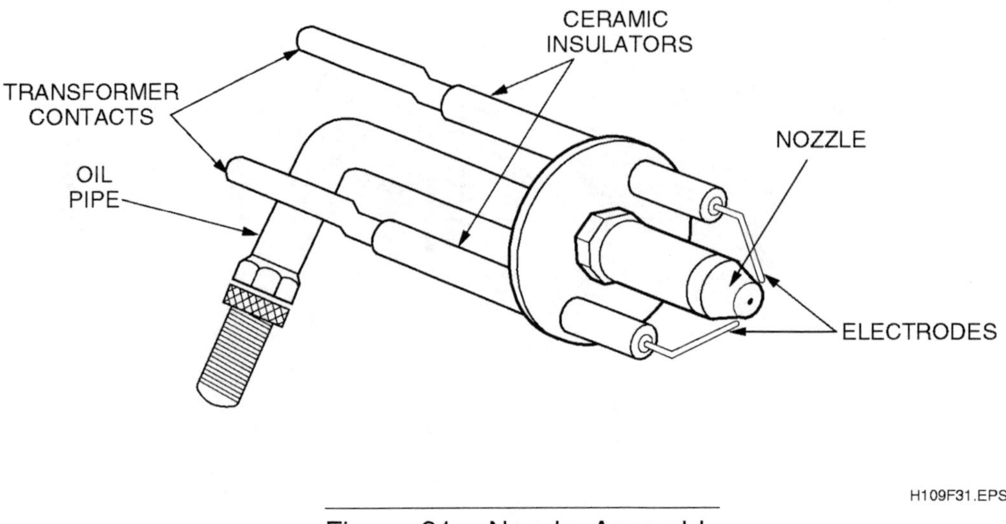

Figure 31. Nozzle Assembly

The nozzle prepares the oil for mixing with the air by atomizing the fuel. Oil passes through a strainer and enters slots that direct the oil to the swirl chamber. The swirl chamber gives the oil a rotary motion when it enters the nozzle orifice, shaping the spray pattern. The nozzle orifice increases the velocity of the oil. The oil leaves the nozzle in the form of a mist or spray and mixes with the air from the blast tube.

Due to the fine tolerances of the nozzle construction, dirty or defective nozzles are usually replaced. Nozzle spray patterns vary with the type of application. The shape of the spray and the angle between the sides of the spray can be varied. There are three spray shapes, as shown in *Figure 32*: hollow (H), semi-hollow (SH) and solid (S). The hollow and semi-hollow are most popular on domestic burners as they are more efficient when used with modern combustion chambers.

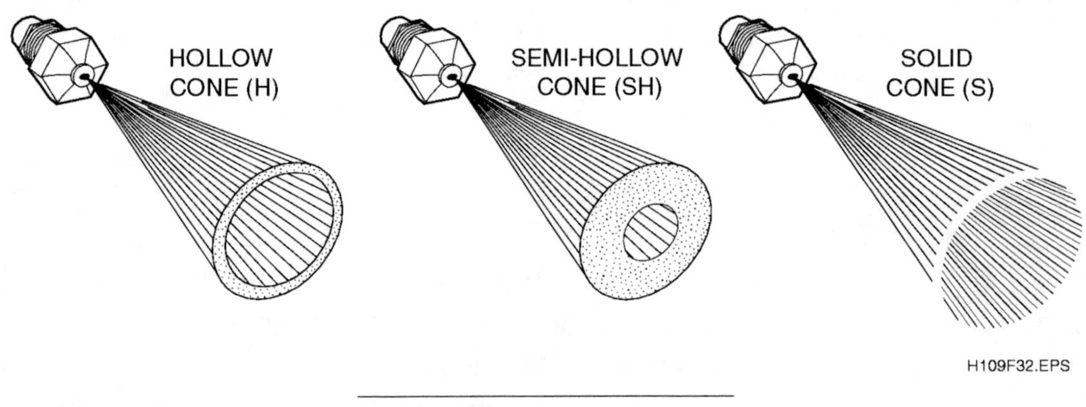

Figure 32. Nozzle Patterns

The angle of the spray must correspond to the type of combustion chamber (*Figure 33*). An angle of 70 to 90 degrees is usually best for square or round chambers; an angle of 30 to 60 degrees is best suited to long, narrow chambers.

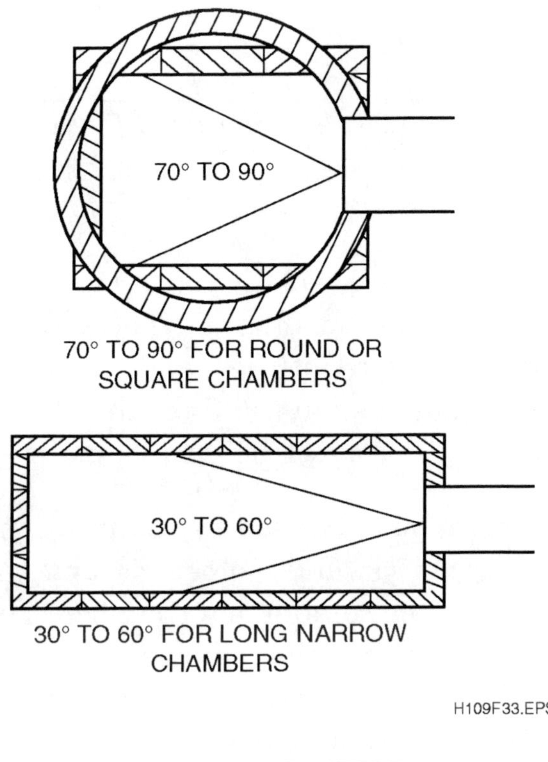

Figure 33.   Spray Angles

5.4.4   Ignition System

The ignition system for high-pressure burners consists of a step-up transformer connected to two electrodes. The transformer supplies high voltage that causes a spark to jump between the two electrodes. The force of the air in the blast tube causes the arc to bend into the fuel-air mixture, igniting it.

Ceramic insulators surround the electrodes and serve to position them. The step-up transformer increases the voltage from 120V to about 10,000V, but reduces the amperage to about 20 milliamps. This amperage is relatively harmless and reduces wear on the electrode tips. The correct position of the electrodes (*Figure 34*) is important. If the electrodes are not centered on the nozzle orifice, the flame will be one-sided and will cause carbon to form on the nozzle.

WARNING!   Due to the high voltage of the transformer and the line voltage delivered to it, caution should be used in checking the transformer.

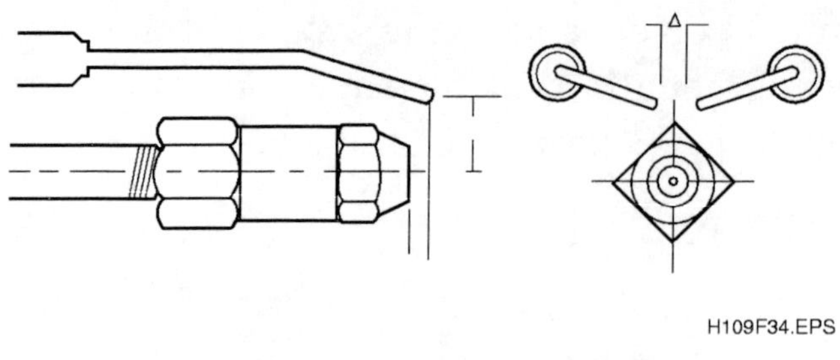

Figure 34. Electrode Position

5.5.0 COMBUSTION CHAMBER

The purpose of the combustion chamber is to protect the heat exchanger from flame damage and to provide reflected heat to the burning oil. The reflected heat warms the tips of the flame, enhancing combustion. No part of the flame should touch the combustion chamber surface. If it does, incomplete combustion could result. The chamber must fit the flame and the nozzle must be located at the correct height above the floor.

Three types of materials are commonly used for combustion chamber construction: metal (stainless steel), insulating fire brick, and molded ceramic. In small furnaces such as those used in residential applications, the combustion chamber is built into the furnace. For large commercial and industrial applications, the combustion chamber may be constructed on site. Such chambers can be made of fire brick or pre-cast ceramic (refractory) material. If pre-cast materials do not fit the job, a combustion chamber may be formed and poured on site using ceramic materials designed for this purpose.

5.6.0 DRAFT REGULATOR

A draft regulator is necessary to control heat and combustion. If too high a draft exists, it causes undue loss of heat through the chimney. If too little draft is available, incomplete combustion could result. The draft regulator should maintain a constant draft over the fire, usually 0.01 to 0.03 in. w.c. A draft regulator (*Figure 35*) consists of a small door in the side of the flue pipe. It is hinged near the center and controlled by adjustable weights.

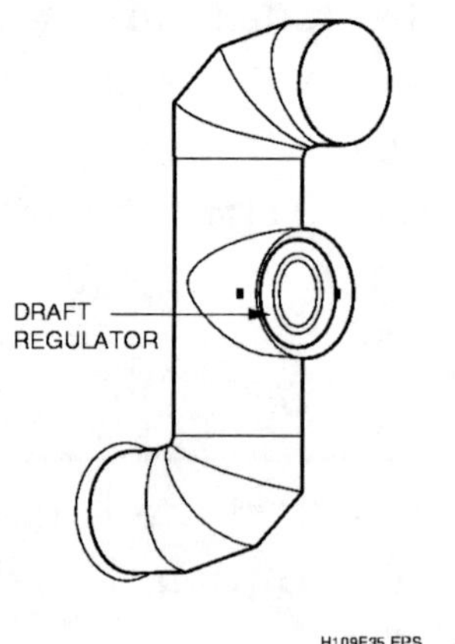

Figure 35. Draft Regulator

HVAC TRAINEE TASK MODULE 03109

5.7.0 OIL STORAGE

Oil furnaces may be supplied from indoor or outdoor oil tanks. Outdoor tanks may be buried. Because many underground oil tanks have leaked, there are strict federal and usually local regulations governing placement, construction, and alarm systems for these tanks. Above-ground tanks over a certain size may require special containment arrangements. No installation should be undertaken without a clear understanding of local and national codes.

5.8.0 OIL FURNACE MAINTENANCE

Like the gas furnace, the oil furnace should be checked at the beginning of each heating season. Filters must be cleaned or replaced and the blower and heat exchangers should be cleaned. Oil burner operation should also be checked. If there is not enough combustion air, the flame may be orange or red instead of yellow. If there is smoke, soot, or odors, it could indicate improper oil pressure, poor draft, or an improper mix of oil and air. A pulsating sound could indicate that the flame is touching the combustion chamber. In any of these cases, some cleaning, adjustment, or repair will be needed.

6.0.0 ELECTRIC HEATING

The electric furnace (*Figure 36*) converts electrical energy directly into heat energy using resistance heaters.

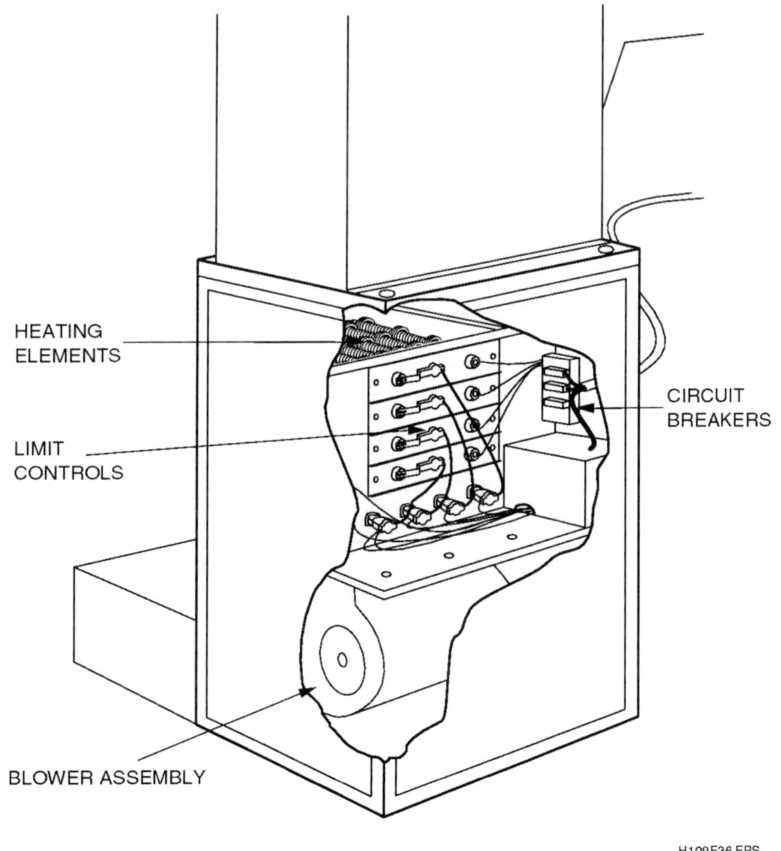

Figure 36. Electric Furnace

Electric furnaces differ from fuel furnaces in that no combustion is required. Because no fuel is burned, there is no need for a chimney or vent to carry the products of combustion outdoors. This feature allows for greater operational safety and more installation flexibility. The return air from the conditioned space passes directly over the resistance heaters and into the supply air plenum. The amount of heat supplied by an electric furnace depends upon the number and size of the resistance heaters used in the application. To avoid overloading the electrical system on start-up, the elements are sequenced on in stages.

The major components of an electric furnace, excluding the controls, are the heating elements, the blower and motor assembly, and the furnace enclosure or cabinet. Accessories such as filters, a humidifier, and a cooling coil may be included.

6.1.0 HEATING ELEMENTS

The function of the heating element (*Figure 37*) is to provide the heat required for the conditioned space. The heating element wires are made of nickel and chromium (Nichrome). The heating element wire is spiraled and threaded through a metal holding rack which has ceramic insulators that prevent the resistance wires from shorting out on the frame of the rack. The heating elements are similar to those found in small household appliances such as electric clothes dryers and toasters.

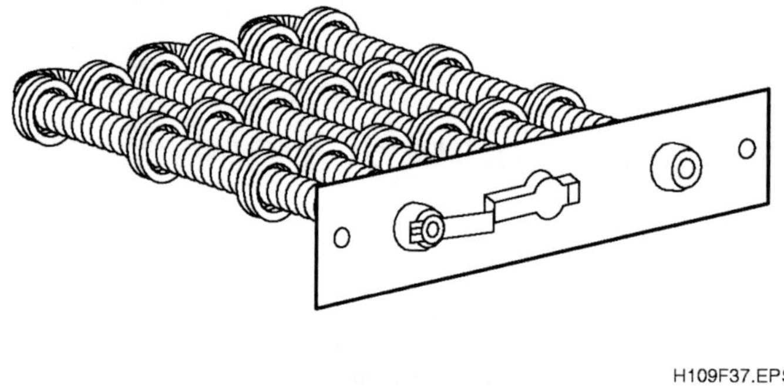

Figure 37. Electric Heating Element

The circuit for each heating element contains a fuse (or circuit breaker) and a safety limit switch. The fuse is a backup for the safety limit switch and is set to open at a temperature slightly higher than the limit switch. The limit switch is usually set to open at approximately 160°F and close when the temperature drops to 125°F.

6.2.0 BLOWER AND MOTOR ASSEMBLY

The fans used in an electric furnace may be either direct-drive or belt-driven. When cooling is added to the heating unit, the use of a multi-speed direct-drive motor may be necessary in order to provide larger air quantities.

6.3.0   FURNACE ENCLOSURE

The casing of the electric furnace is similar to that of a gas or oil furnace but without the vent pipe connection. The interior of the cabinet is designed to permit the air to flow over the heating elements, which are usually insulated from the exterior casing by an air space.

6.4.0   ACCESSORIES

Filters, humidifiers and cooling units are added to an electric furnace in much the same manner as gas and oil furnaces. Therefore, electric heating can provide all of the climate control features found in other types of fuel-fired furnaces.

6.5.0   POWER SUPPLY

An electric furnace power supply is usually 208/240V, single-phase, 60-Hz alternating current. This type of heating unit is supplied by three wires: two "hot" and one grounded. The hot lines leading to the furnace contain fused disconnects. All the wiring should be enclosed in conduit with the proper connectors as specified by the National Electric Code. The NEC also requires that the furnace unit be grounded. The supply ground is provided for this purpose.

SUMMARY

Gas-fired and oil-fired warm air furnaces and boilers provide heat for the majority of buildings in the U.S. Wood and coal are also used as heat sources, but their application is limited because they are less convenient and their combustion products are more damaging to the environment. Electric furnaces provide clean heat, but are usually much more expensive to operate than gas or oil. Oil and gas furnaces operate on the principle of transferring the heat generated by fuel combustion to air or water, which is then circulated through the conditioned space. Furnaces, whether gas or oil, contain a fuel control, a combustion chamber, and heat exchangers. They must also have a means of moving the heat through the conditioned space, as well as a system for venting flue gases to the outside.

References

For more advanced study of topics in this task module, the following books are suggested:

Fundamentals of Gas Heating, The Trane Company, Tyler, Texas.
General Training – Heating (GTH), Carrier Corporation, Syracuse, New York.
Modern Refrigeration and Air Conditioning, The Goodheart-Willcox Company, Inc., South Holland, Illinois.
Refrigeration & Air Conditioning Technology, Second Edition, Delmar Publishers, Inc., Albany, New York.

1. A gas-fired warm air furnace heats by:
    a. radiation.
    b. convection.
    c. conduction.
    d. capillary action.

2. If you burn your finger by touching a hot surface, it is an example of:
    a. radiation.
    b. convection.
    c. conduction.
    d. stupidity.

3. A temperature of 80°F is equal to _____°C.
    a. 12
    b. 202
    c. 27
    d. 176

4. Carbon monoxide (CO) is produced when:
    a. combustion is incomplete.
    b. combustion is complete.
    c. there is water vapor in the flue gas.
    d. there is excess air in the combustion chamber.

5. A gas flame should be:
    a. yellow with a blue tip.
    b. completely blue.
    c. completely yellow.
    d. blue with an orange tip.

6. An oil burner flame should be:
    a. blue with yellow tips.
    b. red with yellow tips.
    c. yellow.
    d. orange.

7. Of the following gases, which has the highest heating value?
    a. Propane
    b. Butane
    c. Natural gas
    d. Manufactured gas

8. In a warm air furnace, conditioned air:
    a. flows through the inside of the heat exchangers.
    b. is heated directly by the burners.
    c. flows over the heat exchangers.
    d. is circulated through the building and the furnace by convection.

9. In a condensing furnace, the condensing heat exchanger:
    a. extracts heat by condensing the conditioned air that passes over the heat exchanger.
    b. is used only in the cooling mode.
    c. extracts heat by condensing the flue gas that flows through the heat exchangers.
    d. none of the above.

10. Humidifiers are used to:
    a. reduce moisture in the air.
    b. add moisture to the air.
    c. keep furniture dry.
    d. store cigars.

11. Direct-venting of flue gases with PVC pipe may be used:
    a. only for the highest efficiency furnaces.
    b. for mid-efficiency and high-efficiency furnaces.
    c. only for natural-draft furnaces.
    d. for any furnace if a chimney is not available.

12. A manometer may be used to measure:
    a. how much water is in the flue gas.
    b. the manifold pressure.
    c. the flame rectification current.
    d. the percentage of primary air used in combustion.

13. In a typical oil burner, the oil is ignited by:
    a. a pilot light.
    b. a cigarette lighter.
    c. electrodes.
    d. a hot surface igniter.

14. The fan in the high-pressure oil burner is used to:
    a. keep the furnace from overheating.
    b. keep the service technician from overheating.
    c. atomize the oil.
    d. provide combustion air.

15. In an electric furnace, the heating coil replaces which of the following items found in a gas furnace?
    a. Burner assembly, heat exchangers, and induced-draft motor
    b. Burner assembly and blower
    c. Induced-draft motor and blower
    d. Condensing coil

ANSWERS TO SELF CHECK REVIEW / PRACTICE QUESTIONS

<u>Answer</u>	<u>Section Reference</u>
1. c	2.1.1
2. c	2.1.1
3. c	2.2.0
4. a	2.4.2
5. d	2.4.4
6. c	2.4.4
7. b	2.5.1
8. c	3.0.0
9. c	3.3.0
10. b	3.7.0
11. a	3.8.4
12. b	4.5.2
13. c	5.0.0
14. d	5.4.1
15. a	6.0.0

ACKNOWLEDGEMENTS

Figure 7 courtesy of Lennox Industries.

HVAC TRAINEE TASK MODULE 03109